AF344441

Developments in Mathematics

Volume 79

Andrej Dujella

Diophantine m-tuples and Elliptic Curves

Springer

Andrej Dujella (ID)
Department of Mathematics
Faculty of Science
University of Zagreb
Zagreb, Croatia

ISSN 1389-2177   ISSN 2197-795X   (electronic)
Developments in Mathematics
ISBN 978-3-031-56723-0   ISBN 978-3-031-56724-7   (eBook)
https://doi.org/10.1007/978-3-031-56724-7

Mathematics Subject Classification: 11D09, 11D25, 11G05, 11J86, 14H52

This work was supported by European Regional Development Fund (KK.01.1.1.01.0004).

This Springer imprint is published by the registered company Springer Nature Switzerland AG
The registered company address is: Gewerbestrasse 11, 6330 Cham, Switzerland

Preface

This book provides an overview of the main historical and recent results and problems concerning Diophantine m-tuples and their various generalizations, with special emphasis on their connections with elliptic curves. A Diophantine m-tuple is a set of m positive integers with the property that the product of any two of its distinct elements plus 1 is a square. Fermat found the first Diophantine quadruple in integers $\{1, 3, 8, 120\}$. If a set of non-zero rationals has the same property, it is called a rational Diophantine m-tuple. The ancient Greek mathematician Diophantus found the first example of a rational Diophantine quadruple.

If we want to extend a Diophantine triple $\{a, b, c\}$ to a quadruple, we have to find an integer or rational x such that $ax + 1$, $bx + 1$ and $cx + 1$ are all squares. By multiplying these three conditions, we obtain a single condition $y^2 = (ax+1)(bx+1)(cx + 1)$, which is, in fact, the equation of an elliptic curve (non-singular cubic curve with a rational point).

Diophantine m-tuples have a very long and exciting history, but it is also a very active research topic today. Some of the famous mathematicians of the past, like Diophantus, Fermat and Euler, as well as some modern ones like Fields Medalist Alan Baker, made important contributions to problems related to Diophantine m-tuples, but many problems remain open. The author's web page [121] contains the full list of references related to Diophantine m-tuples. At the moment, it contains 518 references (38 references before 1990, 95 references before 2000, 245 references before 2010) of authors from all continents. The recent booklet *Wonderland of Families of Diophantine triples* by Deshpande [81] covers a particular subtopic of regular Diophantine triples, and there are some books that treat some aspects of Diophantine m-tuples (e.g. Sections 14.6, 16.7 of *Number Theory* by Dujella [116], Chapter 6 of *Liczby Kwadratowe* by Nowicki [299]). However, until now, no book systematically covered this topic.

Let us mention that the third edition of the well-known book *Unsolved Problems in Number Theory* by Richard Guy [203] contains a new section (Section D29) devoted to Diophantine m-tuples. The problem of the existence of Diophantine quintuples was mentioned in 2001 by Michel Waldschmidt [355] as one of the important problems at the end of the second millennium. A brief survey on

Diophantine m-tuples and their generalizations appeared in "What is …" column in the August 2016 issue of Notices of AMS [114].

The systematic study of connections between Diophantine m-tuples and elliptic curves started around the year 2000. It proved to be very fruitful in both directions: elliptic curves are used in solving some long-standing problems on Diophantine m-tuples, like the existence of infinite families of rational Diophantine sextuples, but also rational Diophantine m-tuples are used in the construction of elliptic curves with interesting Mordell-Weil groups, including some curves with the record (highest known) rank a for given torsion group.

Since it is impossible to cover all aspects appearing in more than 500 publications in one book, we think that focusing on aspects related to connections with elliptic curves is a good and attractive choice. Of course, this book contains fragments of the exciting history of Diophantine m-tuples, which might often have modern interpretations in terms of elliptic curves but which did not originally use this language, as it was unavailable at the time.

In Chap. 1, we present the contributions of Diophantus, Fermat and Euler on this topic, which served as motivation for further investigations. We state the main definitions, results and open problems which will be discussed in the book. We also mention here various generalizations of the notion of Diophantine m-tuples. Later in the book, we will concentrate on "ordinary" (integer or rational) Diophantine m-tuples and discuss only generalizations connected with elliptic curves.

Chapter 2 covers prerequisites on elliptic curves over $\mathbb{Q}$ needed later in the book. We discuss possible torsion groups of elliptic curves over $\mathbb{Q}$, with particular attention to those groups that can appear for curves induced by Diophantine triples. We present methods for computing the rank and constructing elliptic curves with high rank. General methods are illustrated on examples coming from Diophantine tuples. We also explain the functions related to elliptic curves, which are available in the software package PARI/GP.

Chapter 3 is the central part of the book. Here, we introduce elliptic curves induced by Diophantine triples and discuss their properties and applications. One of the main applications is the construction of infinite families of rational Diophantine sextuples (an open problem from the time when Euler found families of such quintuples). There are four known different constructions of such families, which all use elliptic curves in some form, and will be presented in this chapter. Another important application of elliptic curves induced by Diophantine triples comes in constructing high-rank elliptic curves with certain torsion groups. We present details on the construction of some record curves over $\mathbb{Q}(t)$ and $\mathbb{Q}$. We also discuss some other connections between Diophantine m-tuples and elliptic curves.

In Chap. 4, we explain general methods for finding integer points on elliptic curves (transformation to Thue equations, application of elliptic logarithms, solving systems of Pellian equations via linear forms in logarithms and the Baker-Davenport reduction). These methods are then applied to the problem of finding all integer points on elliptic curves induced by (integer) Diophantine triples. We present the proof of an absolute upper bound for the size of Diophantine tuples and sketch the main steps in the results that lead to the proof of the non-existence of Diophantine quintuples.

In Chap. 5, we provide more details on one of the generalizations of the notion of Diophantine m-tuples, namely, that in which the condition that $ab + 1$ is a square is replaced by $ab+n$ is a square for a fixed integer or rational number n. We discuss the problem of the existence of integer $D(n)$-quadruples and rational $D(n)$-quintuples. We will see that the last problem is related to the distribution of ranks in families of twists of certain elliptic curves. Finally, we consider sets which are $D(n)$-triples, quadruples and quintuples for several distinct values of n. Elliptic curves will also play an important role here, but their appearance will differ in those three problems.

The book's primary audience is expected to be researchers and graduate students working in Diophantine equations and elliptic curves. However, this book might be of interest to many other mathematicians interested in number theory and arithmetic geometry. If used as a textbook for a graduate course, the prerequisites would be on the level of a standard first course in elementary number theory (e.g. the Niven-Zuckerman-Montgomery textbook *An Introduction to the Theory of Numbers* [298], or other textbooks that cover the notions of divisibility, congruences and quadratic residues, like *A Concise Introduction to the Theory of Numbers* by Baker [22]). All prerequisites on elliptic curves are provided here (mainly in Chap. 2). For a more systematic introduction to elliptic curves, we can recommend the textbook by Silverman and Tate *Rational Points on Elliptic Curves* [330] (additional recommended books on elliptic curves are given at the beginning of Chap. 2). Most prerequisites related to Diophantine equations and Diophantine approximations are included in the introductory sections of Chap. 4, particularly in Sect. 4.1. Some books that systematically treat aspects of Diophantine equations relevant to this book are *Linear Forms in Logarithms and Applications* by Bugeaud [48], *Number Theory. Volume I: Tools and Diophantine Equations* by Cohen [70], *Diophantine Equations* by Mordell [284] and *The Algorithmic Resolution of Diophantine Equations* by Smart [332].

The author gave a course based on the preliminary version of this book in the academic year 2021/2022 for PhD students at the University of Zagreb. On the course web page [117], additional materials, like homework exercises (mostly included in the book in the exercise sections at the end of each chapter), seminar topics and links to relevant software, can be found. The book could be used as a textbook for a specialized graduate course, and it may also be suitable for a second reading supplement reference in any course on Diophantine equations and/or elliptic curves at the graduate or undergraduate level.

Let me give here some personal comments on how I got involved in Diophantine m-tuples and how they became my main research topic. Since high school mathematics competitions, I have been occupied and inspired by number theory. During my mathematics undergraduate studies, I did not have the opportunity to take courses in number theory. However, I acquired solid knowledge in various areas of mathematics and had the opportunity to listen to the lectures of prominent Croatian mathematicians. I deepened my interest in number theory through the literature I bought in a foreign literature bookstore in Zagreb (mainly Russian editions, accessible to the student pocket). In the last year of my undergraduate studies, I received a financially generous scholarship from the University of Zagreb. That was a very happy circumstance because, among other things, it made it possible for me to buy the proceedings from the 2nd Conference on Fibonacci numbers held in Greece, where in the two papers *On a problem of Diophantus* by Long and Bergum [257] and *More on the problem of Diophantus* by Arkin and Bergum [12], I met for the first time a problem of Diophantus, which I later dealt with for the most part of my scientific career. As I mentioned, my interest in number theory started in high school and is closely related to my participation in math competitions. My PhD dissertation was from that area and contained published results in international journals. Nevertheless, I would say that my work in number theory was on a somewhat amateurish basis until, in July 1996, I participated at the 7th Conference on Fibonacci numbers, which was held in Graz, where I met Professor Attila Pethő. We immediately understood that we had a lot of common mathematical interests. He guided me from that moment with his advice on my scientific career. The results from two of our joint papers are described in Sect. 4.8. He is also "responsible" for suggesting that I study connections between Diophantine m-tuples and elliptic curves. This became a very fruitful research topic, as I hope this book shows.

I would like to thank all my coauthors, collaborators and PhD students. Most of my 13 PhD students had theses related to Diophantine m-tuples (and several of their PhD students, too). Their enthusiasm gave me a lot of inspiration during my research. Many joint results are included in some form in this book. I thank all colleagues who have read some versions of the manuscript of this book and suggested improvements to the text, in particular, Bill Allombert, Nicolae Bonciocat, Yann Bugeaud, Mihai Cipu, Jelena Dujella, Yasutsugu Fujita, Shubham Gupta, Seoyoung Kim, Matija Kazalicki, Franz Lemmermeyer, Takafumi Miyazaki, Filip Najman, Bartosz Naskrecki, Tomislav Pejković, Vinko Petričević, Ákos Pintér, Ivan Soldo, Gökhan Soydan, Maksym Voznyy and Gary Walsh. Nicolae Bonciocat, Mihai Cipu and Tomislav Pejković read all the chapters carefully and provided many very helpful comments and corrections. I would like to thank the referees of this book for their very useful comments, remarks and suggestions. I am grateful to Remi Lodh, Senior Editor for Mathematics Books at Springer Nature, for his constant support and encouragement during the writing process.

I would also like to thank the PhD students at the Croatian Doctoral Program in Mathematics who attended my lectures on the course *Diophantine m-tuples and elliptic curves* in 2021/2022. Because of the situation in that period, most of the activities on the course appeared online, but they regularly attended my online lectures (they are available (in Croatian) on my YouTube channel), solved exercises and gave interesting seminar talks in the frame of our Seminar on Number Theory and Algebra (partly online and partly at the Department of Mathematics in Zagreb).

I also thank my family for their patience, support and understanding while writing this book.

Novigrad and Zagreb, Croatia   Andrej Dujella
January 2024

Contents

Chapter 1
Introduction

1.1 Diophantus of Alexandria

The ancient Greek mathematician Diophantus of Alexandria (third century?) was the first to study the problem of finding numbers with the property that the product of any two of them increased by 1 gives a perfect square. In the fourth part of his book *Arithmetica* (English and Russian translations with comments can be found in [61, 212] and [28], respectively), Problem 20 reads:

> To find four numbers (for Diophantus, it meant positive rational numbers) such that the product of any two of them increased by 1 is a square.

Diophantus solved the problem, i.e. he found four positive rational numbers with the required property:

$$\left\{ \frac{1}{16}, \frac{33}{16}, \frac{17}{4}, \frac{105}{16} \right\}.$$

Indeed,

$$\frac{1}{16} \cdot \frac{33}{16} + 1 = \frac{289}{256} = \left(\frac{17}{16}\right)^2, \quad \frac{1}{16} \cdot \frac{17}{4} + 1 = \frac{81}{64} = \left(\frac{9}{8}\right)^2,$$

$$\frac{1}{16} \cdot \frac{105}{16} + 1 = \frac{361}{256} = \left(\frac{19}{16}\right)^2, \quad \frac{33}{16} \cdot \frac{17}{4} + 1 = \frac{625}{64} = \left(\frac{25}{8}\right)^2,$$

$$\frac{33}{16} \cdot \frac{105}{16} + 1 = \frac{3721}{256} = \left(\frac{61}{16}\right)^2, \quad \frac{17}{4} \cdot \frac{105}{16} + 1 = \frac{1849}{64} = \left(\frac{43}{8}\right)^2.$$

We will show how Diophantus came up with the solution (using today's mathematical notation). Two numbers with the required property can be easily obtained by taking $a = x$ and $b = x + 2$, so $ab + 1 = (x + 1)^2$. Each pair $\{a, b\}$ such that $ab + 1 = r^2$, can be extended to a triple by taking $c = a + b + 2r$. Indeed, we

© The Author(s), under exclusive license to Springer Nature Switzerland AG 2024
A. Dujella, *Diophantine m-tuples and Elliptic Curves*,
Developments in Mathematics 79, https://doi.org/10.1007/978-3-031-56724-7_1

have $ac+1 = a^2 + 2ar + ab + 1 = (a+r)^2$ and $bc+1 = (b+r)^2$. For $a = x$ and $b = x+2$, we get $c = 4x+4$. Now we apply the same construction to the pair $\{a, c\}$ and the equation $ac+1 = (2x+1)^2$. We get $d = x + (4x+4) + 2(2x+1) = 9x+6$. So we obtained the set $\{a, b, c, d\}$ that satisfies five of the six properties from the definition of Diophantine quadruple. The only missing condition is that $bd + 1$ be a square. Therefore, we need to find a rational solution to the equation

$$9x^2 + 24x + 13 = y^2. \tag{1.1}$$

Diophantus knew a method for solving equations of this type. He sought solutions in the form $y = 3x + t$, and after inserting it in the equation, he obtained a linear equation in x. He did not seek a general solution (probably due to a problem with notation), he would just insert some specific value for the parameter t and thus get one solution to the problem. In this case, he put $y = 3x - 4$ and got the linear equation $48x = 3$, whose solution is $x = 1/16$. In this way, he found the first example of a set that we now call a rational Diophantine quadruple.

More generally, by inserting $y = 3x + t$, we get $x = \frac{t^2 - 13}{6(4-t)}$ and an infinite family of rational Diophantine quadruples

$$\left\{ \frac{t^2 - 13}{6(4-t)}, \frac{(t-5)(t-7)}{6(4-t)}, \frac{2(t^2 - 6t + 11)}{3(4-t)}, \frac{3(t-1)(t-3)}{2(4-t)} \right\}.$$

Equation (1.1) is a special case of the equation of the form

$$\alpha^2 x^2 + \beta x + \gamma = y^2. \tag{1.2}$$

Diophantus knew a method for solving such equations so that the solution is sought in the form $y = \alpha x + t$. Equation (1.2) can be understood as an equation of a genus 0 curve with a rational point at infinity. By introducing the projective coordinates $x = X/Z, y = Y/Z$, we obtain the projective equation of the curve

$$\alpha^2 X^2 + \beta X Z + \gamma Z^2 = Y^2,$$

which has a rational point with coordinates $(X, Y, Z) = (1, \alpha, 0)$. The equation of a line through that point has the form $\alpha X - Y + t Z = 0$, which in affine coordinates gives exactly $y = \alpha x + t$.

Triples of the form $\{a, b, a + b + 2r\}$, where $ab + 1 = r^2$, are called *regular Diophantine triples*. An analogous construction of triples is also possible for a problem in which instead of products increased by 1, we consider the products increased by a fixed arbitrary rational number q. Namely, if $ab + q = r^2$, then $a(a + b + 2r) + q = (a + r)^2$ and $b(a + b + 2r) + q = (b + r)^2$, so the triple $\{a, b, a + b + 2r\}$ has the required property and is called a *regular $D(q)$-triple*. In Problems 3 and 4 in Part 5 of *Arithmetica*, Diophantus observed sets of four rational numbers with the property that the product of any two elements increased by q is a perfect square, i.e. rational $D(q)$-quadruples. He found one rational $D(5)$-quadruple

$$\left\{1, \frac{2861}{676}, \frac{7645}{676}, \frac{5084}{169}\right\},$$

and one rational $D(-6)$-quadruple

$$\left\{1, \frac{4993}{784}, \frac{6729}{784}, \frac{5665}{196}\right\},$$

but his construction actually gives a much more general result that for every rational number q there are infinitely many rational $D(q)$-quadruples. This result is particularly interesting because it is not known to date whether for every rational number q there exist infinitely many rational $D(q)$-quintuples (we will discuss this problem in Sect. 5.3). Let us take, as Diophantus, $a = 1$. Then we could take $b = x^2 - q, c = (x+1)^2 - q$, but for somewhat more aesthetically pleasing formulas, we will take (generalizing Diophantus' choice for $q = 5$) $b = (x + (q+1)/2)^2 - q$, $c = (x + 1 + (q + 1)/2)^2 - q$. Then $\{a, b, c\}$ is a regular $D(q)$-triple. Let us now extend the pair $\{b, c\}$ to the regular $D(q)$-triple $\{b, c, d\}$. We get

$$d = 4x^2 + 4xq + 8x + q^2 + 4.$$

It remains only to satisfy the condition that $ad + q$ is a square, which gives

$$4x^2 + (8 + 4q)x + q^2 + q + 4 = (2x + q + 2 + t)^2$$

(again, we introduced $q + 2$ only for more aesthetic formulas), and we get

$$x = -\frac{3q + 2qt + 4t + t^2}{4t}.$$

Inserting the obtained expression for x, gives an infinite family of $D(q)$-quadruples

$$\left\{1, \frac{9q^2 + 12qt - 10qt^2 + 4t^2 + 4t^3 + t^4}{16t^2},\right.$$

$$\left.\frac{9q^2 - 12qt - 10qt^2 + 4t^2 - 4t^3 + t^4}{16t^2}, \frac{(-9q + t^2)(-q + t^2)}{4t^2}\right\}.$$

(Here, t is an arbitrary non-zero rational number that satisfies the condition that elements of the resulting quadruple are distinct non-zero rational numbers.)

1.2　Pierre de Fermat

The first set of four positive integers with the property of Diophantus, the set

$$\{1,\ 3,\ 8,\ 120\},$$

was found by Pierre de Fermat, a French jurist and mathematician (1607–1665). Indeed,

$$1 \cdot 3 + 1 = 2^2, \quad 1 \cdot 120 + 1 = 11^2,$$
$$1 \cdot 8 + 1 = 3^2, \quad 3 \cdot 120 + 1 = 19^2,$$
$$3 \cdot 8 + 1 = 5^2, \quad 8 \cdot 120 + 1 = 31^2.$$

In his construction, Fermat started from the triple $\{1, 3, 8\}$ and tried to find a positive integer d such that $d + 1$, $3d + 1$ and $8d + 1$ are perfect squares. The first condition is satisfied for $d = y^2 + 2y$. The remaining two conditions are now

$$3(y^2 + 2y) + 1 = u^2,$$
$$8(y^2 + 2y) + 1 = v^2.$$

By subtraction, we get

$$5y(y + 2) = (v + u)(v - u),$$

and we are looking for a solution that satisfies $5y = v + u$, $y + 2 = v - u$, i.e. $v = 3y + 1$, $u = 2y - 1$. Now from

$$8(y^2 + 2y) + 1 = (3y + 1)^2,$$

we get $y = 10$, i.e. $d = 120$ (the second solution of the quadratic equation, $y = 0$, gives the trivial extension $d = 0$, which does not satisfy the condition from the definition that d is a positive integer).

Fermat applied this method to other problems leading to a system of equations of the form $ax + p^2 = y^2$, $bx + q^2 = u^2$, $cx + r^2 = v^2$. For example, if we start from the $D(-1)$-triple $\{1, 2, 5\}$ and ask if there exists a positive integer d such that $d + 1$, $2d + 1$ and $5d + 1$ are squares (such a set is said to have the property $D(-1; 1)$), by substituting $d = y^2 + 2y$, we get the system

$$2(y^2 + 2y) + 1 = u^2,$$
$$5(y^2 + 2y) + 1 = v^2.$$

By subtraction, we get

$$3y(y + 2) = (v + u)(v - u),$$

and we are looking for a solution that satisfies $3y = v + u$, $y + 2 = v - u$, i.e. $v = 2y + 1$, $u = y - 1$, so from

$$2(y^2 + 2y) + 1 = (y - 1)^2,$$

we get $y = -6$ and $d = 24$.

1.3 Leonhard Euler

The famous Swiss mathematician Leonhard Euler (1707–1783) gave numerous important contributions to the study of integer and rational Diophantine m-tuples and other sets with similar properties. We will mention some of his results here. Euler generalized Fermat's example of an (integer) Diophantine quadruple and showed that there are infinitely many such sets. More precisely, he showed that if a and b are positive integers such that $ab + 1 = r^2$, then the set

$$\{a,\, b,\, a + b + 2r,\, 4r(r + a)(r + b)\} \tag{1.3}$$

is a Diophantine quadruple.

As we have already seen, Diophantus knew how to extend the pair $\{a, b\}$ to a (regular) triple $\{a, b, a + b + 2r\}$. Let us show how Euler found an extension of this triple with the fourth (integer) element d. We want $ad + 1$, $bd + 1$ and $(a+b+2r)d+1$ to be squares, so their product should also be a square. By multiplying these three conditions, we get the equation

$$(ad + 1)(bd + 1)((a + b + 2r)d + 1) = y^2. \tag{1.4}$$

(In modern terminology, we could say that we obtained the equation of an elliptic curve in the variables d and y.) The idea is to look for a solution to Eq. (1.4) in the form $y = u + vd + wd^2$, where the rational numbers u, v, w are chosen so that both sides of (1.4) coincide in terms of small degrees, i.e. in the constant term and the terms with d and d^2. By comparison, we get: $u^2 = 1$, $2uv = 2(a + b + r)$, $2wu + v^2 = a^2 + 2br + 2ar + 2ab + b^2 + r^2 - 1$. Therefore, we can take $u = 1$, $v = a + b + r$, $w = -1/2$. By inserting this in (1.4), we get

$$ab(a + b + 2r) = -(a + b + r) + \frac{1}{4}d,$$

so

$$d = 4(a + b + r) + 4ab(a + b + 2r)$$
$$= 4(ab + 1)(a + b + r) + 4abr$$
$$= 4r^2(a + b + r) + 4abr$$
$$= 4r(r + a)(r + b).$$

Now direct computation shows that $ad + 1$, $bd + 1$ and $(a + b + 2r)d + 1$ are squares. Indeed,

$$ad + 1 = 4ar(r + a)(r + b) + 1 = (2r^2 + 2ar - 1)^2,$$
$$bd + 1 = 4br(r + a)(r + b) + 1 = (2r^2 + 2br - 1)^2,$$
$$(a + b + 2r)d + 1 = 4r(a + b + 2r)(r + a)(r + b) + 1 = (4r^2 + 2ar + 2br - 1)^2.$$

For $a = 1$ and $b = 3$, we get $a + b + 2r = 8$ and $4r(r + a)(r + b) = 120$, i.e. Fermat's quadruple $\{1, 3, 8, 120\}$.

Quadruples of the form (1.3) are a special case of the *regular Diophantine quadruples*, which are quadruples that satisfy the equation

$$(a + b - c - d)^2 = 4(ab + 1)(cd + 1). \tag{1.5}$$

Indeed, for $c = a + b + 2r$, $d = 4r(r + a)(r + b)$, we have

$$(a + b - c - d)^2 = (2r(1 + 2(r + a)(r + b)))^2$$
$$= 4r^2(4r^2 + 2ar + 2br - 1)^2 = 4(ab + 1)(cd + 1).$$

By writing (1.5) in the form $a^2 + b^2 + c^2 + d^2 - 2ab - 2ac - 2ad - 2bc - 2bd - 2cd - 4abcd - 4 = 0$, we see that it is a symmetric equation. Hence, regular quadruples satisfy also the equations

$$(a - b + c - d)^2 = 4(ac + 1)(bd + 1) \quad \text{and} \quad (a - b - c + d)^2 = 4(ad + 1)(bc + 1). \tag{1.6}$$

Euler found a fifth rational number that extends quadruple (1.3) to a rational Diophantine quintuple. In particular, he extended Fermat's quadruple $\{1, 3, 8, 120\}$ to a quintuple with the rational number $\frac{777480}{8288641}$. The question of whether this extension is unique remained open until 2019, when Stoll [339] proved the uniqueness of the extension. On the other hand, the question of the existence of rational Diophantine sextuples was open until 1999, when the first such sextuple was found by Gibbs [194, 195]. It was the sextuple

$$\left\{ \frac{11}{192}, \frac{35}{192}, \frac{155}{27}, \frac{512}{27}, \frac{1235}{48}, \frac{180873}{16} \right\}.$$

Let us now describe Euler's construction of rational Diophantine quintuples. Let $c = a+b+2r$, $d = 4r(r+a)(r+b)$. We want to find a rational number e such that $ae+1$, $be+1$, $ce+1$ and $de+1$ are squares of rational numbers. If we multiply these four expressions, we get

$$P(e) = 1 + \sigma_1 e + \sigma_2 e^2 + \sigma_3 e^3 + \sigma_4 e^4,$$

where σ_i are elementary symmetric polynomials in the variables a, b, c, d, i.e. $\sigma_1 = a+b+c+d$, $\sigma_2 = ab+ac+ad+bc+bd+cd$, $\sigma_3 = abc+abd+acd+bcd$, $\sigma_4 = abcd$. Let $P(e) = (1 + \frac{1}{2}\sigma_1 e + \rho e^2)^2$. By equating the terms with e^2, we get

$$\rho = \frac{1}{2}\sigma_2 - \frac{1}{8}\sigma_1^2. \tag{1.7}$$

Now we have $\sigma_3 + \sigma_4 e = \sigma_1 \rho + \rho^2 e$, so we get that

$$e = \frac{\sigma_3 - \sigma_1 \rho}{\rho^2 - \sigma_4}. \tag{1.8}$$

Let us show that σ_1, σ_2 and σ_4 satisfy the relation

$$4\sigma_2 = \sigma_1^2 - 4\sigma_4 - 4. \tag{1.9}$$

We calculate:

$$\sigma_1^2 - 4\sigma_2 - 4\sigma_4 - 4$$
$$= (a+b+c+d)^2 - 4(ab+ac+ad+bc+bd+cd) - 4abcd - 4$$
$$= (a+b-c-d)^2 - 4(ab+1)(cd+1) = 0,$$

where we used (1.5). If we insert (1.7) and (1.9) in (1.8), we get

$$e = \frac{4\sigma_3 + 2\sigma_1(\sigma_4 + 1)}{(\sigma_4 - 1)^2}. \tag{1.10}$$

Let us check that $ae + 1$ is a square (the other three conditions are checked analogously). It should be checked that $4\sigma_3 a + 2\sigma_1(\sigma_4 + 1)a + (\sigma_4 - 1)^2$ is a square. Let us now use that $P(-1/a) = 0$, i.e.

$$a^4 - \sigma_1 a^3 + \sigma_2 a^2 - \sigma_3 a + \sigma_4 = 0,$$

which gives $4\sigma_3 a = 4a^4 - 4\sigma_1 a^3 + 4\sigma_2 a^2 + 4\sigma_4$, so

$$4\sigma_3 a + 2\sigma_1(\sigma_4 + 1)a + (\sigma_4 - 1)^2$$
$$= 4a^4 - 4\sigma_1 a^3 + 4\sigma_2 a^2 + 4\sigma_4 + 2\sigma_1(\sigma_4 + 1)a + (\sigma_4 - 1)^2$$
$$= 4a^4 - 4\sigma_1 a^3 + 4\sigma_2 a^2 + 2\sigma_1(\sigma_4 + 1)a + (\sigma_4 + 1)^2$$
$$= (2a^2 - \sigma_1 a - (\sigma_4 + 1))^2,$$

where we also used relation (1.9).

In Sect. 3.1, we will give the interpretation of Euler's construction of quintuples in terms of elliptic curves. We will also generalize Euler's construction to arbitrary quadruples, i.e. not necessarily of the form (1.3).

Euler also studied sets with the property that the product of each of their two elements increased by the sum of these two elements gives a perfect square. Since

$$xy + x + y = (x + 1)(y + 1) - 1,$$

we see that such sets are closely related to $D(-1)$-m-tuples. Here, too, the motivation came from Diophantus, who, in Problem 15 of the third part of *Arithmetica*, found two triples with the given property: $\{4, 9, 28\}$ and $\{\frac{3}{10}, \frac{21}{5}, \frac{7}{10}\}$. Furthermore, Diophantus found in Problem 5 of the fifth part of *Arithmetica* a triple with the same property in which all elements are squares: $\{\frac{25}{9}, \frac{64}{9}, \frac{196}{9}\}$. This triple can be extended to a quadruple by the aforementioned method of Fermat, however Fermat did not explicitly implement his method to this case. The first explicit example of a rational quadruple with this property was given by Euler. It was the set

$$\left\{ \frac{65}{224}, \frac{9}{224}, \frac{9}{56}, \frac{5}{2} \right\}.$$

Euler used the above-mentioned connection with $D(-1)$-quadruples. Let $ab - 1 = r^2$. Then the set $\{a, b, a + b + 2r, a + b - 2r\}$ will be a $D(-1)$-quadruple if $(a + b + 2r)(a + b - 2r) - 1$ is a perfect square. So, we have

$$(a + b)^2 - 4r^2 - 1 = x^2.$$

From here, we have

$$(a - b)^2 = x^2 + 1 + 4r^2 - 4ab = x^2 - 3.$$

From $x^2 - 3 = (x - y)^2$, we get $x = \frac{y^2 + 3}{2y}$. Furthermore, from $4r^2 + 1 + x^2 = (2r + z)^2$, we obtain $r = \frac{x^2 + 1 - z^2}{4z}$. Now Euler took $y = 2$ and $z = \frac{7}{2}$, and he got $x = \frac{7}{4}$, $r = -\frac{131}{224}$, $a - b = \frac{1}{4}$, $a + b = \frac{261}{112}$, $a = \frac{289}{224}$, $b = \frac{233}{224}$, $c = \frac{65}{56}$, $d = \frac{7}{2}$.

From here, diminishing each element of the $D(-1)$-quadruple by 1 gives the above *Eulerian quadruple*.

It is known that there is no Eulerian quadruple in positive integers (Dujella and Fuchs proved in 2005 [124] that every $D(-1)$-quadruple in positive integers must contain the number 1), and also in integers (Bonciocat, Cipu, and Mignotte [44] recently proved that there is no integer $D(-1)$-quadruple). On the other hand, it is known that there are infinitely many rational Eulerian quintuples, e.g. $\{9, \frac{17}{8}, \frac{27}{10}, -\frac{27}{40}, \frac{493}{40}\}$ [97], $\{\frac{29}{24}, \frac{71}{54}, \frac{79}{675}, \frac{1637}{216}, \frac{2911}{200}\}$ [304].

The already mentioned book by Heath [212], apart from the translation of Diophantus' *Arithmetica*, contains also a supplement, which deals with Fermat's and Euler's comments on the work of Diophantus. The topics of Diophantine m-tuples can be found at [212, pp. 162–164, 177–182, 201–202, 344–349]. Another excellent source on the early history of this topic is the second volume of Dickson's *History of the Theory of Numbers* [83, pp. 513–520].

1.4 Definitions, Main Problems and Conjectures

The examples of sets we have given so far motivate the following definitions.

Definition 1.4.1 A set of m positive integers $\{a_1, a_2, \ldots, a_m\}$ is called a *Diophantine m-tuple* if $a_i \cdot a_j + 1$ is a perfect square for all $1 \le i < j \le m$.

For $m = 2, 3, 4, 5, 6, 7$ we usually speak of Diophantine pairs, triples, quadruples, quintuples, sextuples (or hextuples) and septuples, respectively.

Definition 1.4.2 A set of m non-zero rational numbers $\{a_1, a_2, \ldots, a_m\}$ is called a *rational Diophantine m-tuple* if $a_i \cdot a_j + 1$ is the square of a rational number for all $1 \le i < j \le m$.

The question naturally arises about how large these sets can be, in the integer and rational case. This question was recently completely solved in the integer case. On the other hand, in the rational case, we do not even have a generally accepted conjecture on what the answer should be. In particular, no upper bound for the size of rational Diophantine tuples is known.

In the integer case, for a long time, there was an open conjecture that a Diophantine quintuple does not exist, so-called "Diophantine quintuple conjecture".

The first significant result related to this conjecture was proved in 1969 by Baker and Davenport [24]. Using Baker's theory of linear forms in logarithms of algebraic numbers and a reduction method based on continued fractions, they proved that if d is a positive integer such that $\{1, 3, 8, d\}$ is a Diophantine quadruple, then necessarily $d = 120$. This result implies that the Fermat set $\{1, 3, 8, 120\}$ cannot be extended to a Diophantine quintuple. The same result was proven a few years later, using different methods, by Kanagasabapathy and Ponnudurai [227], and Sansone [313]. We will present the proof due to Baker and Davenport in Sect. 4.7.

It is usually stated in the literature that this problem was first posed in 1967 in Gardner's column in *Scientific American* [190], while the first non-trivial results regarding this problem were presented in 1968 by van Lint [349] at a conference in Oberwolfach. However, the problem appeared already in 1957 in the magazine *The Sunday Times*, in a column entitled *A holiday brain teaser* authored by Denton [79]. The numbers $1, 3, 8$ are given in the column, and a reward of 5 pounds is offered for finding the fourth positive integer which, with these three, makes a Diophantine quadruple and a prize of 25 pounds for finding the fifth positive integer which, with the previous four, makes a Diophantine quintuple. The first task is of course much easier, and 131 (mostly correct) solutions were received for it, while a concrete solution was not received for the second task (today we know that there is no solution), and in the second continuation of the column, we can find the solvers' thoughts on whether a possible solution is very large or does not exist at all.

In the already mentioned column in *Scientific American* from 1967, the famous proponent of science, especially mathematics, Martin Gardner, states that numbers $1, 3, 8$ and 120 have an interesting property that the product of any two of them is 1 less than a perfect square, and asks to find a fifth number that can be added to these four numbers so that this property is not violated. The same task is listed in his book *Mathematical Magic Show* from 1977 [191]. The number 0 is given as an answer, but with the comment that it is, of course, a trivial solution intended as a joke and that it is a challenging question whether there is a positive integer with such property. In the book, we can also find the chronology that led to the solution to this problem. One student of Bouwkamp, a professor from Eindhoven, saw the column in *Scientific American* and mentioned it to Bouwkamp, who mentioned it to his colleague van Lint. Van Lint was able to prove that if we want to replace 120 with a positive integer with the same property, that number must have more than $1,700,000$ digits. His proof used the properties of Pellian equations and continued fractions. He presented it at the conference in Oberwolhach in March 1968 and published it the same year in a publication by the Technological University Eindhoven [349]. This motivated Alan Baker and his mentor Harold Davenport to try to solve the problem completely by applying, at that time, the very new Baker's method, which allows to obtain explicit (usually very large) upper bounds for integer solutions for a wide class of Diophantine equations. For that method, Baker was awarded the Fields Medal in 1970. In order to reduce that very large bound to the size for which it could be verified that there are no additional solutions, they introduced the reduction method that is nowadays called the Baker-Davenport reduction.

The main ideas used by Baker and Davenport in their proof of the non-extendibility of the Fermat quadruple $\{1, 3, 8, 120\}$, more precisely, the uniqueness of the extension of the triple $\{1, 3, 8\}$ to a quadruple, were later generalized by other authors who also introduced some new methods that were necessary in studying extensions of parametric families of triples and quadruples (for example, the congruence method and effective estimates for simultaneous rational approximations of quadratic irrationals).

Finally, in 2019, He, Togbé, and Ziegler [211] proved that there is no Diophantine quintuple. In Sect. 4.11, we will say more about this result and some that preceded it.

Let us note that the result by Baker and Davenport says more than that the Fermat quadruple cannot be extended to a quintuple. Namely, it says that the only extension of the triple $\{1, 3, 8\}$ to a quadruple is with the number 120; in other words, the only quadruple containing the triple $\{1, 3, 8\}$ is the regular quadruple $\{1, 3, 8, 120\}$. This, and some analogous results mentioned below, motivated the following conjecture.

Conjecture 1.4.3 *All Diophantine quadruples are regular. In other words, if $\{a, b, c, d\}$ is a Diophantine quadruple, then $d = d_+$ or $d = d_-$, where*

$$d_+ = a + b + c + 2abc + 2\sqrt{(ab + 1)(ac + 1)(bc + 1)}, \tag{1.11}$$

$$d_- = a + b + c + 2abc - 2\sqrt{(ab + 1)(ac + 1)(bc + 1)}. \tag{1.12}$$

That every Diophantine triple $\{a, b, c\}$ can be extended to a quadruple with d_+ and d_- (if $d_- \neq 0$), was first noticed by Gibbs in 1978 [193] and by Arkin, Hoggatt, and Strauss in 1979 [13], and thus they generalized Euler's extension of regular triples to quadruples. Indeed, it is easy to see that

$$ad_+ + 1 = (a\sqrt{bc + 1} + \sqrt{(ab + 1)(ac + 1)})^2,$$

$$bd_+ + 1 = (b\sqrt{ac + 1} + \sqrt{(ab + 1)(bc + 1)})^2,$$

$$cd_+ + 1 = (c\sqrt{ab + 1} + \sqrt{(ac + 1)(bc + 1)})^2,$$

and that analogous equalities hold for d_-.

Let $a < b < c$. It is obvious that $d_+ > c$ and $d_- \geq 0$, while from

$$d_+ d_- = (a + b + c + 2abc)^2 - 4(ab + 1)(ac + 1)(bc + 1)$$

$$= a^2 + b^2 + c^2 - 2ab - 2ac - 2bc - 4 = (c - a - b)^2 - 4(ab + 1)$$

$$= (c - a - b - 2r)(c - a - b + 2r), \tag{1.13}$$

where $ab + 1 = r^2$, it follows that $d_- < c$ and that $d_- = 0$ if and only if $c = a + b \pm 2r$, i.e. if and only if the triple $\{a, b, c\}$ is regular (this explains why for the triple $\{1, 3, 8\}$ we had only one extension to the quadruple with 120, while, for example, with the triple $\{1, 3, 120\}$, we have (at least) two extensions: 8 and 1680). Therefore, we can express Conjecture 1.4.3 by saying that for each triple $\{a, b, c\}$ the only extension to a quadruple $\{a, b, c, d\}$ such that $d > \max\{a, b, c\}$ is given by $d = d_+$. Thus we can call it the *unique extension conjecture*. Apart from the result of Baker and Davenport on the triple $\{1, 3, 8\}$, the conjecture is verified for many other triples. For example, for the triple $\{2, 4, 12\}$ in [351], the triples $\{1, 3, 120\}$, $\{1, 8, 120\}$, $\{1, 8, 15\}$, $\{1, 15, 35\}$, $\{1, 24, 35\}$ and $\{2, 12, 24\}$ in [231], all triples of the form $\{k - 1, k + 1, 4k\}$ in [93] or $\{F_{2k}, F_{2k+2}, F_{2k+4}\}$ (where F_n denotes the nth Fibonacci number) in [98], and all triples containing the pair $\{a, 3a\}$ in [1] or the pair $\{k - 1, k + 1\}$ for $k \geq 2$ (see [152] for $k = 2$, [184] for general k, with the exception of the triples $\{k - 1, k + 1, 16k^3 - 4k\}$, which were handled in [50]). In

Chap. 4, we will give details for some of these results. However, Conjecture 1.4.3 is still open. In 2018, Fujita and Miyazaki [187] proved that any fixed Diophantine triple can be extended to a Diophantine quadruple in at most eleven ways by joining a fourth element exceeding the maximal element in the triple, while Cipu, Fujita, and Miyazaki [67] improved that result by replacing eleven with eight, and this is still the best-known result in this direction. Let us also mention a recent result by Cipu, Dujella, and Fujita [63] that for an arbitrary triple $\{b, c, d\}$, there are at most two extensions to a quadruple $\{a, b, c, d\}$ such that $a < \min(b, c, d)$.

The formulas (1.11) and (1.12) motivate the following question: if the quantity $(ab + 1)(ac + 1)(bc + 1)$ is a square, is $\{a, b, c\}$ necessarily a Diophantine triple? The affirmative answer to this question was given by Kedlaya [230], and we present it in the following proposition.

Proposition 1.4.4 *Let a, b, c be positive integers. Then $(ab + 1)(ac + 1)(bc + 1)$ is a perfect square if and only if $ab + 1$, $ac + 1$ and $bc + 1$ are all perfect squares.*

Proof If $ab + 1$, $ac + 1$ and $bc + 1$ are all perfect squares, then clearly, their product is also a perfect square. Suppose now that there exists a triple a, b, c of positive integers (we may assume that $a \le b \le c$) such that $(ab + 1)(ac + 1)(bc + 1)$ is a perfect square, but not all of $ab + 1$, $ac + 1$, $bc + 1$ are perfect squares. Among all such triples, choose a triple that minimizes $a + b + c$. Define d_- as in (1.12). We know from (1.13) that $d_- < c$. Since $cd_- + 1 \ge 0$, i.e. $d_- \ge -1/c > -1$ (we can exclude the possibility that $c = 1$ because in that case we have $a = b = c = 1$, in which case $(ab + 1)(ac + 1)(bc + 1)$ is not a perfect square), we have $d_- \ge 0$. However, if $d_- = 0$, then, by inserting $d = d_-$ in (1.5) and (1.6), we get that

$$4(ab + 1) = (a + b - c)^2, \quad 4(ac + 1) = (a + c - b)^2, \quad 4(bc + 1) = (b + c - a)^2,$$

contradicting the assumption that not all of $ab + 1$, $ac + 1$ and $bc + 1$ are perfect squares. Therefore d_- is a positive integer.

It remains to check that $(ab + 1)(ad_- + 1)(bd_- + 1)$ is a perfect square, but that not all of $ab + 1$, $ad_- + 1$, $bd_- + 1$ are perfect squares, which will contradict the minimality of $a + b + c$. By multiplying the two equations in (1.6), for $d = d_-$, we get

$$16(ac + 1)(bd_- + 1)(ad_- + 1)(bc + 1) = (a - b + c - d_-)^2(a - b - c + d_-)^2 \tag{1.14}$$

If we multiply both sides of (1.14) by $(ab + 1)(ac + 1)(bc + 1)$ (which is a perfect square by the assumption), we conclude that $(ab + 1)(bd_- + 1)(ad_- + 1)$ is a perfect square. Furthermore, from (1.6), we see that $bd_- + 1$ is a perfect square if and only if $ac + 1$ is a perfect square, while $ad_- + 1$ is a perfect square if and only if $bc + 1$ is a perfect square, so, by the assumption, not all of $ab + 1$, $ad_- + 1$ and $bd_- + 1$ are perfect squares. $\qquad \square$

Let $\{a, b, c\}$ be a Diophantine triple that we want to extend to a quadruple $\{a, b, c, x\}$. Then it should hold that $ax + 1$, $bx + 1$, $cx + 1$ are perfect squares. By multiplying these three conditions, we get the equation of an elliptic curve

$$y^2 = (ax + 1)(bx + 1)(cx + 1). \tag{1.15}$$

We say that this elliptic curve is induced by the triple $\{a, b, c\}$. There will be much more to say about these kinds of elliptic curves later, especially in Chap. 3. For now, let us just state the following conjecture, which is stronger than Conjecture 1.4.3.

Conjecture 1.4.5 *Let $\{a, b, c\}$ be a Diophantine triple. Then all integer solutions of Eq. (1.15) satisfy $x = 0$ or $x = d_+$ or $x = d_-$, and additionally $x = -1$ if $1 \in \{a, b, c\}$.*

The following question, related to Conjecture 1.4.3, appeared recently on several internet forums: Are there distinct positive integers $a_1, a_2, a_3, b_1, b_2, b_3$ such that $a_i b_j + 1$ is a perfect square for each $1 \leq i, j \leq 3$? An extensive computation indicates that the answer might be negative. The negative answer to this question would mean that a version of Conjecture 1.4.3 is valid for arbitrary (not necessarily Diophantine) triples. It will also directly imply the validity of Conjecture 1.4.3 for non-regular Diophantine triples $\{a_1, a_2, a_3\}$ because in that situation, we may take $b_1 = d_-$, $b_2 = d_+$ and the non-existence of b_3 would give the desired conclusion. Note, however, that the situation with Conjecture 1.4.5 is different. Namely, the equation $y^2 = (a_1 x + 1)(a_2 x + 1)(a_3 x + 1)$ may have more than two solutions in positive integers x if $\{a_1, a_2, a_3\}$ is not a Diophantine triple. For example, the equation $y^2 = (x + 1)(55x + 1)(276x + 1)$ has solutions $x = 4, 8, 17, 89, 4870844$ [304].

Let us now say something about the rational case. It has already been mentioned that Euler showed that there exist infinitely many rational Diophantine quintuples. The first examples of rational Diophantine sextuples were found by Gibbs in 1999 [194, 195]. In 2017, Dujella, Kazalicki, Mikić, and Szikszai [137] proved that there are infinitely many rational Diophantine sextuples. Moreover, they proved that there are infinitely many triples (for example, $\{\frac{15}{14}, -\frac{16}{21}, \frac{7}{6}\}$), which can each be extended to sextuples in infinitely many different ways (for example, the previous triple can be extended to the sextuple $\{\frac{15}{14}, -\frac{16}{21}, \frac{7}{6}, -\frac{1680}{3481}, -\frac{910}{1083}, \frac{624}{847}\}$). Furthermore, there are infinitely many rational Diophantine sextuples with positive elements and the same is true for any other choice of signs. This construction essentially uses the properties of elliptic curves induced by rational Diophantine triples and will be described in detail in Sect. 3.2.

In 2017, Dujella and Kazalicki [135], using the idea of Piezas [305], described an alternative construction of a parametric family of rational Diophantine sextuples. That construction also uses elliptic curves, but here in the so called Edwards form. The construction does not start from a triple but from a quadruple $\{a, b, c, d\}$. From the conditions $ab + 1 = t_{12}^2$, $ac + 1 = t_{13}^2$, $ad + 1 = t_{14}^2$, $bc + 1 = t_{23}^2$, $bd + 1 = t_{24}^2$, $cd + 1 = t_{34}^2$, rational points on the curve $(x^2 - 1)(y^2 - 1) = abcd$

are obtained, whose properties are further used in the construction of families of
sextuples $\{a, b, c, d, e, f\}$, the simplest of which being:

$$a = \frac{(t^2 - 2t - 1)(t^2 + 2t + 3)(3t^2 - 2t + 1)}{4t(t^2 - 1)(t^2 + 2t - 1)},$$

$$b = \frac{4t(t^2 - 1)(t^2 - 2t - 1)}{(t^2 + 2t - 1)^3},$$

$$c = \frac{4t(t^2 - 1)(t^2 + 2t - 1)}{(t^2 - 2t - 1)^3},$$

$$d = \frac{(t^2 + 2t - 1)(t^2 - 2t + 3)(3t^2 + 2t + 1)}{4t(t^2 - 1)(t^2 - 2t - 1)},$$

$$e = \frac{-t^5 + 14t^3 - t}{t^6 - 7t^4 + 7t^2 - 1},$$

$$f = \frac{3t^6 - 13t^4 + 13t^2 - 3}{4t(t^4 - 6t^2 + 1)}.$$

A similar construction was used in 2019 by Dujella, Kazalicki, and Petričević
[139] to prove the existence of infinitely many rational Diophantine sextuples in
which the denominators of all elements (in the lowest terms) are perfect squares.
This construction was motivated by the following experimentally found example

$$\left\{\frac{75}{8^2}, \; -\frac{3325}{64^2}, \; -\frac{12288}{125^2}, \; \frac{123}{10^2}, \; \frac{3498523}{2260^2}, \; \frac{698523}{2260^2}\right\}.$$

In 2021, the same authors [140] gave another construction of an infinite
family of rational Diophantine sextuples. The resulting sextuples have a special
structure, namely, they contain two regular quadruples and one regular quintuple (a
quintuple obtained by generalizing the previously described Euler's construction of
quintuples). The starting point in the construction was the following parametrization
of rational Diophantine triples found by Lasić [228]:

$$\left\{\frac{2t_1(1+t_1t_2(1+t_2t_3))}{(-1+t_1t_2t_3)(1+t_1t_2t_3)}, \; \frac{2t_2(1+t_2t_3(1+t_3t_1))}{(-1+t_1t_2t_3)(1+t_1t_2t_3)}, \; \frac{2t_3(1+t_3t_1(1+t_1t_2))}{(-1+t_1t_2t_3)(1+t_1t_2t_3)}\right\}.$$

Some details of the mentioned four constructions of infinite families of rational
Diophantine sextuples will be given in Sects. 3.2–3.5. Of course, the question arises
whether there is any rational Diophantine septuple (7-tuple). That question is still
open, and it is unclear what the answer should be.

There are known examples of "almost" septuples, i.e. sets of seven rationals that
lack only one condition in order to be rational Diophantine septuples. In other words,

there are rational Diophantine quintuples that can be extended to a sextuple in two different ways. For example, the quintuple

$$\left\{\frac{243}{560}, \frac{1147}{5040}, \frac{1100}{63}, \frac{7820}{567}, \frac{95}{112}\right\}$$

can be extended to a sextuple with $\frac{38269}{6480}$ and with $\frac{196}{45}$ (see [196]).

Here we can also mention that Herrmann, Pethő, and Zimmer [214] proved that every rational Diophantine quadruple can be extended in at most finitely many ways to a rational Diophantine quintuple. If we start from a quadruple $\{a, b, c, d\}$ that we want to extend to a quintuple, the first idea might be to look at the curve $y^2 = (ax+1)(bx+1)(cx+1)(dx+1)$. This is a curve of genus 1 (birationally equivalent to an elliptic curve), and it may have infinitely many rational points. But if we first express $e = (x^2 - 1)/a$ from the condition that $ae + 1 = x^2$, and then insert it in $(be+1)(ce+1)(de+1)$, we get the curve $y^2 = a(bx^2+a-b)(cx^2+a-c)(dx^2+a-d)$ of genus 2, which according to the Faltings theorem, has only finitely many rational points.

1.5 Generalizations of Diophantine m-Tuples

There are several natural generalizations of the notion of Diophantine m-tuples. We can replace the squares with k-th powers for some fixed $k \geq 3$. Such sets are called *k-th power Diophantine m-tuples* or *Diophantine m-tuples with the property $D_k(n)$*. Examples of Diophantine triples of the third and fourth power are $\{2, 171, 25326\}$ and $\{1352, 8539880, 9768370\}$, respectively. Bugeaud and Dujella [49] showed in 2003 that there do not exist k-degree Diophantine quadruples for $k \geq 177$ and obtained similar (although somewhat weaker) estimates for $k \leq 176$.

The problem in which the squares are replaced by an arbitrary power was also considered. Thus, here we consider sets that satisfy $a_i a_j + 1 = x^{k_{ij}}$, where $k_{ij} \geq 2$. Luca [259] proved in 2005, under the assumption that the abc-conjecture is valid, that the size of such a set is bounded by an absolute constant. The best known unconditional result was obtained by Stewart [335] in 2008 (see also [205]): if a set $D \subseteq \{1, 2, \ldots, N\}$ has the property that $ab + 1$ is a perfect power for all $a, b \in D$, $a \neq b$, then $|D| \ll (\log N)^{2/3}(\log \log N)^{1/3}$.

Several authors have considered versions of the problem in which the squares are replaced by terms of some recurrence sequence. Let us mention here that in 2008, Luca and Szalay [263] showed that there are no three distinct positive integers a, b, c such that $ab + 1$, $ac + 1$ and $bc + 1$ are all Fibonacci numbers ($F_0 = 0$, $F_1 = 1$, $F_n = F_{n-1} + F_{n-2}$ for $n > 2$), while the same authors in 2009 [264] proved that the only triple of positive integers $a < b < c$ with the property that $ab + 1$, $ac + 1$ and $bc + 1$ are all Lucas numbers ($L_0 = 2$, $L_1 = 1$, $L_n = L_{n-1} + L_{n-2}$ for $n \geq 2$) is the triple $(a, b, c) = (1, 2, 3)$. The analogous problems for more general recurrence sequences were studied in [10, 179–182, 221].

We can replace the number 1 in the condition "$ab + 1$ is a square" with another fixed integer n. Such sets are called $D(n)$-m-*tuples* or m-tuples with property $D(n)$. The case $n = 4$ has many similarities with the classical case $n = 1$. By multiplying all elements of a $D(1)$-m-tuple by 2, we get a $D(4)$-m-tuple. It is easy to see that in a $D(4)$-m-tuple at most two elements can be odd. Indeed, if we had three odd elements, say a_1, a_2, a_3, then from $a_i a_j + 4 \equiv 1 \pmod 8$, it would follow that $a_1 a_2 \equiv 5 \pmod 8$, $a_1 a_3 \equiv 5 \pmod 8$, $a_2 a_3 \equiv 5 \pmod 8$. By multiplying these three congruences, we get $(a_1 a_2 a_3)^2 \equiv 5 \pmod 8$, which is impossible. Thus, every result about the size of $D(1)$-m-tuples has a direct, but somewhat weaker, consequence on the size of $D(4)$-m-tuples. For example, the non-existence of a $D(1)$-quintuple implies the non-existence of a $D(4)$-septuple. However, with careful and technically demanding modification of arguments from the $n = 1$ case, Bliznac Trebješanin and Filipin [41] proved in 2019 that there is no $D(4)$-quintuple.

It is easy to show that there is no $D(n)$-quadruple if $n \equiv 2 \pmod 4$. This result was proved independently in 1985 by Brown [47], Gupta and Singh [200], and Mohanty and Ramasamy [280].

Theorem 1.5.1 *If $n \equiv 2 \pmod 4$, there is no $D(n)$-quadruple.*

Proof From the assumption that $a_i a_j + n$ is a square and the fact that squares of integers, when divided by 4, give remainders 0 or 1, it follows that $a_i a_j \equiv 2$ or 3 $\pmod 4$. We conclude that none of the numbers a_i is divisible by 4. So, we have four numbers, which give some of the three possible remainders: 1, 2, 3. Therefore two of them give the same remainder. Let us say that $a_1 \equiv a_2 \pmod 4$. Then $a_1 a_2 \equiv a_2^2 \equiv 0$ or 1 $\pmod 4$, which contradicts the previously shown $a_1 a_2 \equiv 2$ or 3 $\pmod 4$.

$\square$

On the other hand, it can be shown that if $n \not\equiv 2 \pmod 4$ and $n \notin S = \{-4, -3, -1, 3, 5, 8, 12, 20\}$, then there is at least one $D(n)$-quadruple. Indeed, an integer n that is not of the form $4k + 2$ has one of the following forms:

$$4k + 3, \quad 8k + 1, \quad 8k + 5, \quad 8k, \quad 16k + 4, \quad 16k + 12.$$

For each of these forms, by a method similar to the one used by Diophantus, a formula for a $D(n)$-quadruple was found by Dujella [88]. For example,

$$\{1, \ 9k^2 + 8k + 1, \ 9k^2 + 14k + 6, \ 36k^2 + 44k + 13\}$$

is a $D(4k+3)$-quadruple (we will give similar formulas for other cases in Sect. 5.1). Note that for $a = 9k^2 + 8k + 1$, $b = 9k^2 + 14k + 6$, we have $ab + (4k + 3) = r^2$, where $r = 9k^2 + 11k + 3$, and $a + b - 2r = 1$ and $a + b + 2r = 36k^2 + 44k + 13$, so we know that five conditions from the definition of a $D(4k + 3)$-quadruple are fulfilled. Nevertheless, the sixth condition is easy to check: $1 \cdot (36k^2 + 44k + 13) + (4k + 3) = (6k + 4)^2$. Exceptions from the set S appear because for some (small) values of k, it may happen that these quadruples have two equal elements.

In the case of $n = -1$, we have already mentioned regarding Euler's m-tuples that Dujella and Fuchs [124] proved that there is no $D(-1)$-quintuple, and together with Filipin [122] that there are at most finitely many $D(-1)$-quadruples. Recently, Bonciocat, Cipu, and Mignotte [44] proved the non-existence of $D(-1)$-quadruples. The proof is based on several new ideas and combines in an innovative way techniques proved successful in dealing with the problems of extensions of $D(1)$ and $D(-1)$-tuples (most of them will be explained in Chap. 4) with less usual tools, developed for the study of different problems. Note that this result implies the non-existence of $D(-4)$-quadruples since it is easy to prove that all elements of a $D(-4)$-quadruple must be even (see Proposition 5.1.3), so by dividing the elements of a $D(-4)$-quadruple by 2, we would get a $D(-1)$-quadruple. For $n \in S \setminus \{-4, -1\}$, the question of the existence of $D(n)$-quadruples is still open.

Clearly, if $n = m^2$ is a perfect square, then there are infinitely many $D(n)$-quadruples. Namely, the previously mentioned Euler's result shows that there are infinitely many $D(1)$-quadruples, and it is obvious that by multiplying all the elements of a $D(1)$-quadruple by m, we get a $D(m^2)$-quadruple. It is an open question of how many $D(n)$-quadruples are there in the case when n is not a perfect square and is not one of n for which we know that there is no $D(n)$-quadruple ($n \equiv 2$ (mod 4), $n = -1$, $n = -4$).

Conjecture 1.5.2 *If n is an integer that is not a perfect square, then there are at most finitely many $D(n)$-quadruples.*

We have seen that already Euler proved that there are infinitely many Diophantine quadruples. Martin and Sitar [266] proved that the number of Diophantine quadruples with elements $\leq N$ is asymptotically equal to $C\sqrt[3]{N}\log N$, where $C \approx 0.338285$. It is interesting to mention that the main contribution here comes from quadruples of Euler's form. Similarly, the main contribution to the number of Diophantine triples comes from regular Diophantine triples $\{a, b, a + b + 2r\}$. In [112], it was proved that this number is asymptotically equal to $\frac{3}{\pi^2}N\log N$ (see Exercise 18 in Sect. 4.12). Let $D_{m,n}(N) = |\{S \subseteq \{1, 2, \ldots, N\} : S \text{ is a } D(n) - m\text{-tuple}\}|$. It was conjectured in [111] that $D_{3,n}(N) \sim C(n)N\log(N)$ if n is a perfect square, while $D_{3,n}(N) \sim C(n)N$ otherwise. This conjecture has been verified recently for the case that n is prime in [3].

Let

$$M_n = \sup\{|\mathcal{S}| : \mathcal{S} \text{ has the property } D(n)\}$$

and

$$M = \sup\{M_n : n \in \mathbb{Z} \setminus \{0\}\}.$$

It is known that M_n is finite for every integer $n \neq 0$. The conjecture is that the number M is also finite, i.e. the numbers M_n are bounded from above by a constant independent of n. However, this is an open problem. It is known that M_p, for p prime, is bounded by a constant independent of p (Dujella and Luca [142] proved in 2005 that $M_p < 3 \cdot 2^{168}$, see Sect. 5.2.4), but for general n only results like $M_n \leq 31$ for $|n| \leq 400$, $M_n < 15.476 \log |n|$ for $|n| > 400$, $M_n < 2.6071 \log |n|$ for sufficiently large $|n|$ are known [29, 106, 108], see Sect. 5.2.

Combining the two mentioned generalizations of Diophantine m-tuples, we can get arbitrarily large sets. Namely, for every positive integer m, there is a positive integer n and a set of positive integers A such that $|A| \geq m$ and $ab + n$ is a perfect power for all $a, b \in A$, $a \neq b$. More precisely, for sufficiently large x, we can take $m = \lfloor \left(\frac{\log \log x}{2 \log \log \log x} \right)^{1/3} \rfloor$ and (using the Chinese remainder theorem) construct the set $A_m = \{a_1, \dots, a_m\} \subset [1, x]$ and a positive integer $n_m \in [1, x]$ such that $a_i a_j + n_m = x_{ij}^{k_{ij}}$ for $1 \leq i < j \leq m$, where x_{ij} are positive integers, and exponents k_{ij} are the first $\binom{m}{2}$ prime numbers (see [31]).

Let us mention that in [32, 311], so-called (F, m)-Diophantine sets were considered, i.e. sets of integers with the property that for any two distinct elements a and b, $F(a, b)$ is an m-th power, where F a bivariate polynomial with integer coefficients and $m \geq 2$ an integer. The choice $F(x, y) = xy + 1$ and $m = 2$ gives the classical case.

When studying $D(n)$-m-tuples, the integer n is usually fixed in advance. However, we can ask whether the same set can have the property $D(n)$ for several distinct values of n. This question was asked in 2001 by A. Kihel and O. Kihel [235]. For example, the set $\{8, 21, 55\}$ is a $D(1)$-triple and also a $D(4321)$-triple, while $\{1, 8, 120\}$ is both a $D(1)$-triple and a $D(721)$-triple. In 2018, Adžaga, Dujella, Kreso, and Tadić [4] proved that there are infinitely many triples $\{a, b, c\}$ which are simultaneously $D(1)$, $D(n_2)$ and $D(n_3)$ triples for $1 < n_2 < n_3$. One example of such an infinite family of triples is

$$a = 2(i + 1)i, \quad b = 2(i + 2)(i + 1), \quad c = 4(2i^2 + 4i + 1)(2i + 3)(2i + 1),$$

with

$$n_2 = 32i^4 + 128i^3 + 172i^2 + 88i + 16,$$

$$n_3 = 256i^8 + 2048i^7 + 6720i^6 + 11648i^5 + 11456i^4 + 6400i^3 + 1932i^2 + 280i + 16,$$

for an arbitrary positive integer i. The construction uses integer points on the elliptic curve

$$y^2 = (x + ab)(x + ac)(x + bc).$$

This curve is isomorphic to the above-mentioned curve $y^2 = (ax + 1)(bx + 1)(cx + 1)$ (via transformations $x \mapsto abcx$, $y \mapsto abcy$), so the rational points on one and the other curve are directly connected by these transformations. However, with integer points, there is no direct connection, and unlike the curve $y^2 = (ax + 1)(bx + 1)(cx + 1)$, the number of integer points on $y^2 = (x + ab)(x + ac)(x + bc)$ essentially depends on the rank of the elliptic curve.

It is known that there are sets that are simultaneously $D(n_1)$- and $D(n_2)$-quadruples for $n_1 \neq n_2$. For example, $\{27, 115, 160, 1755\}$ is a $D(-2016)$-quadruple and a $D(37296)$-quadruple, while $\{1458, 66248, 5000, 14112\}$ is a $D(16769025)$- and a $D(406425600)$-quadruple and also a $D(0)$-quadruple since all its elements are squares multiplied by 2. However, $n = 0$ is often excluded from the definition of $D(n)$-m-tuples because it is trivial to see that there are infinite sets with the $D(0)$-property—squares or squares multiplied by the same number, but when this condition is combined with other non-trivial conditions, interesting problems arise, so in such a context it makes sense also to allow $n = 0$ in the definition. Moreover, there are infinitely many quadruples with these properties (see [156, 157]). Recently, Dujella, Kazalicki, and Petričević [141] proved that there exist infinitely many $D(n)$-quintuples whose elements are squares (so they are also $D(0)$-quintuples). One such example is the $D(480480^2)$-quintuple $\{225^2, 286^2, 819^2, 1408^2, 2548^2\}$. Triples, quadruples and quintuples with property $D(n)$ for several n will be discussed in detail in Sect. 5.4.

The problem of Diophantus can be considered over any commutative ring with unity. Let us mention the results obtained by Franušić and Soldo [173, 174, 177, 333] over rings of integers in some quadratic fields, which indicate that there might be a close connection between the existence of $D(n)$-quadruples and representability of the element n as a difference of two squares in the observed ring. Notice that the integers $n \equiv 2 \pmod 4$ from Theorem 1.5.1 are exactly those integers which cannot be represented as a difference of two squares of integers. There are analogous results for certain cubic and quartic fields [175, 176]. However, in 2023, Chakraborty, Gupta, and Hoque [59] showed that in certain rings of the form $\mathbb{Z}[\sqrt{4k+2}]$, there are elements z which are not difference of two squares but there exists a $D(z)$-quadruple (explicit examples are given for $z = 26 + 6\sqrt{10}$ in $\mathbb{Z}[\sqrt{10}]$ and $z = 18 + 2\sqrt{58}$ in $\mathbb{Z}[\sqrt{58}]$). Let us also mention that Adžaga [2] proved that in the ring of integers in an imaginary quadratic field, there is no $D(1)$-m-tuple for $m \geq 43$, while Gupta [199] proved the analogous result for $D(-1)$-m-tuple for $m \geq 37$ (by the recent result due to Cipu and Fujita [66], there are no such m-tuples for $m \geq 25$).

A simple connection between the problem of the existence of a $D(n)$-quadruple and the representation of the number n as a difference of two squares in the observed ring appears if we allow two elements of the quadruple to be equal. Namely, if $n = k^2 - a^2$, then the four numbers $a, a, 2a + 2k, 5a + 4k$ have the property that

$$a \cdot a + n = k^2,$$

$$a \cdot (2a + 2k) + n = (a + k)^2,$$

$$a \cdot (5a + 4k) + n = (2a + k)^2,$$

$$(2a + 2k) \cdot (5a + 4k) + n = (3a + 3k)^2.$$

Of course, this does not solve the problem of the existence of $D(n)$-quadruples, but it gives a hint that for the numbers which can be represented as a difference of two squares, either there are $D(n)$-quadruples or the proof of their non-existence might be a difficult problem (because we will not be able to solve it by simply testing the congruences to some modulus as in the case of numbers of the form $4k + 2$ from Theorem 1.5.1).

Various versions of the problem of Diophantus have also been studied in polynomial rings. In 2004, Dujella and Fuchs [123] proved that every Diophantine quadruple in $\mathbb{Z}[x]$ is regular. Recently, Filipin and Jurasić [168] proved that the same statement also holds in $\mathbb{R}[x]$. On the other hand, Dujella and Jurasić [133] showed in 2010 that this statement is not valid in $\mathbb{C}[x]$ because for every $p \in \mathbb{C}[x]$,

$$\left\{ \frac{\sqrt{-3}}{2}, \ -\frac{2\sqrt{-3}}{3}(p^2 - 1), \ \frac{-3 + \sqrt{-3}}{3}p^2 + \frac{2\sqrt{-3}}{3}, \ \frac{3 + \sqrt{-3}}{3}p^2 + \frac{2\sqrt{-3}}{3} \right\}$$

is a Diophantine quadruple that is not regular.

Dujella and Luca [143] considered the higher power variant of the problem of Diophantus for polynomials. Let K be an algebraically closed field of characteristic 0. They proved that for every $k \geq 3$ there exists a constant $P(k)$, depending only on k, such that if $\{a_1, a_2, \ldots, a_m\}$ is a set of polynomials, not all of them constant, with coefficients in K, with the property that $a_i a_j + 1$ is a k-th power of an element of $K[X]$ for $1 \leq i < j \leq m$, then $m \leq P(k)$. More precisely, they proved that: $m \leq 5$ if $k = 3$, $m \leq 4$ if $k = 4$, $m \leq 3$ for $k \geq 5$, $m \leq 2$ for k even and $k \geq 8$. Furthermore, Dujella, Fuchs, and Luca [126] proved that $m \leq 10$ if $k = 2$. They also obtained an absolute upper bound for the size of a set of polynomials with the property that the product of any two elements plus 1 is a perfect power.

Notice that in the definition of a (rational) Diophantine m-tuple, we excluded the requirement that the product of an element with itself increased by 1 gives a square. It is obvious that this condition cannot be satisfied in integers (the equation $a^2 + 1 = r^2$ has no solutions in positive integers). But in rationals, there is no apparent reason why such sets (which we could call *strong Diophantine m-tuples*) would not exist. Each element a of such a set should satisfy that $a^2 + 1$ is a square, and therefore $a = X/Y$, where (X, Y, Z) is a Pythagorean triple, i.e. $X^2 + Y^2 = Z^2$. In 2008, Dujella and Petričević [154] proved that there are infinitely many strong rational Diophantine triples (for example,

$$\left\{ \frac{1976}{5607}, \ \frac{3780}{1691}, \ \frac{14596}{1197} \right\}$$

is one such triple). However, it is not known if there are any strong Diophantine quadruples. It is known that there are quadruples that lack only one condition to be strong Diophantine quadruples. For example, for

$$\left\{ \frac{140}{51}, \ \frac{2223}{30464}, \ \frac{278817}{33856}, \ \frac{3182740}{17661} \right\},$$

the only missing condition is that the product of the third and fourth elements increased by 1 is a square. Strong $D(-1)$-triples were also studied (see [130]), and more generally, strong $D(q)$-triples (see [146]), and it is known that for infinitely many square-free integers q, there are infinitely many strong $D(q)$-triples. For example,

$$\left\{1, \frac{5}{4}, \frac{14645}{484}\right\}$$

is a strong $D(-1)$-triple, while

$$\left\{\frac{7769}{1638}, \frac{38893009}{50902488}, \frac{50817649}{35348950}\right\}$$

is a strong $D(2)$-triple.

1.6 Exercises

1. Find a positive rational number x with the property that $\{x, 4x+4, 9x+6, 25x+20\}$ is a rational Diophantine quadruple.
2. Find two positive rational numbers a and b such that $a+1, b+1, a+b+1$ and $a-b+1$ are all squares (Problem 13 in the fourth part of *Arithmetica* by Diophantus).
3. Find three positive rational numbers such that the product of squares of any two of them increased by 1 is a square. (Problem 24 in the fifth part of *Arithmetica* by Diophantus).
4. Determine all values of a rational parameter t for which the set

$$\left\{\frac{t^2 - 13}{6(4 - t)}, \frac{(t-5)(t-7)}{6(4 - t)}, \frac{2(t^2 - 6t + 11)}{3(4 - t)}, \frac{3(t - 1)(t - 3)}{2(4 - t)}\right\}$$

   is a regular rational Diophantine quadruple.
5. Let $a = \frac{25}{9}, b = \frac{64}{9}, c = \frac{196}{9}$. Find a rational number $d \neq 0$ with the property that $ad + a + d, bd + b + d$ and $cd + c + d$ are squares of rational numbers.
6. Let $\{a, b, c\}$ be a regular Diophantine triple such that $a + b + c \equiv 0 \pmod 3$. Show that there exists a positive integer k such that the set $\{k - a, k - b, k - c\}$ is a Diophantine triple (see [81, Chapter 3]).
7. Find a positive integer d such that the set $\{1, 5, 10\}$ has the property $D(-1; 1)$.
8. Let $a = 8, b = 120, c = 190$. Find a positive integer d and a rational number $e \neq 0$ with the property that $\{a, b, c, d\}$ is a Diophantine quadruple, and $\{a, b, c, d, e\}$ is a rational Diophantine quintuple.
9. Let x_1, x_2, x_3 be rational numbers such that the denominator of

$$x_4 = \frac{8(x_3 - x_1 - x_2)(x_1 + x_3 - x_2)(x_2 + x_3 - x_1)}{(x_1^2 + x_2^2 + x_3^2 - 2x_1x_2 - 2x_1x_3 - 2x_2x_3)^2}$$

is non-zero. Prove that $x_1x_4 + 1$, $x_2x_4 + 1$ and $x_3x_4 + 1$ are squares of rational numbers. If $x_1 = F_{2n+1}$, $x_2 = F_{2n+3}$, $x_3 = F_{2n+5}$, prove that x_4 is a positive integer.

10. Find a non-zero rational function $f(t)$ with the property that

$$\{t - 1,\ t + 1,\ 4t,\ 16t^3 - 4t,\ f(t)\}$$

is a Diophantine quintuple in $\mathbb{Q}(t)$. (It is shown in [339] that $f(t)$ is unique.)

11. Find four positive integers with the property that the product of any two of them increased by 1 is a triangular number, i.e. number of the form $k(k + 1)/2$ for a positive integer k (see [169]).

12. Find all Diophantine quadruples with $\max\{a, b, c, d\} \leq 10^6$ and check that all of them are regular.

13. Find two Diophantine quadruples with the same largest element.

14. Let a, b, c be positive integers such that $(ab + 4)(ac + 4)(bc + 4)$ is a perfect square. Prove that $ab + 4$, $ac + 4$ and $bc + 4$ are all perfect squares.

15. Find three distinct positive integers a, b, c such that $(ab + 2)(ac + 2)(bc + 2)$ is a perfect square, but $ab + 2$ is not a perfect square.

16. Find five positive integers x such that $(x + 1)(84x + 1)(10333x + 1)$ is a perfect square.

17. Find integers $a_3 > a_2 > a_1 \geq 2$ with the property that the number $(a_1x + 1)(a_2x + 1)(a_3x + 1)$ is a perfect square for at least five positive integers x.

18. Let a, b, c be odd integers such that $\{a, b, c\}$ is a $D(n)$-triple for an integer n. Prove that then $a \equiv b \equiv c \pmod 4$.

19. Prove that for any positive integer $n \geq 2$, the set

$$\{1,\ n^4 - 3n^2 + 1,\ n^2(n^2 - 1),\ 4n^2(n^2 - 1)(n^2 - 2)\}$$

is a $D(n^2)$-quadruple.

20. For each $k \in \{4, 5, 6, 7, 8, 9, 10\}$, find positive integers $a < b < c < d < e$ with the property that $\{a, b, c, d\}$ and $\{a, b, c, e\}$ are $D(k^2)$-quadruples.

21. Prove that a number $a + bi \in \mathbb{Z}[i]$ cannot be represented as a difference of squares of two elements in $\mathbb{Z}[i]$ if and only if b is odd or $a \equiv b \equiv 2 \pmod 4$.

22. Let $\{a, b\}$ be a strong Diophantine pair with the property that $1 - ab$ is a perfect square. Show that $\left\{a, b, \frac{a+b}{1-ab}\right\}$ is a strong Diophantine triple.

Chapter 2
Elliptic Curves over the Rationals

2.1 Introduction to Elliptic Curves

Elliptic curves play an important role in several areas of mathematics (number theory, algebraic geometry, complex analysis), and recently they have become significant for applications in cryptography. An elliptic curve can be defined over an arbitrary field. In number theory, the most important case is the field of rational numbers $\mathbb{Q}$, while for cryptographic applications, the most important are finite fields. In problems related to (rational) Diophantine m-tuples elliptic curves over $\mathbb{Q}$ naturally appear, so in this overview of the properties of elliptic curves, we will mainly talk about them.

In this chapter, we provide the prerequisites on elliptic curves needed in the following chapter where we will discuss connections between Diophantine m-tuples and elliptic curves. Special emphasis will be given to the problems and algorithms related to the torsion group and rank. In the exposition, we will follow closely [116, Chapter 15]. For more systematical introductions to elliptic curves, we can recommend the textbooks [220, 239, 258, 330, 356] and more advanced books [328, 329], while the details on most of the presented algorithms can be found in [73, 74, 319]. We will not discuss applications of elliptic curves to cryptography, primality testing and factorization (except briefly mentioning the relevance of elliptic curves with large torsion and positive rank for the Elliptic curve factorization method), so we direct readers interested in such applications to [40, 208, 240].

Let K be a field. An *elliptic curve* over K is a non-singular projective cubic curve over K with at least one point over K. It has an (affine) equation of the form

$$F(x, y) = ax^3 + bx^2 y + cxy^2 + dy^3 + ex^2 + fxy + gy^2 + hx + iy + j = 0,$$

where the coefficients $a, b, c, \ldots, j$ are in K, and the non-singularity means that at every point on the curve, considered in the projective plane $\mathbb{P}^2(\overline{K})$ over an algebraic closure of K, at least one partial derivative of the function F is non-

© The Author(s), under exclusive license to Springer Nature Switzerland AG 2024   23
A. Dujella, *Diophantine m-tuples and Elliptic Curves*,
Developments in Mathematics 79, https://doi.org/10.1007/978-3-031-56724-7_2

zero. It can be shown that every such equation, by a birational transformation (rational transformation, the inverse of which is also a rational transformation), can be reduced to the form

$$y^2 + a_1 xy + a_3 y = x^3 + a_2 x^2 + a_4 x + a_6, \tag{2.1}$$

which is called the *(long) Weierstrass form* of an elliptic curve.

Furthermore, if the characteristic of the field K is different from 2 and 3 (so we can complete to a square and a cube by dividing by 2 and 3 if necessary), then this equation can be transformed into the form

$$y^2 = x^3 + ax + b, \tag{2.2}$$

which is called the *short Weierstrass form*. The condition of non-singularity now becomes that the cubic polynomial $f(x) = x^3 + ax + b$ does not have multiple roots (in the algebraic closure $\overline{K}$), and this is equivalent to the condition that the *discriminant* $\Delta = -16(4a^3 + 27b^2)$ is non-zero.

We often understand the set of K-rational points on an elliptic curve E over the field K (with characteristic different from 2 and 3), as the set of all points $(x, y) \in K \times K$ which satisfy the equation $y^2 = x^3 + ax + b$, where $a, b \in K$ and $4a^3 + 27b^2 \neq 0$, together with the "point at infinity" O. We denote this set by $E(K)$.

The point at infinity appears naturally if we consider the elliptic curve in the projective plane. The *projective plane* $\mathbb{P}^2(K)$ is obtained by introducing the equivalence relation $(X, Y, Z) \sim (kX, kY, kZ)$, for $k \in K \setminus \{0\}$, on the set $K^3 \setminus \{(0, 0, 0)\}$. If we introduce, in the (affine) equation of the elliptic curve, the substitutions $x = \frac{X}{Z}$, $y = \frac{Y}{Z}$, we get the projective equation

$$Y^2 Z = X^3 + aXZ^2 + bZ^3.$$

If $Z \neq 0$, then the equivalence class of (X, Y, Z) has a representative $(x, y, 1)$, so we can identify that class with (x, y). However, there is also one equivalence class which contains points for which $Z = 0$. It has a representative $(0, 1, 0)$, and we identify that class with the point at infinity O.

One of the most important properties of elliptic curves is that on the set $E(K)$, of their K-rational points, there is a natural way to define a binary operation for which the set of points $E(K)$ becomes an Abelian group. To explain this, let us assume that $K = \mathbb{R}$, the field of real numbers. Then the elliptic curve $E(\mathbb{R})$ (without the point at infinity) can be represented as a subset of the plane. The polynomial $f(x)$ can have either one (if $\Delta < 0$) or three (if $\Delta > 0$) real roots. Depending on this, the graph of the corresponding elliptic curve has one or two components, as shown in Figs. 2.1 and 2.2.

We will define the addition operation on $E(\mathbb{R})$. Let $P, Q \in E(\mathbb{R})$. Let us draw a line through the points P and Q. Counting multiplicity, it intersects the curve E in three points. We denote the third point by $P * Q$. Now define $P + Q$ to be the

Fig. 2.1 One component

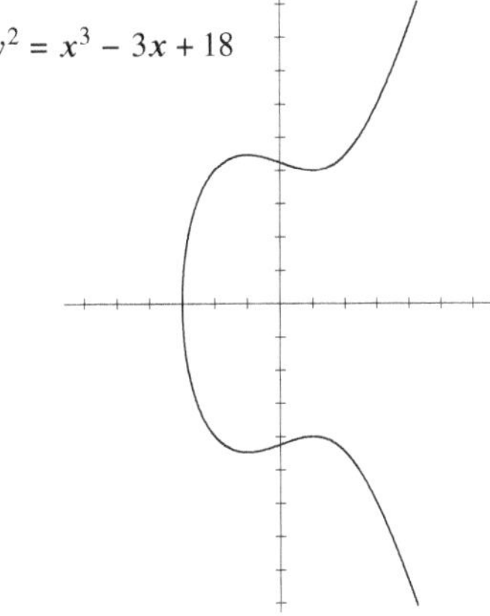

Fig. 2.2 Two components

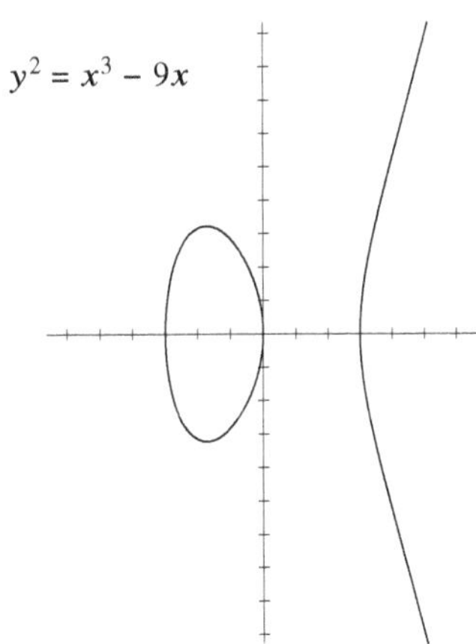

symmetric point of $P * Q$ with respect to the x-axis. If $P = Q$, then instead of the secant, we draw the tangent through the point P (see Figs. 2.3 and 2.4). If P is an inflexion point, then we take that $P * P = P$. Furthermore, we assume that the line intersects the curve at the point at infinity O if and only if it is perpendicular to the x-axis, so $P + O = O + P = P$ for every $P \in E(\mathbb{R})$.

Hence, the operation (addition) on the set $E(\mathbb{R})$ is introduced "geometrically", so that the sum of three distinct points on the curve is equal to the neutral element if and only if they are collinear. Of course, this geometric law can also be described by explicit formulas for coordinates of the sum of points. Formulas obtained in that way can then be used to define the addition of points on an elliptic curve over an arbitrary field (with a slight modification if the field characteristic is 2 or 3). Let us now state those formulas.

Let $P = (x_1, y_1)$, $Q = (x_2, y_2)$. Then

(1) $-O = O$;
(2) $-P = (x_1, -y_1)$;

Fig. 2.3 Secant line

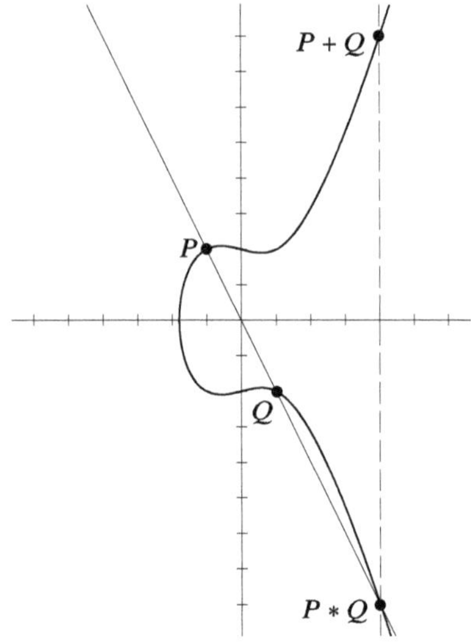

Fig. 2.4 Tangent line

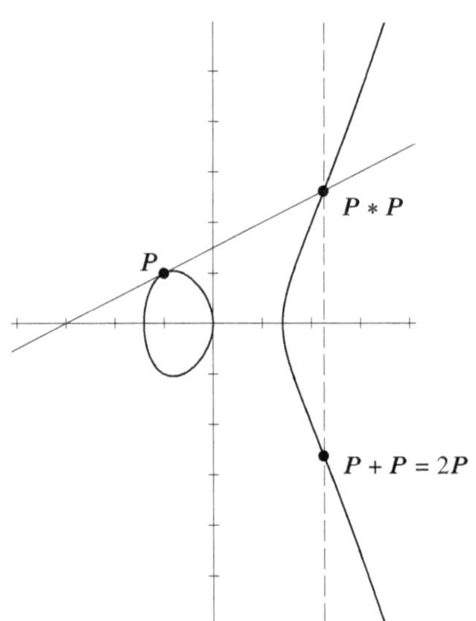

(3) $O + P = P$;

(4) if $Q = -P$, then $P + Q = O$;

(5) if $Q \neq -P$, then $P + Q = (x_3, y_3)$, where

$$x_3 = \lambda^2 - x_1 - x_2, \quad y_3 = -y_1 + \lambda(x_1 - x_3),$$

$$\lambda = \begin{cases} \frac{y_2 - y_1}{x_2 - x_1}, & \text{if } x_2 \neq x_1, \\ \frac{3x_1^2 + a}{2y_1}, & \text{if } x_2 = x_1. \end{cases}$$

The number λ is either the slope of the line through P and Q, or of the tangent line at the point P in the case $P = Q$. By inserting the equation of the line through the point P, i.e. $y = \lambda(x - x_1) + y_1$, in the equation of the elliptic curve, and by equating the coefficients of x^2 (Viète formula) in

$$x^3 + ax + b - (\lambda(x - x_1) + y_1)^2 = (x - x_1)(x - x_2)(x - x_3),$$

we get the above formula for the coordinate x_3.

It turns out that $(E(K), +)$ is an Abelian group. All properties of Abelian groups are evident, except for the associativity, whose proof is more complicated. One way to prove associativity is by listing all possible cases (depending on which of the cases from the definition of addition appears) and comparing the expressions obtained on the left and right-hand sides of $(P + Q) + R = P + (Q + R)$. There are also more elegant proofs (some of which we will mention shortly), but they require knowledge of some facts from complex analysis or projective geometry (the proof using Bézout's theorem on the number of intersection points of algebraic curves can be found in [279, 330]). Elementary proofs can be found in [178, 363].

We may ask where does the name "elliptic curve" come from. The name is clearly associated with an ellipse, but so far, we have not seen any connection between those two concepts. The connection between elliptic curves and ellipses comes through the problem of calculating the perimeter of an ellipse. Let an ellipse be given by the equation $q^2 x^2 + p^2 y^2 = p^2 q^2$. Then its perimeter is equal to the value of the integral

$$4 \int_0^1 \frac{p^2 - (p^2 - q^2)t^2}{\sqrt{(1 - t^2)(p^2 - (p^2 - q^2)t^2)}} \, dt.$$

By rational substitutions, this integral can be transformed into a similar integral where the square root of a cubic function appears. In general, integrals in which square roots of polynomials of the third or fourth degree (without multiple roots) appear are called *elliptic integrals*. They cannot be expressed in terms of elementary functions. However, they can be expressed in terms of the *Weierstrass $\wp$-function*.

This function satisfies the differential equation

$$\left(\frac{\wp'}{2}\right)^2 = \wp^3 + a\wp + b.$$

Here, its role is analogous to the role of the sine (or cosine) function in the calculation of integrals, where under the root appears a quadratic function. Namely, the function $y = \sin x$ satisfies the differential equation $y^2 + (y')^2 = 1$. Similarly to the unit circle, which can be parametrized by $(\cos t, \sin t)$, the complex points on the elliptic curve $y^2 = x^3 + ax + b$ can be parametrized by $(\wp(t), \frac{1}{2}\wp'(t))$. Moreover, it can be shown that if $P = (\wp(t), \frac{1}{2}\wp'(t))$ and $Q = (\wp(u), \frac{1}{2}\wp'(u))$, then $P + Q = (\wp(t + u), \frac{1}{2}\wp'(t + u))$. Thus, the addition of points on $E(\mathbb{C})$

corresponds to the addition of complex numbers. Knowing this fact provides an elegant proof of the associativity for addition of points on an elliptic curve.

When viewed over the field $\mathbb{R}$, the elliptic curve is indeed a "curve", i.e. a 1-dimensional object. However, observed over $\mathbb{C}$, it becomes a 2-dimensional object ("surface") in the 4-dimensional (real) space. Let us try to visualize that surface.

This is where the $\wp$-function can help us. It has many important properties. One of them is that it is doubly periodic, i.e. there are complex numbers ω_1 and ω_2 (such that $\omega_1/\omega_2 \notin \mathbb{R}$) with the property $\wp(z + m\omega_1 + n\omega_2) = \wp(z)$ for all integers m, n. (Here we have an analogy with the (single) periodicity of the trigonometric functions sine and cosine.)

Let us denote by L the "lattice" of all points of the form $m\omega_1 + n\omega_2$. The function $\wp$ is analytic in all points of the complex plane, except in the points from the lattice L in which there is a pole of the second order (i.e. $\wp$ is a meromorphic function). Namely,

$$\wp(z) = \frac{1}{z^2} + \sum_{w \in L,\, w \neq 0} \left(\frac{1}{(z-w)^2} - \frac{1}{w^2} \right).$$

In general, meromorphic doubly periodic functions are called *elliptic functions*.

The above parametrization of the points on the elliptic curve using the function $\wp$ is an isomorphism of the groups $E(\mathbb{C})$ and $\mathbb{C}/L$ (see [321, Chapter 7]). The function $\wp$ is completely determined by its values in the "fundamental parallelogram" which consists of all complex numbers of the form $\alpha\omega_1 + \beta\omega_2, 0 \leq \alpha, \beta < 1$.

The difference between points located opposite to each other on parallel sides of that parallelogram is an element from L. Therefore, these points are identified in the set $\mathbb{C}/L$. To visualize that set, we can imagine that we first "glue" two opposite sides of the parallelogram. That is how we get a cylinder. After that, we "glue" the bases of that cylinder, and that is how we get a torus.

We can imagine the torus as a sphere with a "hole". It turns out that plane algebraic curves can be identified in 3-dimensional space as spheres with finitely many holes. This number of holes can be understood as an informal definition of the *genus* of a curve. An alternative (more general) definition of an elliptic curve is that it is a non-singular projective algebraic curve of genus equal to 1 with at least one point defined over the base field. This definition includes not only non-singular cubic curves, but also all those curves that are birationally equivalent to them. Birational transformations preserve the genus of the curve, but they do not preserve its degree.

If the curve has degree n, then its genus is $\leq (n-1)(n-2)/2$, and if the curve is non-singular, then its genus is exactly equal to $(n-1)(n-2)/2$. It is known that the *hyperelliptic curves* with the equation $y^2 = f(x)$, where $f(x)$ is a polynomial of degree $n \geq 3$ without multiple roots, have genus $\lfloor (n-1)/2 \rfloor$ (see [336, Chapter 6]). In particular, this means that, apart from the case when $n = 3$, in the case when $n = 4$ we also have an elliptic curve if the curve has at least one rational point (unlike the case $n = 3$, where there is always a rational point (the point at infinity), for $n = 4$ such a point does not always have to exist).

Let us illustrate this observation with an example. Let C be the curve given by the equation

$$y^2 = x^4 + 3x^2 + 2x.$$

It has an obvious rational point $(0, 0)$. By substitutions $x = \frac{2}{v}$, $y = \frac{2t}{v^2}$ we get the curve $t^2 = v^3 + 3v^2 + 4$ and finally, by substitution $v + 1 = s$ we get elliptic curve E in the short Weierstrass form

$$t^2 = s^3 - 3s + 6.$$

The transformation that transfers C into E is $x = \frac{2}{s-1}$, $y = \frac{2t}{(s-1)^2}$. The inverse transformation is $s = \frac{x+2}{x}$, $t = \frac{2y}{x^2}$. Therefore, this is a birational transformation.

The genus of a curve plays an important role in the classification of Diophantine equations. Namely, the number of integer or rational solutions of the equation and the structure of the set of these solutions depends on it.

Curves of genus 0 (with at least one rational point) are exactly those that possess parametrization with rational functions. Every curve of the second degree (conic) has genus 0. For example, the curve $x^2 + y^2 = 1$ has a rational parametrization

$$x = \frac{1 - t^2}{1 + t^2}, \qquad y = \frac{2t}{1 + t^2},$$

which we can conclude from the formulas for Pythagorean triples (see, e.g. [116, Section 10.2]) and the obvious connections between integer solutions of the equation $X^2 + Y^2 = Z^2$ and rational solutions of the equation $x^2 + y^2 = 1$, so we put $t = n/m$ in $X = m^2 - n^2$, $Y = 2mn$, $Z = m^2 + n^2$, $x = X/Z$, $y = Y/Z$; or as in the examples below, by drawing a line with a rational slope t through the rational point $(-1, 0)$ on the curve.

Singular cubic curves also have genus 0. For example, the curve $y^2 = x^3$ has the singular point $(0, 0)$ (cusp, see Fig. 2.5). Therefore, this cubic curve is not elliptic. Its rational parametrization is $x = t^2$, $y = t^3$. As another example, let us consider the curve $y^2 = x^3 + 2x^2$. It also has the singular point $(0, 0)$ (node, see Fig. 2.6) and has a rational parametrization $x = t^2 - 2$, $y = t^3 - 2t$.

It is evident that these two cubic curves have infinitely many integer points. Pell's equation $x^2 - dy^2 = 1$ (d is a positive integer that is not a perfect square) is an example of a second degree curve that has infinitely many integer points (see Sect. 4.1.1). A curve of genus 1 can have only finitely many integer points. There can be an infinite number of rational points, but they are "finitely generated" (all points can be obtained from finitely many points by applying the group operation on the elliptic curve). A curve of genus greater than 1 can have only finitely many

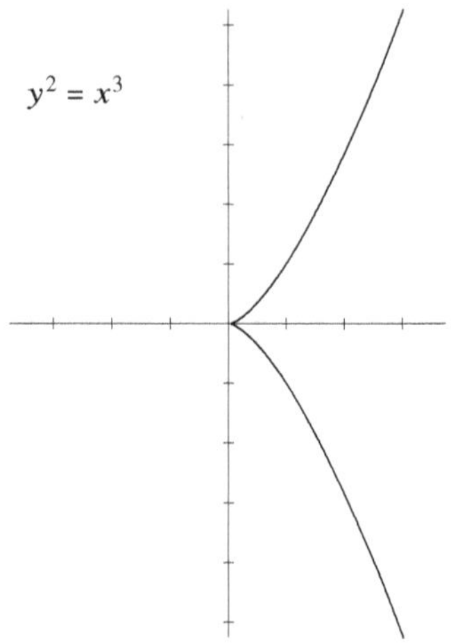

Fig. 2.5 Singular curve—cusp

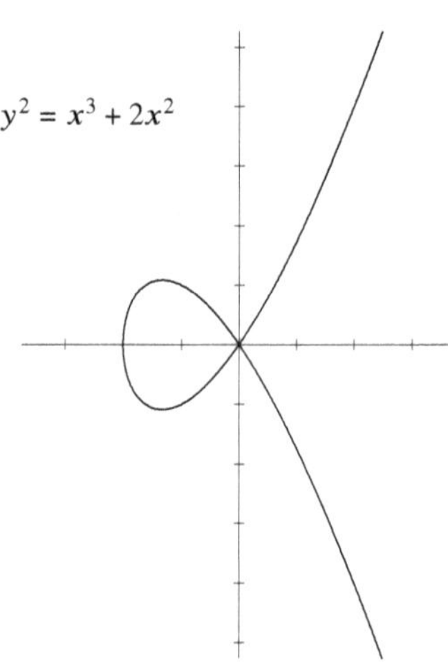

Fig. 2.6 Singular curve—node

rational points. This statement is the famous Mordell conjecture, which was proved by Faltings in 1983 (the proof can be found in [215]).

2.2   Equations of Elliptic Curves

Although the most interesting for number theory are elliptic curves over the field $\mathbb{Q}$, we will start with considerations that are valid in any field with characteristic different from 2 and 3 (for those characteristics, analogous results can be obtained with slight modifications). We will first be interested in how various equations of elliptic curves (and, more generally, curves of genus 1) can be transformed into the Weierstrass form.

Let K be a field of characteristic different from 2, and let us observe a cubic curve with coefficients from the field K that has at least one K-rational point (p, q). Following [73, Chapter 1.4], we will describe Nagell's algorithm by which the equation of the curve can be transformed into the Weierstrass form (or establish that the curve is singular, so it is not elliptic).

By replacing u by $u + p$ and v by $v + q$, we can assume that the K-rational point is exactly $(0, 0)$. So we have the equation

$$f(u, v) = s_1 u^3 + s_2 u^2 v + s_3 u v^2 + s_4 v^3 + s_5 u^2 + s_6 u v + s_7 v^2 + s_8 u + s_9 v = 0. \tag{2.3}$$

If $s_8 = s_9 = 0$, then $(0, 0)$ would be singular. Therefore (replacing u and v if necessary), we can assume that $s_9 \neq 0$. Let us introduce the projective coordinates: $u = \frac{U}{W}, v = \frac{V}{W}$ and write the equation in the form

$$F = F_3 + F_2 W + F_1 W^2 = 0,$$

where $F_i, i = 1, 2, 3$, are homogeneous polynomials of degree i:

$$F_3 = s_1 U^3 + s_2 U^2 V + s_3 U V^2 + s_4 V^3,$$

$$F_2 = s_5 U^2 + s_6 U V + s_7 V^2,$$

$$F_1 = s_8 U + s_9 V.$$

The rational point $P = (u, v) = (0, 0)$ now has coordinates $(U, V, W) = (0, 0, 1)$. The tangent line at the point P is given by the equation $F_1 = 0$ and intersects the curve at point $Q = (-e_2 s_9, e_2 s_8, e_3)$, where $e_i = F_i(s_9, \; s_8)$ for $i = 2, 3$. Note that e_2 and e_3 cannot both be 0, because then the tangent line would be a component of the curve, so we would not have an elliptic curve ($e_2 = 0$ means that $P = Q$ is the point of inflexion, while $e_3 = 0$ means that Q is the point at infinity). If $e_3 \neq 0$, we introduce the substitution

$$U = U' - \frac{s_9 e_2}{e_3} W', \quad V = V' + \frac{s_8 e_2}{e_3} W', \quad W = W',$$

and if $e_3 = 0$, then we introduce the substitution

$$U = U' - s_9 W', \quad V = V' + s_8 W', \quad W = U'.$$

In both cases, the point Q is at the origin $(U', V', W') = (0, 0, 1)$ of the new coordinate system, and the tangent line at the point P has the equation $s_8 U' + s_9 V' = 0$.

Now we can return to the affine coordinates $u' = \frac{U'}{W'}, v' = \frac{V'}{W'}$ because we only needed the projective coordinates to resolve the case when Q was a point at infinity in the original coordinates.

We write the equation in u' and v' as $f' = f'_1 + f'_2 + f'_3 = 0$, where $f'_i = f'_i(u', v')$ are homogeneous parts of f' of degree i. So we have

$$u'^2 f'_3(1, t) + u' f'_2(1, t) + f'_1(1, t) = 0, \tag{2.4}$$

where $t = \frac{v'}{u'}$. Equation (2.4) can be understood as a quadratic equation in u'. Its solutions are

$$u' = \frac{-\phi_2 \pm \sqrt{\delta}}{2\phi_3}, \tag{2.5}$$

where $\phi_i = f'_i(1, t)$ and $\delta = \phi_2^2 - 4\phi_1\phi_3$. The values of t for which $\delta = 0$ are the slopes of the tangent lines to the curve which pass through the point $Q = (0, 0)$ (because the lines through Q have the equation $v' = tu'$). One of these values is $t_0 = -\frac{s_8}{s_9}$. We see that δ is a fourth degree polynomial, which has one zero t_0. We put $t = t_0 + \frac{1}{\tau}$, so $\rho = \tau^4 \delta$ is a cubic polynomial in τ.

Finally, if

$$\rho = c\tau^3 + d\tau^2 + e\tau + k,$$

then it must be $c \neq 0$ (because for $c = 0$, we would not have an elliptic curve), so the substitutions $\tau = \frac{x}{c}$, $\rho = \frac{y^2}{c^2}$ give us the Weierstrass equation

$$y^2 = x^3 + dx^2 + cex + c^2 k.$$

The connection of the original variables u, v with x, y can be obtained via (2.5), where $t = t_0 + \frac{c}{x}$, $\delta = \frac{c^2 y^2}{x^4}$.

Example 2.2.1 Let us illustrate the construction just described on the example of the curve given by the equation $u^3 + v^3 = 1$, which is closely related to the Fermat equation for the exponent 3: $x^3 + y^3 = z^3$.

The curve $u^3 + v^3 - 1 = 0$ has an obvious rational point $(u, v) = (0, 1)$. To translate it to the point $(0, 0)$, we replace v by $v + 1$. Thus we get the equation $u^3 + (v + 1)^3 - 1 = 0$, i.e.

$$u^3 + v^3 + 3v^2 + 3v = 0.$$

Here $s_8 = 0$, $s_9 = 3 \neq 0$. Furthermore, we have $e_2 = 0$, $e_3 = 27$, so $P = Q$. With the substitution $v = tu$, we get the quadratic equation in terms of u

$$(t^3 + 1)u^2 + 3ut^2 + 3t = 0,$$

whose discriminant is $\delta = -3t^4 - 12t$. We see (and we know from the construction) that the obtained polynomial of the fourth degree has a rational zero $t_0 = 0$, so we introduce the substitution $t = 1/\tau$. Thus we get $\rho = -12\tau^3 - 3$. Finally, by substituting $\tau = -x/12$, we get the equation of the elliptic curve

$$y^2 = x^3 - 432.$$

Often elliptic curves are written in (long) Weierstrass form

$$y^2 + a_1 xy + a_3 y = x^3 + a_2 x^2 + a_4 x + a_6.$$

That form is "good" over any field (regardless of the characteristic), and it is very suitable for describing elliptic curves over $\mathbb{Q}$ with points of given finite order. Let us show how we can get a short Weierstrass form from it (in a field of characteristic different from 2 and 3). By substituting $y \mapsto \frac{1}{2}(y - a_1 x - a_3)$, we eliminate all the terms that contain y, except y^2 (we can do this if the characteristic is not 2, so we can divide by 2). We get

$$y^2 = 4x^3 + b_2 x^2 + 2b_4 x + b_6, \tag{2.6}$$

where

$$b_2 = a_1^2 + 4a_2,$$

$$b_4 = a_1 a_3 + 2a_4,$$

$$b_6 = a_3^2 + 4a_6.$$

We also define $b_8 = \frac{1}{4}(b_2 b_6 - b_4^2) = a_1^2 a_6 - a_1 a_3 a_4 + 4a_2 a_6 + a_2 a_3^2 - a_4^2$. We notice that if we assign to each a_i the "weight" i, then each of the b_i is a homogeneous expression of weight i; if we assign the weight 2 to x and the weight 3 to y, then all summands in (2.6) have weight 6.

If the characteristic of the field is different from 3, then we can introduce substitutions $x \mapsto \frac{x - 3b_2}{36}$, $y \mapsto \frac{y}{108}$ and get the equation in the short Weierstrass form

$$y^2 = x^3 - 27c_4 x - 54c_6,$$

where

$$c_4 = b_2^2 - 24b_4,$$

$$c_6 = -b_2^3 + 36b_2 b_4 - 216b_6.$$

In certain situations (for example, when constructing curves of high rank), we encounter curves of the form

$$v^2 = au^4 + bu^3 + cu^2 + du + e, \quad a \neq 0. \tag{2.7}$$

Let us assume that the polynomial on the right-hand side of (2.7) has no multiple roots and that the curve has at least one point (p, q) with coordinates from K. Then (2.7) can be transformed into the Weierstrass form by a birational transformation (with coefficients in K). By replacing u by $u + p$, we can assume that $p = 0$, i.e. that the K-rational point on (2.7) is $(0, q)$.

Let us first assume that $q = 0$. Then $e = 0$, so it must be $d \neq 0$ because otherwise $u = 0$ would be a multiple root. Multiplying (2.7) by $\frac{d^2}{u^4}$, we get

$$\left(\frac{dv}{u^2}\right)^2 = \left(\frac{d}{u}\right)^3 + c\left(\frac{d}{u}\right)^2 + bd\left(\frac{d}{u}\right) + ad^2,$$

i.e. the Weierstrass equation in $\frac{d}{u}$ and $\frac{dv}{u^2}$. The point $(0, 0)$ corresponds to the point at infinity O.

The case $q \neq 0$ is more complicated, but with a direct calculation, we can check that the following proposition holds (see [57, Chapter 5.8], [73, Chapter 1.2], [356, Chapter 2.5.3]).

Proposition 2.2.2 *Let K be a field of characteristic different from 2. Consider the curve given by the equation*

$$v^2 = au^4 + bu^3 + cu^2 + du + q^2,$$

where $a, b, c, d, q \in K$. Let

$$x = \frac{2q(v + q) + du}{u^2}, \quad y = \frac{4q^2(v + q) + 2q(du + cu^2) - \frac{d^2u^2}{2q}}{u^3},$$

and define

$$a_1 = \frac{d}{q}, \quad a_2 = c - \frac{d^2}{4q^2}, \quad a_3 = 2qb, \quad a_4 = -4q^2a, \quad a_6 = a_2a_4.$$

Then

$$y^2 + a_1xy + a_3y = x^3 + a_2x^2 + a_4x + a_6.$$

The inverse transformation is given by

$$u = \frac{2q(x + c) - \frac{d^2}{2q}}{y}, \quad v = -q + \frac{u(ux - d)}{2q}.$$

The point $(u, v) = (0, q)$ corresponds to the point at infinity O, and the point $(u, v) = (0, -q)$ corresponds to the point $(x, y) = (-a_2, a_1a_2 - a_3)$.

In 2007, Edwards [160] described a new interesting form of the equation of a curve birationally equivalent to an elliptic curve. A significant property of this form is that it allows unique formulas for adding points, i.e. they do not distinguish the cases $P + Q$ for $P \neq Q$ and $P + P$. Furthermore, it turned out (see [34]) that this form also offers some advantages in implementation (smaller number of multiplications in the field for calculating the sum of points). Therefore, *Edwards curves* and their variants have been a subject of great interest and research in the last fifteen years.

Here is also necessary to distinguish the case of characteristic 2. Therefore, we will give Edwards equation only for the case of a field with characteristic different from 2.

Proposition 2.2.3 *Let K be a field of characteristic different from* 2. *Let $c, d \in K \setminus \{0\}$ and assume d is not a square in K. Then the curve*

$$C : \quad u^2 + v^2 = c^2(1 + du^2v^2)$$

is birationally equivalent to the elliptic curve

$$E : \quad y^2 = (x - c^4d - 1)(x^2 - 4c^4d),$$

and the corresponding substitutions are given by

$$x = \frac{-2c(w - c)}{u^2}, \quad y = \frac{4c^2(w - c) + 2c(c^4d + 1)u^2}{u^3},$$

where $w = (c^2du^2 - 1)v$.

The point $(0, c)$ is the neutral element for the addition on the curve C. The opposite element of (u, v) is $-(u, v) = (-u, v)$, and the following formula gives the addition law

$$(u_1, v_1) + (u_2, v_2) = \left(\frac{u_1v_2 + u_2v_1}{c(1 + du_1u_2v_1v_2)}, \frac{v_1v_2 - u_1u_2}{c(1 - du_1u_2v_1v_2)} \right),$$

for all points $(u_i, v_i) \in C(K)$.

A typical shape of an Edwards curve can be seen in Fig. 2.7.

The equation of the curve C can be written in the form

$$u^2 - c^2 = (c^2du^2 - 1)v^2 = \frac{w^2}{c^2du^2 - 1},$$

or

$$w^2 = c^2du^4 - (c^4d + 1)u^2 + c^2.$$

Fig. 2.7 Edwards curve
$u^2 + v^2 = 9(1 - u^2 v^2)$

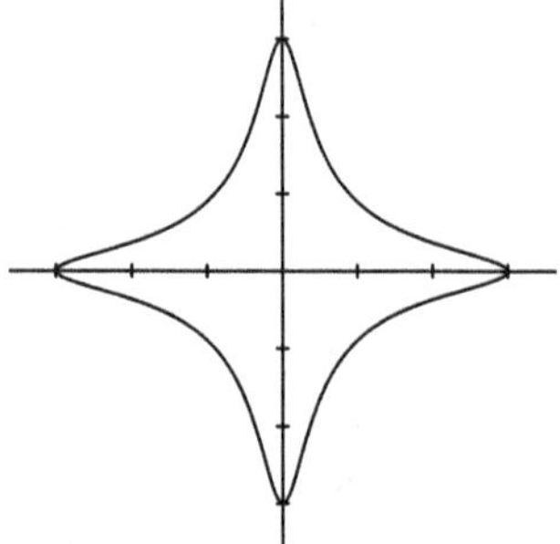

This quartic, with the rational point $(0, c)$, can now be transformed in the manner described above into the Weierstrass equation.

Let us check that the denominators in the addition formula are different from zero. Assume that $du_1 u_2 v_1 v_2 = -1$ (the case $du_1 u_2 v_1 v_2 = 1$ is similar). Then $u_1 v_1 = -\frac{1}{du_2 v_2}$ and by inserting it into the equation of C, we get

$$u_1^2 + v_1^2 = c^2\left(1 + \frac{1}{du_2^2 v_2^2}\right) = \frac{u_2^2 + v_2^2}{du_2^2 v_2^2}.$$

From here, it follows that

$$(u_1 + v_1)^2 = \frac{1}{d}\left(\frac{u_2^2 + v_2^2 - 2u_2 v_2}{u_2^2 v_2^2}\right) = \frac{1}{d}\frac{(u_2 - v_2)^2}{(u_2 v_2)^2}.$$

Since, by the assumption, d is not a square, it follows that $u_1 + v_1 = u_2 - v_2 = 0$. Analogously, from

$$(u_1 - v_1)^2 = \frac{1}{d}\frac{(u_2 + v_2)^2}{(u_2 v_2)^2},$$

it follows that $u_1 - v_1 = u_2 + v_2 = 0$. We get that $u_1 = v_1 = u_2 = v_2 = 0$, which is a contradiction with $du_1 u_2 v_1 v_2 = -1$.

Let us now return to the Weierstrass equation

$$y^2 + a_1 xy + a_3 y = x^3 + a_2 x^2 + a_4 x + a_6.$$

Using its coefficients a_1, a_2, a_3, a_4, a_6, we have defined the quantities b_2, b_4, b_6, b_8, c_4, c_6. Using them, we can now define two other important quantities:

- *discriminant* $\Delta = -b_2^2 b_8 - 8b_4^3 - 27b_6^2 + 9b_2 b_4 b_6 = \frac{c_4^3 - c_6^2}{1728}$,
- *j-invariant* $j = \frac{c_4^3}{\Delta}$.

A curve is non-singular if and only if $\Delta \neq 0$. The discriminant defined in this way equals $1/16 \times \mathrm{Disc}(4x^3 + b_2 x^2 + 2b_4 x + b_6)$. In general, the discriminant of a polynomial f of degree n with the leading coefficient a_n and roots $x_1, \ldots, x_n$ (from the algebraic closure $\overline{K}$) can be defined as

$$\mathrm{Disc}(f) = a_n^{2n-2} \prod_{i<j}^{n} (x_i - x_j)^2,$$

and it is equal to 0 if and only if f has multiple roots. If an elliptic curve is given by the equation $y^2 = x^3 + ax + b$, then $\Delta = -16(4a^3 + 27b^2)$. For curves defined over $\mathbb{R}$, the sign of the discriminant tells us how many components the graph of the curve has: if $\Delta < 0$, then there is one component and if $\Delta > 0$, then there are two components.

The name j-invariant comes from the fact that isomorphic curves (over $\overline{\mathbb{Q}}$) have the same j-invariants. The most general form of an isomorphism between two elliptic curves over $\mathbb{Q}$, given by the long Weierstrass forms, is

$$\begin{aligned}
x &= u^2 x' + r, \\
y &= u^3 y' + su^2 x' + t,
\end{aligned} \tag{2.8}$$

where $r, s, t \in \mathbb{Q}$ and $u \in \mathbb{Q}^*$. The effects of these substitutions on the coefficients a_i are

$$\begin{aligned}
ua_1' &= a_1 + 2s, \\
u^2 a_2' &= a_2 - sa_1 + 3r - s^2, \\
u^3 a_3' &= a_3 + ra_1 + 2t, \\
u^4 a_4' &= a_4 - sa_3 + 2ra_2 - (t + rs)a_1 + 3r^2 - 2st, \\
u^6 a_6' &= a_6 + ra_4 + r^2 a_2 + r^3 - ta_3 - t^2 - rta_1,
\end{aligned}$$

from which we obtain

$$u^4 c_4' = c_4, \quad u^6 c_6' = c_6, \quad u^{12} \Delta' = \Delta, \quad j' = j.$$

In the case of short Weierstrass forms, the only admissible substitutions are

$$x = u^2 x', \quad y = u^3 y', \quad u \in \mathbb{Q}^*.$$

Two elliptic curves are isomorphic over the algebraic closure $\overline{\mathbb{Q}}$ if and only if they have equal j-invariants (see [329, Chapter 3.1 and Appendix A]).

If we consider elliptic curves over an algebraically closed field, then the j-invariant tells us whether the curves are isomorphic. But if the field is not algebraically closed, for example, if we consider curves over $\mathbb{Q}$, then two curves may have equal j-invariants, while they may not be transformed one into the other by rational functions with coefficients from $\mathbb{Q}$. For example, the curves $y^2 = x^3 - 4x$ and $y^2 = x^3 - 25x$ both have $j = 1728$. The first has finitely many rational points, while the second has an infinite number of them (the point $(-4, 6)$ is of infinite order). Therefore, these curves are not isomorphic over $\mathbb{Q}$, but they are isomorphic over $\mathbb{Q}(\sqrt{10})$ (the substitutions are $(x, y) \mapsto (u^2 x, u^3 y)$, $u = \frac{\sqrt{10}}{2}$).

For $j \neq 0, 1728$, the curve

$$y^2 = x^3 - \frac{3j}{j - 1728}x + \frac{2j}{j - 1728}$$

has j-invariant equal to j. The curves of the form $y^2 = x^3 + b$ have j-invariant equal to 0, while the curves of the form $y^2 = x^3 + ax$ have j-invariant equal to $1728 = 12^3$.

Let E be an elliptic curve defined over $\mathbb{Q}$. By changing the variables, if necessary, we may assume that E is given by the equation

$$y^2 = x^3 + ax + b, \tag{2.9}$$

where $a, b \in \mathbb{Z}$ (we call such an equation an integral model for E). For a prime number $p > 3$, we can consider the equation $y^2 = x^3 + ax + b \bmod p$. If this equation defines an elliptic curve over the field $\mathbb{F}_p$, then we say that E has *good reduction* modulo p.

Let us notice that for the curve E, there are many choices for $a, b \in \mathbb{Z}$ in expression (2.9). In the definition of good (bad) reduction, we assume that a, b are chosen so that E has the "best possible" properties. So, for each p, we look for a, b with the property that the cubic polynomial $P(x) = x^3 + ax + b$ has as many distinct roots modulo p as possible, and that the discriminant $\Delta = -16(4a^3 + 27b^2)$ is divisible by the smallest power of p. We say that such an equation is minimal for p. It turns out that it is possible to choose a, b that have this property for all p. The corresponding equation is called the (global) *minimal Weierstrass equation* of E. To define analogous terms for $p = 2$ and $p = 3$, we need to look at the general (long) Weierstrass form.

For prime numbers with bad reduction, the cubic polynomial $P(x) = x^3 + ax + b$ has a multiple root modulo p. If the polynomial P has a triple root, we say that E has *additive reduction*, and if the polynomial P has a double root, then we say that E has *multiplicative reduction*. In addition, we distinguish between split and non-split multiplicative reduction. We have *split reduction* if the slope of the tangent line at the singular point is from $\mathbb{F}_p$; otherwise it is *non-split*. The latter can be determined looking at the equation of the curve written in the form $y^2 = x^2(x + c)$. The equations of the tangent lines at the singular point $(0, 0)$ are $y = \pm\sqrt{c}x$, so we see that we have multiplicative split reduction if and only if c is a square in $\mathbb{F}_p$ (i.e. if and only if c is a quadratic residue modulo p). Let us mention that the set of non-singular points has a group structure, and becomes isomorphic to the additive group (of order p) in the case of additive reduction, to the multiplicative group of the base field (of order $p - 1$) or to the kernel of the norm map from the quadratic extension of the base field (of order $p + 1$) in the two cases of multiplicative reduction (see [356, Section 2.10]).

Example 2.2.4 Consider the elliptic curve over $\mathbb{Q}$ given by the equation

$$y^2 = x^3 - 1037232x + 965662992.$$

Its discriminant is $\Delta = -2^{12}3^{12}7^{12}11$. We conclude that E has good reduction everywhere, except maybe at $p = 2, 3, 7, 11$. Also, we immediately see that bad reduction at $p = 11$ will not be possible to remove (because the discriminants of the isomorphic curves differ by a factor of the form u^{12}), while for $p = 2, 3, 7$, it might be possible. By substitutions

$$x = 7^2 x_1, \quad y = 7^3 y_1$$

we get the equation

$$y_1^2 = x_1^3 - 432x_1 + 8208,$$

The discriminant is $-2^{12}3^{12}11$, so E has good reduction at 7.

When considering the reduction at $p = 2$ and $p = 3$, we will have to give up the short Weierstrass form. For $p = 3$ we use substitutions of the form $x_1 = 3^2 x_2 + r$, $y_1 = 3^3 y_2$, and from the condition that the coefficients of the Weierstrass equation are integers, we get the condition $r \equiv -3 \pmod 9$. Let us take $r = -12$, and by substitutions

$$x_1 = 9x_2 - 12, \quad y_1 = 27y_2$$

we get the equation

$$y_2^2 = x_2^3 - 4x_2^2 + 16, \tag{2.10}$$

with discriminant equal to $-2^{12}11$, so E has good reduction at $p = 3$.

For $p = 2$, we use substitutions of the form $x_2 = 2^2 x_3$, $y_2 = 2^3 y_3 + t$, where from the condition that the coefficients of the Weierstrass equation are integers, we get that $t \equiv 4 \pmod 8$. Let us take $t = 4$. Then by substitutions

$$x_2 = 4x_3, \quad y_2 = 8y_3 + 4$$

we get the equation

$$y_3^2 + y_3 = x_3^3 - x_3^2. \tag{2.11}$$

This curve is non-singular for $p = 2$ because its discriminant is equal to -11 (another way to see it is by considering the partial derivative with respect to y_3, i.e. $2y_3 + 1 = 1$, in $\mathbb{F}_2$, which is always different from zero). Therefore, E has good reduction at $p = 2$.

We conclude that E has good reduction at all prime numbers, except for $p = 11$, where it has bad reduction. Equation (2.11) is the minimal Weierstrass equation of E.

Let us look at the situation for $p = 11$, using Eq. (2.10). In $\mathbb{F}_{11}$, we have

$$x_2^3 - 4x_2^2 + 16 = (x_2 + 1)^2(x_2 + 5),$$

so we see that E has multiplicative reduction at $p = 11$. The tangent lines at the singular point $(x_2, y_2) = (-1, 0)$ have slopes $\pm 2 \in \mathbb{F}_{11}$ (because the equation $\alpha^2 = 4$ has a solution in $\mathbb{F}_{11}$), so E has multiplicative split reduction at $p = 11$.

The global minimal equation of E has the property that its $|\Delta|$ is minimal among all integral models of E. In an isomorphism (2.8) between two minimal equations, we must have $u = \pm 1$, while $r, s, t \in \mathbb{Z}$. By choosing the parameters r, s, t, it is always possible to achieve that $a_1, a_3 \in \{0, 1\}$, $a_2 \in \{-1, 0, 1\}$. An equation that satisfies these conditions is called *reduced*. It is not difficult to see that the only transformation (except the identity map) that maps one reduced equation into another reduced equation is the transformation $(r, s, t, u) = (0, -a_1, -a_3, -1)$. It is the transformation that maps (x, y) into its inverse $-(x, y) = (x, -y - a_1 x - a_3)$ and does not change the equation. We conclude that every elliptic curve has a unique reduced minimal Weierstrass equation, and this fact enables very easy distinguishing of curves. Thus, in various tables with data on concrete elliptic curves, we will mostly find only data for the reduced minimal equations.

Let an elliptic curve be given by an equation with integer coefficients. If $v_p(\Delta) < 12$ or $v_p(c_4) < 4$ or $v_p(c_6) < 6$, then that equation is minimal for the prime number p. If $p \neq 2, 3$, then the converse also holds: if $v_p(\Delta) \geq 12$ and $v_p(c_4) \geq 4$, then the equation is not minimal for p. The situation with $p = 2$ and $p = 3$ is more complicated. There, the minimal equation for p can be calculated by Tate's algorithm. It is an algorithm that calculates minimal equations for each p, global minimal equation, conductor and some other data (see [74, Chapter 3.2]).

For an elliptic curve defined over $\mathbb{Q}$, we define a quantity associated with the discriminant called the *conductor*:

$$N = \prod_p p^{f_p}.$$

If $p \neq 2, 3$, then f_p can be easily determined from the minimal Weierstrass model for E:

- $f_p = 0$ if $p \nmid \Delta$;
- $f_p = 1$ if $p \mid \Delta$ and $p \nmid c_4$;
- $f_p \geq 2$ if $p \mid \Delta$ and $p \mid c_4$; if $p \neq 2, 3$, then $f_p = 2$.

In various tables and databases of elliptic curves over $\mathbb{Q}$, e.g. [75, 256], mostly only the reduced minimal Weierstrass equation curves are listed, while curves are usually sorted according to their conductors. The smallest N for which there are elliptic curves over $\mathbb{Q}$ with the conductor equal to N is $N = 11$. The following curves have conductor 11:

$$y^2 + y = x^3 - x^2,$$

$$y^2 + y = x^3 - x^2 - 10x - 20,$$

$$y^2 + y = x^3 - x^2 - 7820x - 263580.$$

Their discriminants are -11, -11^5, -11, respectively. The first two have points of order 5, while the third has no non-trivial rational points.

The next possibilities for the conductor are $N = 14, 15, 17, 19, 20, 21, 24, \ldots$ All curves with a conductor less than 37 have only finitely many rational points (they have rank equal to 0). One of the curves with conductor 37,

$$y^2 + y = x^3 - x,$$

has a point of infinite order $(0, 0)$ (and rank equal to 1).

2.3 Elliptic Curves in the Software Package PARI/GP

In the software package PARI/GP [301], many important functions related to elliptic curves have been implemented. We will explain only some of them here, while we will mention others at the places in the book where they naturally appear (the list of all functions related to elliptic curves can be obtained by typing `?12`).

We assume that the given curve is in the Weierstrass form $y^2 + a_1 xy + a_3 y = x^3 + a_2 x^2 + a_4 x + a_6$, and in PARI, it is represented as a five-component vector $e = [a_1, a_2, a_3, a_4, a_6]$. Points on E are represented as two-component vectors $[x, y]$, except for the point at infinity, which is represented as the one-component vector $[0]$.

Before applying any other functions, we "initialize" an elliptic curve using the function `ellinit`.

$E = \texttt{ellinit}(e)$: computes the following data for an elliptic curve over $\mathbb{Q}$:

$$a_1, \ a_2, \ a_3, \ a_4, \ a_6, \ b_2, \ b_4, \ b_6, \ b_8, \ c_4, \ c_6, \ \Delta, \ j.$$

E.g. the discriminant of E can be obtained as $E[12]$ or $E.\texttt{disc}$, while the j-invariant is $E[13]$ or $E.\texttt{j}$. The coefficients c_4 and c_6 are obtained as $E.\texttt{c4}$ and $E.\texttt{c6}$.

Additional data depends on the field over which the curve is defined. For curves over the field $\mathbb{C}$, the following data are also available:

- E.omega[1] is the real, and E.omega[2] is the complex period of E. In other words, $\omega_1 = E$.omega[1] and $\omega_2 = E$.omega[2] form a basis of the complex lattice for E.
- E.area is the area of the fundamental parallelogram of E.

Let us now list some of the functions related to elliptic curves available in PARI. We will mention more functions later, for example, those for computing the torsion group and rank.

elladd$(E, P1, P2)$: computes the sum of the points $P1$ and $P2$ on the elliptic curve E.

ellsub$(E, P1, P2)$: computes the difference $P1 - P2$ of points on the elliptic curve E.

ellmul(E, P, n): computes the multiple nP of the point P on the elliptic curve E.

ellordinate(E, x): gives a 0, 1 or 2-component vector containing the y-coordinates of the points on the elliptic curve E having x as x-coordinate.

ellisoncurve(E, P): gives 1 (i.e. "true") if P is a point on E, and 0 (i.e. "false") otherwise.

ellchangecurve(E, v): gives the elliptic curve obtained from E with substitutions determined by the vector $v = [u, r, s, t]$, i.e. the connection of the old coordinates x, y and the new x', y' is given by $x = u^2 x' + r$, $y = u^3 y' + s u^2 x' + t$.

ellminimalmodel$(E, \{\&v\})$: gives the reduced minimal model for an elliptic curve over $\mathbb{Q}$. To the optional variable v the vector $[u, r, s, t]$ is assigned, which gives the corresponding change of variables so that the curve obtained by this function is identical to that of ellchangecurve(E, v).

ellglobalred(E): computes the conductor, the global minimal model of E and the global Tamagawa number c. The result of this function is a five-component vector $[N, v, c, F, L]$, where N is the conductor, v gives the change of variables which transforms E to the minimal integral model (ellminimalmodel), while c is the product of local Tamagawa numbers c_p, a quantity which appears in the explicit version of the Birch and Swinnerton-Dyer conjecture, F is the factorization of N, while L represents data from local reductions modulo prime factors of N ($L[i]$ is elllocalred$(E, F[i, 1])$).

ellwp$(E, \{z = x\})$: computes the value at z of the Weierstrass function $\wp$ attached to the elliptic curve E (given by ellinit or as the lattice $[\omega_1, \omega_2]$).

ellpointtoz(E, P): computes the complex number t (modulo the lattice determined by E), which corresponds to the point P (its parameter), i.e. $\wp(t) = P[1]$, $\wp'(t) = P[2]$.

ellztopoint(E, z): computes the coordinates $[x, y]$ of the point on the elliptic curve E corresponding to the complex number z. Hence this is the inverse function of ellpointtoz. The point $[x, y]$ represents the value of the Weierstrass $\wp$-function and its derivative at the point z. If z is a point of the lattice which defines E over $\mathbb{C}$, then the result of this function is the point at infinity $[0]$.

Example 2.3.1 We will illustrate the use of some functions from PARI on the elliptic curve from Example 2.11.

```
? E=ellinit([0,0,0,-1037232,965662992])
%1 = [0, 0, 0, -1037232, 965662992, 0, -2074464, 3862651968,
-1075850221824, 49787136, -834332825088, -331424164353036496896,
-4096/11]
? factor(E.disc)
%2 = [-1, 1; 2, 12; 3, 12; 7, 12; 11, 1]
? F=ellminimalmodel(E, &v);
? F[1..5]
%3 = [0, -1, 1, 0, 0]
? v
%4 = [42, -588, 0, 37044]
? ellordinate(F,0)
%5 = [0, -1]
? P = [0, 0]; for(n=2,5,print(ellmul(F,P,n)))
[1, -1] [1, 0] [0, -1] [0]
```

Example 2.3.2 In this example, we will take

$$y^2 + y = x^3 - x,$$

the curve with the smallest conductor which has infinitely many rational points.

```
? E=ellinit([0,0,1,-1,0])
%1 = [0, 0, 1, -1, 0, 0, -2, 1, -1, 48, -216, 37, 110592/37]
? P = [0,0]
%2 = [0,0]
? ellisoncurve(E,P)
%3 = 1
? for(n=2,12,print(ellmul(E,P,n)))
[1, 0] [-1, -1] [2, -3] [1/4, -5/8] [6, 14] [-5/9, 8/27]
[21/25, -69/125] [-20/49, -435/343] [161/16, -2065/64]
[116/529, -3612/12167] [1357/841, 28888/24389]
? t1 = ellpointtoz(E,P)
%4 = 0.92959271528539567744051993445 + 1.2256946909933950304271124l6*I
? t2 = ellpointtoz(E,[2, -3])
%5 = 0.72491221490962306778878739B4
? t2 - 4*t1
%6 = -2.9934586462319596298320099B0 - 4.9027787639735B0121708449664*I
? E.omega[1]
%7 = 2.9934586462319596298320099B0
? 2*E.omega[2]
-4.9027787639735B0121708449664*I
? G=ellglobalred(E)
%8 = [37, [1, 0, 0, 0], 1, Mat([37, 1]), [[1, 5, 0, 1]]]
? cond = G[1]
%9 = 37
```

Example 2.3.3 The algorithm for transforming a general cubic equation with a rational point into the (long) Weierstrass form is implemented in PARI, and called via the function `ellfromeqn`. Let us illustrate its application on the curve from Example 2.2.1. To make sure that the elliptic curve we get by applying this function is equivalent to the elliptic curve obtained in Example 2.2.1, we will compare their reduced minimal equations. The function `ellfromeqn` can also be applied to other forms of curves of genus 1, such as Edwards curves and hyperelliptic curves of degree 4 (in general, it gives a Weierstrass model for the Jacobian of a genus 1 curve, see, e.g. [14, 171]).

```
? ellfromeqn(x^3+y^3-1)
%1 = [0, 0, -9, 0, -27]
? e1=ellinit([0, 0, -9, 0, -27]);
? e2=ellinit([0, 0, 0, 0, -432]);
? f1=ellminimalmodel(e1);
? f1[1..5]
%2 = [0, 0, 1, 0, -7]
? f2=ellminimalmodel(e2);
? f2[1..5]
%3 = [0, 0, 1, 0, -7]
? ellfromeqn(u^2+v^2-9*(1-u^2*v^2))
%4 = [0, 80, 0, 324, 25920]
? ellfromeqn(y^2-(x+1)*(3*x+1)*(8*x+1)*(120*x+1))
%5 = [0, 1475, 0, 546048, 51031296]
```

2.4 Torsion Group

The most important fact about elliptic curves over $\mathbb{Q}$ is the Mordell-Weil theorem.

Theorem 2.4.1 (Mordell-Weil) *The group $E(\mathbb{Q})$ is a finitely generated Abelian group.*

This theorem was proved in 1922 by the American-born British mathematician Louis Joel Mordell (1888–1972), while in 1928, the French mathematician André Weil (1906–1998) generalized it to Abelian varieties over number fields. In other words, the Mordell-Weil theorem tells us that there exists a finite set of rational points $\{P_1, \ldots, P_k\}$ on E from which all other rational points on E can be obtained by drawing secant and tangent lines. As every finitely generated Abelian group is isomorphic to a product of cyclic groups (more precisely, to a product of the form $\mathbb{Z}^n \times \mathbb{Z}_{k_1} \times \cdots \times \mathbb{Z}_{k_m}$ with $k_1 \mid k_2 \mid \cdots \mid k_m$, where $\mathbb{Z}_k$ denotes the quotient ring $\mathbb{Z}/k\mathbb{Z}$; [219, Chapter 2.2]), we get the following immediate consequence of the Mordell-Weil theorem.

Corollary 2.4.2

$$E(\mathbb{Q}) \cong E(\mathbb{Q})_{\text{tors}} \times \mathbb{Z}^r$$

The subgroup $E(\mathbb{Q})_{\text{tors}}$ of $E(\mathbb{Q})$, which consists of all points of finite order, is called the *torsion group* of E, and the non-negative integer r is called the *rank* of E and is denoted by rank(E) (more precisely, rank$(E(\mathbb{Q}))$). The corollary tells us that there are r rational points $P_1, \ldots, P_r$ of infinite order on the curve E with the property that every rational point P on E can be written in the form

$$P = T + m_1 P_1 + \cdots + m_r P_r,$$

where T is a point of finite order, and $m_1, \ldots, m_r$ are integers. Here $m_1 P_1$, for a positive integer m_1, denotes the sum $P_1 + \cdots + P_1$ of m_1 summands, which is often denoted by $[m_1]P_1$, while $[-m_1]P_1 = -[m_1]P_1$.

The question naturally arises as to what values $E(\mathbb{Q})_{\text{tors}}$ and $\text{rank}(E)$ can take. Furthermore, the question is how to compute them for a specific curve E. It turns out that it is much easier to answer these questions for the torsion group than for the rank.

Let us consider for a moment the points of finite order over $\mathbb{C}$ and $\mathbb{R}$. We said that an elliptic curve over $\mathbb{C}$ can be identified with the quotient group $\mathbb{C}/L$, where $L = \{m_1\omega_1 + m_2\omega_2 : m_1, m_2 \in \mathbb{Z}\}$. Therefore, $nP = O$ if and only if the parameter of P (z from the fundamental parallelogram such that $\wp(z) = x(P)$) is of the form $\frac{m_1}{n}\omega_1 + \frac{m_2}{n}\omega_2$, $0 \leq m_1, m_2 < n$. Thus, the solutions of the equation $nP = O$ form a group isomorphic to $\mathbb{Z}/n\mathbb{Z} \times \mathbb{Z}/n\mathbb{Z}$.

In the case of a curve with real coefficients, the lattice L has a base in which one of the periods, say ω_1, is real, and in the case when $\Delta > 0$ (i.e. the graph of E has two components), the second period, ω_2, is purely imaginary (thus the fundamental parallelogram is actually a rectangle). For the case when $\Delta < 0$, a detailed description can be found in [345, Chapter 3.4]. Points from $E(\mathbb{R})$ correspond to parameters $t \in [0, \omega_1)$ (the lower base of the parallelogram) and, in case that the graph of E has two components, $t - \frac{1}{2}\omega_2 \in [0, \omega_1)$ (the midsegment of the rectangle parallel to the base). Therefore, the group $E(\mathbb{R})$ is isomorphic to the circle group S^1 (when $\Delta < 0$) or $\mathbb{Z}/2\mathbb{Z} \times S^1$ (when $\Delta > 0$). The solutions of the equation $nP = O$ form the group isomorphic to $\mathbb{Z}/n\mathbb{Z}$ or $\mathbb{Z}/2\mathbb{Z} \times \mathbb{Z}/n\mathbb{Z}$.

Let us now return to curves over $\mathbb{Q}$. From what we have said, it follows that the group $E(\mathbb{Q})_{\text{tors}}$ should be a finite subgroup of S^1 or $\mathbb{Z}/2\mathbb{Z} \times S^1$. It is known that all finite subgroups of S^1 are cyclic. Therefore, $E(\mathbb{Q})_{\text{tors}}$ is isomorphic to one of the groups of the form $\mathbb{Z}/k\mathbb{Z}$ or $\mathbb{Z}/2\mathbb{Z} \times \mathbb{Z}/2k\mathbb{Z}$ (note that if k is odd, then we have $\mathbb{Z}/2\mathbb{Z} \times \mathbb{Z}/k\mathbb{Z} \cong \mathbb{Z}/2k\mathbb{Z}$).

In 1977, Mazur [270, 271] proved that there are exactly 15 possible torsion groups for elliptic curves over $\mathbb{Q}$. These groups are:

$$\mathbb{Z}/k\mathbb{Z}, \quad \text{for } k = 1, 2, 3, 4, 5, 6, 7, 8, 9, 10, 12,$$

$$\mathbb{Z}/2\mathbb{Z} \times \mathbb{Z}/2k\mathbb{Z}, \quad \text{for } k = 1, 2, 3, 4.$$

Points of order 2 on the curve $y^2 = x^3 + ax^2 + bx + c$ are precisely those with y-coordinate equal to 0. We can have 0, 1 or 3 such points, which depends on the number of rational roots of the polynomial $x^3 + ax^2 + bx + c$. Those points, together with the point O, form a subgroup of $E(\mathbb{Q})_{\text{tors}}$ which is either trivial or isomorphic to $\mathbb{Z}/2\mathbb{Z}$ or isomorphic to $\mathbb{Z}/2\mathbb{Z} \times \mathbb{Z}/2\mathbb{Z}$. We can find other points of finite order with the help of the Lutz-Nagell theorem. The idea is to find a curve model where all torsion points will have integer coordinates. This is precisely the equation $y^2 = x^3 + ax^2 + bx + c$ which we get from the Weierstrass equation (2.1) by eliminating the terms with xy and y (by substitutions (2.8) with $u = 2$ if necessary). If $a_1 = a_3 = 0$, then the Weierstrass equation already has the desired form; otherwise, we put $a = b_2$, $b = 8b_4$, $c = 16b_6$. Then, in order to obtain an estimate for y, use the fact that, for a torsion point $P = (x, y)$, both P and $2P$ have integer coordinates. The Lutz-Nagell theorem is often stated for an equation of the form $y^2 = x^3 + ax + b$, which is not a loss of generality because the term with x^2 can be eliminated by completing to a perfect cube. However, this elimination includes additional scaling of coordinates, and the result is (unnecessarily) larger estimate for y. Note that for a curve with the general Weierstrass equation, for a point $P(x, y)$ of finite order, $4x$ and $8y$ are integers.

Theorem 2.4.3 (Lutz-Nagell) *Let E be an elliptic curve given by the equation*

$$y^2 = f(x) = x^3 + ax^2 + bx + c, \qquad (2.12)$$

where $a, b, c \in \mathbb{Z}$. If $P = (x_1, y_1)$ is a point of finite order in $E(\mathbb{Q})$, then $x_1, y_1 \in \mathbb{Z}$.

A proof of the Lutz-Nagell theorem can be found in [356, Chapter 8.1]. The theorem was named after the French mathematician Élisabeth Lutz (1914–2008) and the Norwegian mathematician Trygve Nagell (1895–1988), who proved it independently in the 1930s.

Proposition 2.4.4 *Let E be an elliptic curve given by Eq. (2.12), where $a, b, c \in \mathbb{Z}$. If $P = (x_1, y_1)$ is a point of finite order in $E(\mathbb{Q})$, then either $y_1 = 0$ or $y_1^2 \mid \Delta_0$, where $\Delta_0 = -\Delta/16 = 27c^2 + 4a^3c + 4b^3 - a^2b^2 - 18abc$.*

Proof If $2P = O$, then $P = -P = (x_1, -y_1)$, so $y_1 = 0$. Otherwise, $2P = (x_2, y_2)$, where, according to the Lutz-Nagell theorem, $x_2, y_2 \in \mathbb{Z}$. From the addition formula on E, we have $2x_1 + x_2 = \lambda^2 - a$, where $\lambda = \frac{f'(x_1)}{2y_1}$ is the slope of the tangent line to E at the point P. We see that $\lambda \in \mathbb{Z}$, which implies that $y_1 \mid f'(x_1)$. Now from the formula

$$\Delta_0 = (-27f(x) + 54c + 4a^3 - 18ab)f(x) + (f'(x) + 3b - a^2)f'(x)^2 \qquad (2.13)$$

and $y_1^2 = f(x_1)$, it follows that $y_1^2 \mid \Delta_0$. Formula (2.13) can be obtained by applying the (extended) Euclidean algorithm to the polynomials $f(x)$ and $(f'(x))^2$. $\qquad \square$

The Lutz-Nagell theorem gives us a finite list of candidates for torsion points. More precisely, it gives us candidates for the y-coordinates of the points. For a given y, it is not difficult to find integer solutions to the equation $x^3 + ax^2 + bx + c - y^2 = 0$ (either by examining the factors of $y^2 - c$ or through Cardan's formula for solving cubic equations). If P is a torsion point, then for every positive integer n, the point nP must be either O or one of the points from the list. Since the list is finite, either we will get that $nP = mP$ for some $m \neq n$, in which case $(n-m)P = O$ and the point P is a torsion point, or some multiple nP will be out of the list, showing that P is not a torsion point. Alternatively, we can use Mazur's theorem, according to which the order of each torsion point is ≤ 12. Therefore, if $nP \neq O$ for all $n \leq 12$, then P is not a torsion point.

Let us assume that we have found all the torsion points and, after that, we want to determine the structure of the torsion group. According to Mazur's theorem, the only cases when the order of the group does not completely determine the structure of the group are cases $|E(\mathbb{Q})_{\mathrm{tors}}| = 4$, 8 and 12, when we have two possibilities: $\mathbb{Z}/4k\mathbb{Z}$ or $\mathbb{Z}/2\mathbb{Z} \times \mathbb{Z}/2k\mathbb{Z}$, $k = 1, 2, 3$. If we have exactly one point of order 2, then $E(\mathbb{Q})_{\mathrm{tors}} \cong \mathbb{Z}/4k\mathbb{Z}$, and if we have three points of order two, then $E(\mathbb{Q})_{\mathrm{tors}} \cong \mathbb{Z}/2\mathbb{Z} \times \mathbb{Z}/2k\mathbb{Z}$.

Example 2.4.5 Let us determine the torsion group of the elliptic curve

$$E : \quad y^2 = (-x + 4)(3x + 4)(4x + 4)$$

induced by the $D(4)$-triple $\{-1, 3, 4\}$.

Let us first transform the curve into the form from the Lutz-Nagell theorem (multiply the equation by $(-3)^2$, replace x by $x/(-3)$ and y by $2y/(-3)$):

$$y^2 = x^3 + 5x^2 - 72x + 144 = (x - 3)(x - 4)(x + 12).$$

Here $\Delta_0 = -57600$. If $y = 0$, then $x = 3, 4$ or -12, so we have three points $(3, 0)$, $(4, 0)$ and $(-12, 0)$ of order 2. If $y \neq 0$, then $y^2 \mid 576000$, i.e. $y \mid 240$. By testing all possibilities, e.g. using this code in PARI:

```
fordiv(240,y,f=factor(x^3+5*x^2-72*x+144-y^2);
if(length(f~)>1,print(y," ",f)))
```

we find the following points with integer coordinates: $P_1 = (0, 12)$, $P_2 = (8, 20)$, $-P_1 = (0, -12)$, $-P_2 = (8, -20)$ (corresponding to $y = 12$ and $y = 20$). Calculating multiples, we get $2P_1 = (4, 0)$, $2P_2 = (4, 0)$. Thus, the points $\pm P_1$ and $\pm P_2$ are points of order 4. Since we have a total of 8 points of finite order (together with the point at infinity), of which 3 are points of order 2, we conclude that

$$E(\mathbb{Q})_{\mathrm{tors}} = \left\{ O, (-1, 0), \left(-\frac{4}{3}, 0 \right), (4, 0), (0, 8), (0, -8), \right.$$
$$\left. \left(-\frac{8}{3}, \frac{40}{3} \right), \left(-\frac{8}{3}, -\frac{40}{3} \right) \right\}$$
$$\cong \mathbb{Z}/2\mathbb{Z} \times \mathbb{Z}/4\mathbb{Z}.$$

Let us mention that $\{-1, 3, 4\}$ is the only $D(4)$-triple that induces an elliptic curve with the torsion group $\mathbb{Z}/2\mathbb{Z} \times \mathbb{Z}/4\mathbb{Z}$. All other $D(4)$-triples induce curves with the torsion group $\mathbb{Z}/2\mathbb{Z} \times \mathbb{Z}/2\mathbb{Z}$ or $\mathbb{Z}/2\mathbb{Z} \times \mathbb{Z}/6\mathbb{Z}$ (see Sect. 3.8 and [144]). The conjecture is that the group $\mathbb{Z}/2\mathbb{Z} \times \mathbb{Z}/6\mathbb{Z}$ cannot appear here.

A problem with applying the Lutz-Nagell theorem may occur if it is difficult to factorize the discriminant Δ or if it has many quadratic factors. Then the following fact can be helpful.

Proposition 2.4.6 *Let E be an elliptic curve given by the equation*

$$y^2 = x^3 + ax^2 + bx + c,$$

where $a, b, c \in \mathbb{Z}$. Let p be an odd prime such that $p \nmid \Delta_0$, and let

$$\rho_p : E(\mathbb{Q}) \to E(\mathbb{F}_p)$$

denote the reduction modulo p. If the point $P \in E(\mathbb{Q})$ is of finite order and $\rho_p(P) = O$, then $P = O$.

Proof According to the Lutz-Nagell theorem, all torsion points (except O) have integer coordinates, so when reducing modulo p they are not reduced to O. $\square$

According to Proposition 2.4.6, the kernel (set of domain elements that are mapped into the neutral element of the codomain) of the restriction ρ_p to $E(\mathbb{Q})_{\text{tors}}$ is trivial. The image of this restriction is a subgroup of $E(\mathbb{F}_p)$, and since the order of the subgroup divides the group order, we conclude that $|E(\mathbb{Q})_{\text{tors}}|$ divides $|E(\mathbb{F}_p)|$. If we take several values of p, then the greatest common divisor g of corresponding values of $|E(\mathbb{F}_p)|$ must be a multiple of $|E(\mathbb{Q})_{\text{tors}}|$.

Computing the order of $E(\mathbb{F}_p)$ for large p is not easy. But in applications for the computation of torsion groups, p is usually very small (we choose a few smallest odd primes p that do not divide the discriminant), so for calculation of $|E(\mathbb{F}_p)|$ the following formula, which uses the Legendre symbol, is quite satisfactory:

$$|E(\mathbb{F}_p)| = p + 1 + \sum_{x \in \mathbb{F}_p} \left(\frac{x^3 + ax^2 + bx + c}{p} \right).$$

In PARI, $|E(\mathbb{F}_p)|$ can be obtained as `ellcard(E, p)`.

Example 2.4.7 Let us determine the torsion group of the elliptic curve

$$E : \quad y^2 = (x + 1)(3x + 1)(8x + 1)$$

induced by the Diophantine triple $\{1, 3, 8\}$.

The minimal Weierstrass equation of E is

$$y^2 = x^3 - x^2 - 120x + 432.$$

Here $\Delta_0 = -2822400 = -2^8 3^2 5^2 7^2$, so we should, using the Lutz-Nagell theorem, test all divisors y of 1680. Instead, we can check that $|E(\mathbb{F}_{11})| = 12$ and $|E(\mathbb{F}_{13})| = 20$. Since $\gcd(12, 20) = 4$ and the curve has three points of order 2, it follows that $E(\mathbb{Q})_{\text{tors}} = \left\{ O, (-1, 0), (-\frac{1}{3}, 0), (-\frac{1}{8}, 0) \right\} \cong \mathbb{Z}/2\mathbb{Z} \times \mathbb{Z}/2\mathbb{Z}$.

In the software package PARI, the torsion group of an elliptic curve over $\mathbb{Q}$ can be calculated via the function `elltors`. The result is a 3-component vector $[t, v_1, v_2]$, where t is the order of the torsion group, v_1 gives the structure of the torsion group as a product of cyclic groups, while v_2 gives generators of those cyclic groups.

We have already mentioned that in 1977 Mazur proved the following theorem.

Theorem 2.4.8 *There are exactly 15 possible torsion groups for elliptic curves over* $\mathbb{Q}$. *These groups are:*

$$\mathbb{Z}/k\mathbb{Z}, \quad \text{for } k = 1, 2, 3, 4, 5, 6, 7, 8, 9, 10, 12,$$

$$\mathbb{Z}/2\mathbb{Z} \times \mathbb{Z}/2k\mathbb{Z}, \quad \text{for } k = 1, 2, 3, 4.$$

The difficult part of this result lies in proving that groups which are not listed in the theorem cannot appear as torsion groups of elliptic curves over $\mathbb{Q}$.

On the other hand, it is not difficult to show that for each of the 15 groups listed in Mazur's theorem, it is possible to construct infinitely many elliptic curves with that torsion (sub)group. We will do this in detail for groups of the form $\mathbb{Z}/2\mathbb{Z} \times \mathbb{Z}/2k\mathbb{Z}$ because these groups occur for elliptic curves induced by rational Diophantine triples (and quadruples).

Nevertheless, let us say something briefly about torsion groups of the form $\mathbb{Z}/k\mathbb{Z}$ (the details can be found in [116, Section 15.3] or [243]). It turns out that for this purpose, it is convenient to consider curves in the long Weierstrass form

$$y^2 + a_1 xy + a_3 y = x^3 + a_2 x^2 + a_4 x + a_6. \tag{2.14}$$

Therefore, we give the formulas for adding points on the curve given by (2.14): if $P_1 = (x_1, y_1)$, $P_2 = (x_2, y_2)$, then $P_1 + P_2 = (x_3, y_3)$, where

$$x_3 = \lambda^2 + a_1 \lambda - a_2 - x_1 - x_2,$$

$$y_3 = -(\lambda + a_1)x_3 - \mu - a_3,$$

$$\lambda = \begin{cases} \dfrac{y_2 - y_1}{x_2 - x_1}, & \text{if } x_2 \neq x_1, \\ \dfrac{3x_1^2 + 2a_2 x_1 + a_4 - a_1 y_1}{2y_1 + a_1 x_1 + a_3}, & \text{if } P_2 = P_1, \end{cases}$$

$$\mu = \begin{cases} \dfrac{y_1 x_2 - y_2 x_1}{x_2 - x_1}, & \text{if } x_2 \neq x_1, \\ \dfrac{-x_1^3 + a_4 x_1 + 2a_6 - a_3 y_1}{2y_1 + a_1 x_1 + a_3}, & \text{if } P_2 = P_1. \end{cases}$$

Furthermore, $-P_1 = (x_1, -y_1 - a_1 x_1 - a_3)$.

Let P be a point in $E(\mathbb{Q})$ of order k. Without loss of generality we can assume that $P = (0, 0)$ (by the substitution $(x, y) \mapsto (x - x_P, y - y_P)$). Then in Eq. (2.14) we have $a_6 = 0$, and due to non-singularity, one of the numbers a_3 and a_4 is different from zero. If $k \geq 4$, then we can assume that both a_2 and a_3 are different from zero. Then the equation of the curve can be transformed into *Tate's normal form*

$$y^2 + (1 - c)xy - by = x^3 - bx^2, \tag{2.15}$$

where b and c are rationals such that the curve is non-singular. In this equation, the first few multiples of the point P have very simple coordinates, e.g.

$$- P = (0, b), \quad 2P = (b, bc), \quad -2P = (b, 0), \quad 3P = (c, b - c), \quad -3P = (c, c^2),$$

$$4P = \left(\frac{b(b - c)}{c^2}, \frac{-b^2(b - c - c^2)}{c^3} \right), \quad -4P = \left(\frac{b(b - c)}{c^2}, \frac{b(b - c)^2}{c^3} \right).$$

Now we write the condition $2kP = O$ in the form $kP = -kP$, while the condition $(2k + 1)P = O$ is written in the form $(k + 1)P = -kP$. For example, we get:

- The point P is of order 4, i.e. $2P = -2P$, if and only if $c = 0$. So, the general form of the curve with the torsion subgroup $\mathbb{Z}/4\mathbb{Z}$ is

$$y^2 + xy - by = x^3 - bx^2, \quad b \in \mathbb{Q} \setminus \left\{ -\frac{1}{16}, 0 \right\}.$$

  We excluded the values $b = -1/16$ and $b = 0$ (and the same will happen with exceptions in the following formulas) because they produce singular curves.
- The point P is of order 5, i.e. $3P = -2P$, if and only if $b = c$. So, the general form of the curve with the torsion group $\mathbb{Z}/5\mathbb{Z}$ is

$$y^2 + (1 - b)xy - by = x^3 - bx^2, \quad b \in \mathbb{Q} \setminus \{0\}.$$

- The point P is of order 6, i.e. $3P = -3P$, if and only if $b = c + c^2$. Thus, the general form of the curve with the torsion subgroup $\mathbb{Z}/6\mathbb{Z}$ is

$$y^2 + (1 - c)xy - (c + c^2)y = x^3 - (c + c^2)x^2, \quad c \in \mathbb{Q} \setminus \left\{ -\frac{1}{9}, -1, 0 \right\}.$$

- The point P is of order 7, i.e. $4P = -3P$, if and only if $b(b - c) = c^3$. The equation $b^2 - bc = c^3$ can be understood as an equation of a singular cubic, with a singularity at $(b, c) = (0, 0)$. Inserting $b = cd$ into the equation, we get the parametrization $c = d^2 - d, b = d^3 - d^2$. Thus, the general form of the curve with the torsion group $\mathbb{Z}/7\mathbb{Z}$ is

$$y^2 + (1 - c)xy - by = x^3 - bx^2, \quad b = d^3 - d^2, \ c = d^2 - d, \ d \in \mathbb{Q} \setminus \{0, 1\}.$$

We will now consider torsion groups $\mathbb{Z}/2\mathbb{Z} \times \mathbb{Z}/2k\mathbb{Z}$ for $k = 1, 2, 3, 4$. All such curves have three points of order 2. Therefore, we will consider here curves with the equation

$$y^2 = (x - \alpha)(x - \beta)(x - \gamma), \tag{2.16}$$

where α, β, γ are three distinct rational numbers. Curve (2.16) has three rational points of order 2, and therefore has a torsion subgroup isomorphic to $\mathbb{Z}/2\mathbb{Z} \times \mathbb{Z}/2\mathbb{Z}$.

In the construction of curves with torsion group $\mathbb{Z}/2\mathbb{Z} \times \mathbb{Z}/4\mathbb{Z}$, we use the following fact (see also [220, Chapter 1.4] and [239, Chapter 4.2]).

Theorem 2.4.9 *Let E be an elliptic curve over a field K, $\mathrm{char}(K) \neq 2$. Let*

$$E : \quad y^2 = (x - \alpha)(x - \beta)(x - \gamma), \quad \alpha, \beta, \gamma \in K.$$

For a point $Q = (x_2, y_2) \in E(K)$ there is a point $P = (x_1, y_1) \in E(K)$ such that $2P = Q$ if and only if $x_2 - \alpha$, $x_2 - \beta$ and $x_2 - \gamma$ are squares in K.

Proof Let us first prove that if there is a point P such that $2P = Q$, then $x_2 - \alpha$, $x_2 - \beta$ and $x_2 - \gamma$ are squares.

Let $P = (x_1, y_1)$ be a point with the desired property and let $y = \lambda x + \mu$ be the tangent line at P. Consider the polynomial

$$(x - \alpha)(x - \beta)(x - \gamma) - (\lambda x + \mu)^2.$$

Its roots are x_1 (root with multiplicity 2) and x_2 (because $-Q = (x_2, -y_2)$ lies on the tangent line). Hence,

$$(x - \alpha)(x - \beta)(x - \gamma) - (\lambda x + \mu)^2 = (x - x_1)^2(x - x_2). \tag{2.17}$$

By inserting $x = \alpha$ into (2.17), we get

$$-(\lambda \alpha + \mu)^2 = (\alpha - x_1)^2(\alpha - x_2),$$

from which we conclude that $x_2 - \alpha$ is a square. By inserting $x = \beta$, resp. $x = \gamma$, we get that $x_2 - \beta$ and $x_2 - \gamma$ are squares too.

Now suppose that $x_2 - \alpha$, $x_2 - \beta$ and $x_2 - \gamma$ are squares, and we want to find (x_1, y_1) such that $2(x_1, y_1) = (x_2, y_2)$. By substituting variables, we can assume that $x_2 = 0$. So, $\alpha = -\alpha_1^2$, $\beta = -\beta_1^2$, $\gamma = -\gamma_1^2$, and we can choose the signs of $\alpha_1, \beta_1, \gamma_1$ so that $y_2 = \alpha_1 \beta_1 \gamma_1$.

The tangent line at the point (x_1, y_1) passes through the point $(0, -y_2)$, so it has an equation of the form $y = kx - y_2$. The cubic polynomial

$$p(x) = (x - \alpha)(x - \beta)(x - \gamma) - (kx - y_2)^2$$

has the root 0 and the double root x_1, so it is equal to $x(x - x_1)^2$. Therefore, the quadratic polynomial

$$p(x)/x = x^2 - (\alpha + \beta + \gamma + k^2)x + \alpha\beta + \alpha\gamma + \beta\gamma + 2ky_2$$

has a double root, and so its discriminant is equal to 0. Thus, we get an equation of the fourth degree in k. If we show that this equation has a root k_0 in K, then $x_1 = \frac{1}{2}(\alpha + \beta + \gamma + k_0^2)$ will be a double root of $p(x)$, so the point $(x_1, y_1) = (x_1, k_0 x_1 - y_2)$ will have the required property that $2(x_1, y_1) = (0, y_2)$.

Therefore, it remains to show that the equation

$$k^4 + (2\alpha + 2\beta + 2\gamma)k^2 - 8y_2 k + \alpha^2 + \beta^2 + \gamma^2 - 2\alpha\beta - 2\alpha\gamma - 2\beta\gamma = 0 \qquad (2.18)$$

has a K-rational root. If we compare the polynomial on the left-hand side of Eq. (2.18) with $(k^2 - \alpha + \beta + \gamma)^2$, and express $\alpha, \beta, \gamma, y_2$ using $\alpha_1, \beta_1, \gamma_1$, we get the equation

$$(k^2 + \alpha_1^2 - \beta_1^2 - \gamma_1^2)^2 = (2(\alpha_1 k + \beta_1 \gamma_1))^2,$$

i.e.

$$k^2 + \alpha_1^2 - \beta_1^2 - \gamma_1^2 = \pm 2(\alpha_1 k + \beta_1 \gamma_1),$$

whose solutions are $k = \alpha_1 \pm (\beta_1 + \gamma_1)$ (for the sign $+$), resp. $k = -\alpha_1 \pm (\beta_1 - \gamma_1)$ (for the sign $-$). So, we proved that Eq. (2.18) has four K-rational roots. $\square$

We will later (in Sects. 2.5 and 3.1) also need the following theorem [239, Proposition 4.6], which can be seen as a generalization of one part of Theorem 2.4.9.

Theorem 2.4.10 *Let E be an elliptic curve over $\mathbb{Q}$ given by the equation*

$$E \ : \ y^2 = (x - \alpha)(x - \beta)(x - \gamma), \quad \alpha, \beta, \gamma \in \mathbb{Q}.$$

Define the function $\varphi : E(\mathbb{Q}) \to \mathbb{Q}^*/\mathbb{Q}^{*2}$ by

$$\varphi(P) = \begin{cases} (x - \alpha)\mathbb{Q}^{*2} & \text{if } P = (x, y) \neq O, (\alpha, 0), \\ (\alpha - \beta)(\alpha - \gamma)\mathbb{Q}^{*2} & \text{if } P = (\alpha, 0), \\ \mathbb{Q}^{*2} & \text{if } P = O. \end{cases}$$

Then φ is a group homomorphism.

Proof Let $P_1 + P_2 = P_3$. We need to prove that $\varphi(P_1)\varphi(P_2)\varphi(P_3)^{-1} \in \mathbb{Q}^{*2}$. Since $\varphi(P_3) = \varphi(-P_3)$ and $\varphi(P_3) = \varphi(P_3)^{-1}$, it is sufficient to prove that $P_1 + P_2 + P_3 = O$ implies $\varphi(P_1)\varphi(P_2)\varphi(P_3) \in \mathbb{Q}^{*2}$.

If one of P_i, say P_1, is equal to O, then the statement easily follows because $P_2 + P_3 = O$ implies $\varphi(P_2) = \varphi(P_3)$, so $\varphi(P_1)\varphi(P_2)\varphi(P_3) \in \mathbb{Q}^{*2}$. Therefore, we can further assume that $P_i = (x_i, y_i) \neq O$ for $i = 1, 2, 3$.

Assume that $P_1 = (\alpha, 0)$. Then $P_2, P_3 \neq (\alpha, 0)$ because otherwise one of them would be O. From the assumption that $P_1 + P_2 + P_3 = O$, it follows that the points P_1, P_2, P_3 lie on a line. Let the equation of that line be $y = kx + \ell$. Then $x_1 = \alpha, x_2, x_3$ are roots of the polynomial $(x - \alpha)(x - \beta)(x - \gamma) - (kx + \ell)^2$, so we have

$$(x - \alpha)(x - \beta)(x - \gamma) - (kx + \ell)^2 = (x - \alpha)(x - x_2)(x - x_3). \qquad (2.19)$$

From (2.19), it follows that $x - \alpha$ divides $kx + \ell$, so $kx + \ell = k(x - \alpha)$. After cancelling by $x - \alpha$, (2.19) becomes

$$(x - \beta)(x - \gamma) - k^2(x - \alpha) = (x - x_2)(x - x_3). \tag{2.20}$$

If we put $x = \alpha$ in (2.20), we get that

$$(\alpha - \beta)(\alpha - \gamma) = (x_2 - \alpha)(x_3 - \alpha),$$

which implies that $\varphi(P_1) = \varphi(P_2)\varphi(P_3)$, so $\varphi(P_1)\varphi(P_2)\varphi(P_3) \in \mathbb{Q}^{*2}$.

It remains to consider the case when all $P_i \neq (\alpha, 0)$. Let $y = kx + \ell$ be the line passing through the points P_1, P_2, P_3 again. As before, we conclude that

$$(x - \alpha)(x - \beta)(x - \gamma) - (kx + \ell)^2 = (x - x_1)(x - x_2)(x - x_3). \tag{2.21}$$

By inserting $x = \alpha$ in (2.21), we get $(x_1 - \alpha)(x_2 - \alpha)(x_3 - \alpha) = (k\alpha + \ell)^2$ and we conclude that $\varphi(P_1)\varphi(P_2)\varphi(P_3) \in \mathbb{Q}^{*2}$. $\qquad\square$

Let us now return to the construction of curves with the torsion group $\mathbb{Z}/2\mathbb{Z} \times \mathbb{Z}/4\mathbb{Z}$. Without loss of generality, we can assume that the point $P = (0, 0)$ is one of the points of order 2, and precisely that point for which there exists $Q \in E(\mathbb{Q})$ such that $2Q = P$. This means that the curve has an equation

$$y^2 = x(x - \alpha)(x - \beta)$$

and that the numbers $-\alpha$ and $-\beta$ are squares in $\mathbb{Q}$. So, the general form of the curve with the torsion subgroup $\mathbb{Z}/2\mathbb{Z} \times \mathbb{Z}/4\mathbb{Z}$ is

$$y^2 = x(x + r^2)(x + s^2), \quad r, s \in \mathbb{Q} \setminus \{0\}, \ r \neq \pm s, \tag{2.22}$$

or (dividing the equation by r^6 and with the notation $u = s/r$)

$$y^2 = x(x + 1)(x + u^2), \quad u \in \mathbb{Q} \setminus \{0, \pm 1\}. \tag{2.23}$$

A point of order 4 on (2.22) is the point $(rs, rs(r + s))$, while on (2.23) it is the point $Q = (u, u(u + 1))$. To obtain a curve with torsion group $\mathbb{Z}/2\mathbb{Z} \times \mathbb{Z}/8\mathbb{Z}$, there should be a point R (of order 8) such that $2R = Q$. According to Theorem 2.4.9, a necessary and sufficient condition for the existence of such a point is that $u, u + 1$ and $u(u + 1)$ are squares of rational numbers. So, we have $u = v^2$ and $v^2 + 1 = w^2$. Hence, $v = \frac{2t}{t^2 - 1}$ for some $t \in \mathbb{Q}$. Therefore, the general form of a curve with the torsion subgroup $\mathbb{Z}/2\mathbb{Z} \times \mathbb{Z}/8\mathbb{Z}$ is

$$y^2 = x(x + 1)\left(x + \left(\frac{2t}{t^2 - 1}\right)^4\right), \quad t \in \mathbb{Q} \setminus \{-1, 0, 1\}, \tag{2.24}$$

or

$$y^2 = x(x + (t^2 - 1)^4)(x + 16t^4), \quad t \in \mathbb{Q} \setminus \{-1, 0, 1\}, \tag{2.25}$$

We present the following interesting (and perhaps unexpected) result.

Theorem 2.4.11 *Every elliptic curve over $\mathbb{Q}$ with torsion group $\mathbb{Z}/2\mathbb{Z} \times \mathbb{Z}/8\mathbb{Z}$ is birationally equivalent to a curve of the form*

$$y^2 = (ax + 1)(bx + 1)(cx + 1), \tag{2.26}$$

where $\{a, b, c\}$ is a rational Diophantine triple.

Proof Multiplying Eq. (2.26) by $a^2 b^2 c^2$ and introducing the substitution $x \mapsto (x - ab)/abc$, we get the equation

$$y^2 = x(x + ac - ab)(x + bc - ab).$$

If $ab = -1$, i.e. $b = -1/a$, then $ac - ab$ and $bc - ab$ are squares, so we get an elliptic curve whose torsion group contains $\mathbb{Z}/2\mathbb{Z} \times \mathbb{Z}/4\mathbb{Z}$. One of the points of order 4 is $Q = (\sigma\tau, \sigma\tau(\sigma + \tau))$, where $ac + 1 = \sigma^2$, $bc + 1 = \tau^2$. Let us assume that the triple $\{a, b, c\}$ is regular, i.e. it is of the form

$$\left\{ a, -\frac{1}{a}, a - \frac{1}{a} \right\}.$$

Then $\sigma = a$, $\tau = \frac{1}{a}$, so $Q = (1, a + \frac{1}{a})$, and the curve has the equation

$$y^2 = x(x + a^2)\left(x + \frac{1}{a^2}\right).$$

To obtain the point of order 8 and the torsion group $\mathbb{Z}/2\mathbb{Z} \times \mathbb{Z}/8\mathbb{Z}$, we want the point Q to be of the form $Q = 2R$ for some point $R \in E(\mathbb{Q})$. According to Theorem 2.4.9, a necessary and sufficient condition for this is that $a^2 + 1$ be the square of a rational number, i.e. $a = \frac{2t}{t^2 - 1}$. We get the equation

$$y^2 = x\left(x + \frac{4t^2}{(t^2 - 1)^2}\right)\left(x + \frac{(t^2 - 1)^2}{4t^2}\right).$$

After multiplying this equation by $(4t^2(t^2 - 1)^2)^3$ and introducing the substitution $x \mapsto x/(4t^2(t^2 - 1)^2)$, we get exactly Eq. (2.24).   □

We are left with the torsion group $\mathbb{Z}/2\mathbb{Z} \times \mathbb{Z}/6\mathbb{Z}$. To obtain it, we should have a point P of order 3 on a curve (without loss of generality, we can assume that its first coordinate is equal to 0) for which there is a point Q of order 6 such that $2Q = P$. According to Theorem 2.4.9, then in (2.16) we must have $\alpha = -r^2$, $\beta = -s^2$,

$\gamma = -t^2$. So we get the curve

$$y^2 = (x + r^2)(x + s^2)(x + t^2), \tag{2.27}$$

which, in addition to three points of order 2, also has one additional obvious rational point $P = (0, rst)$. If the point P were of order 3, we would get the desired torsion group. So, we must satisfy the condition $-P = 2P$, which gives

$$\frac{(r^2 s^2 + r^2 t^2 + s^2 t^2)^2}{4 r^2 s^2 t^2} - r^2 - s^2 - t^2 = 0,$$

i.e.

$$(sr + ts + tr)(-sr + ts + tr)(-sr + ts - tr)(sr + ts - tr) = 0.$$

We can take $t = \frac{rs}{r-s}$, so we get that the general form of the curve with the torsion group $\mathbb{Z}/2\mathbb{Z} \times \mathbb{Z}/6\mathbb{Z}$ is

$$y^2 = (x + r^2)(x + s^2)\left(x + \frac{r^2 s^2}{(r-s)^2}\right), \quad r, s \in \mathbb{Q} \setminus \{0\}, \ r \neq \pm s, \frac{1}{2}s, 2s,$$

or, with the notation $u = s/r$,

$$y^2 = (x + 1)(x + u^2)\left(x + \frac{u^2}{(u-1)^2}\right), \quad u \in \mathbb{Q} \setminus \left\{0, \pm 1, \frac{1}{2}, 2\right\}.$$

2.5 Rank of Elliptic Curves

Questions related to the rank of an elliptic curve over $\mathbb{Q}$ are much more difficult than questions related to torsion groups, and satisfactory answers are still not known. For a long time, it was believed that the rank could be arbitrarily large, i.e. that for any $M \in \mathbb{N}$, there is an elliptic curve E over $\mathbb{Q}$ such that $\mathrm{rank}(E) \geq M$. However, recently, several papers have appeared that give certain heuristic arguments that the rank might still be bounded. Today we only know that there is an elliptic curve of rank ≥ 28. That curve was found in 2006 by Elkies [161]. Klagsbrun, Sherman, and Weigandt [238] proved in 2019 that the rank of that curve is equal to 28, assuming that the generalized Riemann hypothesis holds. Its (minimal) equation is:

$$y^2 + xy + y = x^3 - x^2 -$$

$$20067762415575526585033208209338542750930230312178956502x+$$

$$34481611795030556467032985690390720374855944359319180361266008296291939448732243429.$$

Table 2.1 History of elliptic curves rank records

Rank $\geq$	Year	Authors
3	1938	Billing
4	1945	Wiman
6	1974	Penney and Pomerance
7	1975	Penney and Pomerance
8	1977	Grunewald and Zimmert
9	1977	Brumer and Kramer
12	1982	Mestre
14	1986	Mestre
15	1992	Mestre
17	1992	Nagao
19	1992	Fermigier
20	1993	Nagao
21	1994	Nagao and Kouya
22	1997	Fermigier
23	1998	Martin and McMillen
24	2000	Martin and McMillen
28	2006	Elkies

An overview of the findings of the record curves is given in Table 2.1 (details about the record curves can be found on the web page [115]):

Strictly speaking, no algorithm for computing the rank is known. Namely, for "algorithms" which are used to compute the rank, some of which we will describe here, there is no guarantee that they will give a result in all cases. An important part of these algorithms includes the decision whether there are rational points on a certain genus 1 curve which is known to have points everywhere locally (i.e. over $\mathbb{R}$ and the p-adic fields $\mathbb{Q}_p$ for all primes p). However, there is no known algorithm that would answer that question. Furthermore, even if we ignore this problem (because it may not appear for a specific curve we are observing), for curves that do not have rational points of order 2 and have large coefficients, known algorithms are mostly not efficient enough for practical use.

Assume that E has a point of order 2. In that case, computing the rank is usually easier than in the general case. We will describe the method for computing the rank (following [330, Chapter 3]), called "descent via 2-isogeny". By changing the coordinates, we can assume that the point of order 2 is exactly the point $(0, 0)$ and that E has the equation

$$y^2 = x^3 + ax^2 + bx, \tag{2.28}$$

where $a, b \in \mathbb{Z}$. If the starting curve was given by the equation $y^2 = x^3 + a_2x^2 + a_4x + a_6$ and if x_0 is a root of polynomial $x^3 + a_2x^2 + a_4x + a_6$, then we put $a = 3x_0 + a_2$, $b = (a + a_2)x_0 + a_4$. If the initial curve was given by the Weierstrass equation, then for x_0 we take a root of the cubic polynomial $x^3 + b_2x^2 + 8b_4x + 16b_6$

and put $a = 3x_0 + b_2, b = (a + b_2)x_0 + 8b_4$. The non-singularity condition for the curve E is $\Delta = 16b^2(a^2 - 4b) \neq 0$.

For the curve E' which has the equation

$$y^2 = x^3 + a'x^2 + b'x, \tag{2.29}$$

where $a' = -2a$ and $b' = a^2 - 4b$, we say that it is 2-isogenous to the curve E. The non-singularity condition for both curves E and E' is the same and can be expressed in the form $bb' \neq 0$.

In general, an isogeny is a homomorphism between two elliptic curves which is given by rational functions. In our case, we have the function $\phi : E \to E'$, $\phi(P) = (\frac{y^2}{x^2}, \frac{y(x^2-b)}{x^2})$ for $P = (x, y) \neq O, (0, 0)$, and $\phi(P) = O$ otherwise. Analogously, we define $\psi : E' \to E$ with $\psi(P') = (\frac{y'^2}{4x'^2}, \frac{y'(x'^2-b')}{8x'^2})$ for $P' = (x', y') \neq O, (0, 0)$, and $\psi(P') = O$ otherwise. It holds $(\psi \circ \phi)(P) = 2P$ for all $P \in E$ and $(\phi \circ \psi)(P') = 2P'$ for all $P' \in E'$.

We also define the functions $\alpha : E(\mathbb{Q}) \to \mathbb{Q}^*/\mathbb{Q}^{*2}$, $\beta : E'(\mathbb{Q}) \to \mathbb{Q}^*/\mathbb{Q}^{*2}$ by $\alpha(O) = 1 \cdot \mathbb{Q}^{*2}$, $\alpha(0, 0) = b \cdot \mathbb{Q}^{*2}$, $\alpha(x, y) = x \cdot \mathbb{Q}^{*2}$ for $P = (x, y) \neq O, (0, 0)$, and analogously for β. Similarly as in Theorem 2.4.10, we can see that α and β are group homomorphisms. It is clear that $\mathrm{Ker}(\phi) = \{O, (0, 0)\}$, $\mathrm{Ker}(\psi) = \{O, (0, 0)\}$, and it can be shown that $\mathrm{Im}(\phi) = \mathrm{Ker}(\beta)$ and $\mathrm{Im}(\psi) = \mathrm{Ker}(\alpha)$. The number 2 in the name of 2-isogeny comes from the fact that the kernels of ϕ and ψ have two elements.

These functions are used in the first step in the proof of the Mordell-Weil theorem, i.e. in the proof that the subgroup $2E(\mathbb{Q})$ has a finite index in the group $E(\mathbb{Q})$. Namely, it is easy to see that this statement follows from the finiteness of the indices $[E(\mathbb{Q}) : \psi(E'(\mathbb{Q}))]$ and $[E'(\mathbb{Q}) : \phi(E(\mathbb{Q}))]$, and according to the group isomorphism theorem (see [245, Chapter 1.3]), this follows from the finiteness of the groups $\mathrm{Im}(\alpha)$ and $\mathrm{Im}(\beta)$. The connection of these functions with the rank is even more explicit. Namely,

$$2^r = \frac{[E(\mathbb{Q}) : \psi(E'(\mathbb{Q}))] \cdot [E'(\mathbb{Q}) : \phi(E(\mathbb{Q}))]}{4} = \frac{|\mathrm{Im}(\alpha)| \cdot |\mathrm{Im}(\beta)|}{4},$$

where $r = \mathrm{rank}(E(\mathbb{Q}))$. The details can be found in [330, Chapter 3].

It is also true that $r = \mathrm{rank}(E'(\mathbb{Q}))$. However, torsion groups of E and E' do not have to be isomorphic in general, but $|E(\mathbb{Q})_{\mathrm{tors}}| = 2^i |E'(\mathbb{Q})_{\mathrm{tors}}|$, where $i \in \{-1, 0, 1\}$.

We want to get a description of the elements of $\mathrm{Im}(\alpha)$. By $\tilde{x}$ we will denote the class of x in $\mathbb{Q}/\mathbb{Q}^{*2}$.

Let $(x, y) \in E(\mathbb{Q})$. If $x = 0$, then $(x, y) = (0, 0)$ and $\alpha(x, y) = \tilde{b}$. If $x \neq 0$, we write x and y in the form $x = \frac{m}{e^2}$, $y = \frac{n}{e^3}$, $\gcd(m, e) = \gcd(n, e) = 1$ and insert them in the equation of E. We get

$$n^2 = m(m^2 + ame^2 + be^4).$$

We put $b_1 = \pm \gcd(m, b)$, where the sign is chosen so that $mb_1 > 0$. Then $m = b_1 m_1$, $b = b_1 b_2$, $n = b_1 n_1$, and we get

$$n_1^2 = m_1 (b_1 m_1^2 + a m_1 e^2 + b_2 e^4).$$

Since the factors on the right-hand side of the last equation are relatively prime, and $m_1 > 0$, we conclude that there are integers M and N so that $m_1 = M^2$ and $b_1 m_1^2 + a m_1 e^2 + b_2 e^4 = N^2$. Thus, we finally get the equation

$$N^2 = b_1 M^4 + a M^2 e^2 + b_2 e^4, \tag{2.30}$$

in which the unknowns are M, e and N. Now $\alpha(x, y) = (\frac{b_1 M^2}{e^2}) \cdot \mathbb{Q}^{*2} = \widetilde{b_1}$.

We conclude that $\operatorname{Im}(\alpha)$ consists of $\widetilde{1}$, $\widetilde{b}$ and all $\widetilde{b_1}$ where b_1 is a divisor of the number b for which Eq. (2.30), where $b_1 b_2 = b$, has solutions $N, M, e \in \mathbb{Z}$, $e \neq 0$. Then $(\frac{b_1 M^2}{e^2}, \frac{b_1 M N}{e^3}) \in E(\mathbb{Q})$. We notice that Eq. (2.30) always has a solution for $b_1 = 1$, that is $(M, e, N) = (1, 0, 1)$ and for $b_1 = b$, that is $(M, e, N) = (0, 1, 1)$.

In examining the solvability of Eq. (2.30), we can assume that $\gcd(M, e) = 1$. Also, it is no loss of generality if we only consider the divisors of b_1 which are square-free. Alternatively, if we consider all divisors of b_1, we can search only for solutions that satisfy $\gcd(N, e) = \gcd(M, N) = 1$.

We have the following algorithm for computing the rank of an elliptic curve E which has a rational point of order 2, i.e. it has an equation of the form (2.28). For each factorization $b = b_1 b_2$, where b_1 is a square-free integer, we write Eq. (2.30). We are trying to determine whether this equation has a non-trivial integer solution (note that the local-global principle of Hasse and Minkowski does not have to hold for such equations, which means that we do not have an algorithm that would certainly answer this question). Every solution (M, e, N) of Eq. (2.30) induces a point on curve E with coordinates $x = \frac{b_1 M^2}{e^2}$, $y = \frac{b_1 M N}{e^3}$. Let r_1 be the number of factorizations for which the corresponding Eq. (2.30) has a solution, and let r_2 be the number defined in the same way for curve E'. Then there exist nonnegative integers e_1 and e_2 such that $r_1 = 2^{e_1}$, $r_2 = 2^{e_2}$, and we have

$$\operatorname{rank}(E) = e_1 + e_2 - 2.$$

Example 2.5.1 Consider the set $\{1, 2, 5\}$. It is a $D(-1)$-triple. Namely, $1 \cdot 2 - 1$, $1 \cdot 5 - 1$ and $2 \cdot 5 - 1$ are perfect squares. The question arises, whether this set can be extended to a $D(-1)$-quadruple, i.e. does there exist $x \in \mathbb{Z}$ such that

$$1 \cdot x - 1, \quad 2 \cdot x - 1, \quad 5 \cdot x - 1$$

are squares of integers? We will show that the only solution is $x = 1$, which implies that the set $\{1, 2, 5\}$ cannot be extended to a $D(-1)$-quadruple. In fact, we will solve the more general problem of finding all rational points on the elliptic curve

$$y^2 = (x - 1)(2x - 1)(5x - 1), \tag{2.31}$$

and prove that $\{1, 2, 5\}$ cannot be extended even to a rational $D(-1)$-quadruple.

The above curve can be transformed into Weierstrass form by multiplying both sides by 10^2, substituting $10y$ and $10x$ by y and x respectively, which leads to the curve

$$y^2 = x^3 - 17x^2 + 80x - 100.$$

By the map $x \mapsto x + 5$, we transform the curve into the form suitable for computing the rank:

$$E : \quad y^2 = x^3 - 2x^2 - 15x.$$

Its 2-isogenous curve is

$$E' : \quad y^2 = x^3 + 4x^2 + 64x.$$

For the curve E, the possibilities for the number b_1 are ± 1, ± 3, ± 5, ± 15. The corresponding Diophantine equations are $N^2 = M^4 - 2M^2 e^2 - 15e^4$, $N^2 = -M^4 - 2M^2 e^2 + 15e^4$, $N^2 = 3M^4 - 2M^2 e^2 - 5e^4$, $N^2 = -3M^4 - 2M^2 e^2 + 5e^4$, $N^2 = 5M^4 - 2M^2 e^2 - 3e^4$, $N^2 = -5M^4 - 2M^2 e^2 + 3e^4$, $N^2 = 15M^4 - 2M^2 e^2 - e^4$, $N^2 = -15M^4 - 2M^2 e^2 + e^4$. By the symmetry, it suffices to examine the solvability of the first four equations. The first equation has a solution $(M, e, N) = (1, 0, 1)$, and the fourth equation has a solution $(M, e, N) = (1, 1, 0)$. The second equation is equivalent to $N^2 = (3e^2 - M^2)(5e^2 + M^2)$. It is easy to see that $\gcd(3e^2 - M^2, 5e^2 + M^2) \in \{1, 2\}$, so we have two possibilities: either both factors are squares, or both factors are double squares. But $3e^2 - M^2 = s^2$ is impossible modulo 3 because $(\frac{-1}{3}) = -1$, while $5e^2 + M^2 = 2t^2$ is impossible modulo 5 because $(\frac{2}{5}) = -1$. The third equation is equivalent to $N^2 = (M^2 + e^2)(3M^2 - 5e^2)$. We have again the same two possibilities for the factors in the last expression, and again, both possibilities fail: $3M^2 - 5e^2 = t^2$ is impossible modulo 5 because $(\frac{3}{5}) = -1$, while $3M^2 - 5e^2 = 2t^2$ is impossible modulo 8 because $3M^2 - 5e^2 \equiv 6 \pmod 8$ and $2t^2 \equiv 2 \pmod 8$. Therefore, $e_1 = 2$.

For E', $b'_1 \in \{\pm 1, \pm 2\}$ (square-free divisors of 64), so the corresponding Diophantine equations are $N^2 = M^4 + 4M^2 e^2 + 64e^4$, $N^2 = -M^4 + 4M^2 e^2 - 64e^4$, $N^2 = 2M^4 + 4M^2 e^2 + 32e^4$ i $N^2 = -2M^4 + 4M^2 e^2 - 32e^4$. The first equation has a solution $(M, e, N) = (1, 0, 1)$. The second and fourth equations are equivalent to $N^2 = -(M^2 - 2e^2)^2 - 60e^4$ and $N^2 = -2(M^2 - e^2)^2 - 30c^2$, respectively, and obviously have no non-trivial solutions. The third equation is equivalent to $2 \cdot (N/2)^2 = (M^2 + e^2)^2 + 15e^4$ and there is no solution modulo 5 because $(\frac{2}{5}) = -1$. Therefore, $e_2 = 0$. We conclude that $\mathrm{rank}(E) = 2 + 0 - 2 = 0$.

It remains to find the torsion points on E. It has three points of order 2: $(0, 0)$, $(-3, 0)$, $(5, 0)$. Since $7 \nmid \Delta = 2^{10}3^2 5^2$ and $|E(\mathbb{F}_7)| = 4$, we conclude that the torsion group of E is $\mathbb{Z}/2\mathbb{Z} \times \mathbb{Z}/2\mathbb{Z}$ and that the only rational points on E are $(0, 0)$, $(-3, 0)$, $(5, 0)$. Hence, the only rational points on curve (2.31) are $(1, 0)$, $(\frac{1}{2}, 0)$, $(\frac{1}{5}, 0)$. Since $1 \cdot \frac{1}{2} - 1$ and $1 \cdot \frac{1}{5} - 1$ are not squares of rational numbers, we get that the only rational number x with the property that $1 \cdot x - 1$, $2 \cdot x - 1$ and $5 \cdot x - 1$ are squares is the number $x = 1$.

Example 2.5.2 Let us calculate the rank of the elliptic curve

$$y^2 = (x + 1)(3x + 1)(8x + 1)$$

induced by the Diophantine triple $\{1, 3, 8\}$.

As in the previous example, we transform the curve into the form suitable for calculating the rank, so first, we transform it to the Weierstrass form

$$y^2 = (x + 3)(x + 8)(x + 24),$$

and then by the map $x \mapsto x - 8$ we get

$$E : \quad y^2 = x^3 + 11x^2 - 80x.$$

Its 2-isogenous curve is

$$E' : \quad y^2 = x^3 - 22x^2 + 441x.$$

For the curve E, the possibilities for the number b_1 are $\pm 1, \pm 2, \pm 5, \pm 10$. We claim that all eight associated equations have solutions: $N^2 = M^4 + 11M^2 e^2 - 80e^4$, $N^2 = -M^4 + 11M^2 e^2 + 80e^4$, $N^2 = 2M^4 + 11M^2 e^2 - 40e^4$, $N^2 = -2M^4 + 11M^2 e^2 + 40e^4$, $N^2 = 5M^4 + 11M^2 e^2 - 16e^4$, $N^2 = -5M^4 + 11M^2 e^2 + 16e^4$, $N^2 = 10M^4 + 11M^2 e^2 - 8e^4$, $N^2 = -10M^4 + 11M^2 e^2 + 8e^4$. Indeed, the solutions are $(M, e, N) = (1, 0, 1)$, $(4, 1, 0)$, $(2, 1, 6)$, $(1, 1, 7)$, $(1, 1, 0)$, $(0, 1, 4)$, $(2, 1, 14)$, $(1, 1, 3)$, respectively. Hence, $e_1 = 3$.

For E', we have $b'_1 \in \{\pm 1, \pm 3, \pm 7, \pm 21\}$. The equation for $b'_1 = 1$ obviously has a solution, while the equations for $b'_1 < 0$ obviously have no solutions. We will show that the remaining three equations have no solution. For $b'_1 = 3$, we get the equation

$$3N^2 = (3M^2 - 11e^2)^2 + 320e^4,$$

which has no solution modulo 5 because $(\frac{3}{5}) = -1$, i.e. 3 is not a quadratic residue modulo 5. Similarly, for $b'_1 = 7$, the equation

$$7N^2 = (7M^2 - 11e^2)^2 + 320e^4$$

has no solutions modulo 5 because $(\frac{7}{5}) = (\frac{2}{5}) = -1$. Finally, let us consider the equation for $b_1' = 21$:

$$21N^2 = (21M^2 - 11e^2)^2 + 320e^4. \tag{2.32}$$

If one of the numbers M, e is even, then, since the square of an odd number when divided by 8 gives the remainder 1, the left-hand side of (2.32) is $\equiv 5 \pmod 8$, while the right-hand side is $\equiv 1 \pmod 8$. If the numbers M, e are both odd, then $21M^2 - 11e^2 \equiv 2 \pmod 8$, so the right-hand side of (2.32) is divisible by 4 but not by 8. Thus, $N \equiv 2 \pmod 4$. After dividing by 4, the left-hand side of (2.32) is $\equiv 5 \pmod 8$, while the right-hand side is $\equiv 1 \pmod 8$. Therefore, $e_2 = 0$. We conclude that rank$(E) = 3 + 0 - 2 = 1$.

Let us notice that in the previous example, in eliminating b_1' for which the corresponding Diophantine equation has no solutions, we used the fact that a negative number cannot be a square in $\mathbb{R}$ and that the numbers 2 and 3 are not squares in $\mathbb{Z}/5\mathbb{Z}$. However, with Diophantine equations of degree greater than 2, it may happen that they have solutions in $\mathbb{R}$ and have solutions in $\mathbb{Z}/m\mathbb{Z}$ for every integer m but still do not have non-trivial solutions in $\mathbb{Q}$. One such example is the equation

$$N^2 = 17M^4 - 4e^4,$$

which appears when computing the rank of the elliptic curve $y^2 = x^3 + 17x$ (see [56, Chapter 18]). In such cases, determining the rank is significantly more demanding.

Let $\omega(b)$ denote the number of distinct prime factors of b. Then b has $2^{\omega(b)+1}$ (positive and negative) square-free factors. Now from the formula $2^r = \frac{|\text{Im}(\alpha)| \cdot |\text{Im}(\beta)|}{4}$, it follows directly that $r \leq \omega(b) + \omega(b')$. However, from Eq. (2.30), it follows that if $a \leq 0$ and $b > 0$, then b_1 must be positive. Analogously, if $a' \leq 0$ and $b' > 0$, then b_1' must be positive. Also from

$$N^2 = b_1 \left(M^2 + \frac{ae^2}{2b_1} \right)^2 - \frac{b'e^4}{4b_1},$$

it follows that if $b' < 0$, then b_1 must be positive and analogously if $b < 0$, then b_1' must be positive. Notice that b and b' cannot be both negative because $4b + b' = a^2$. Obviously $a \leq 0$ or $a' \leq 0$. Therefore, negative divisors cannot appear in at least one of the sets $\text{Im}(\alpha)$, $\text{Im}(\beta)$. We conclude that

$$r \leq \omega(b) + \omega(b') - 1. \tag{2.33}$$

We say that an elliptic curve of the form (2.28) is of *maximal rank* if in (2.33) the equality holds, i.e. if $r = \omega(b) + \omega(b') - 1$. In [7], examples of curves of maximal rank are given for curves of rank r, where $r = 1, 2, \ldots, 12$.

Now let us say something about checking the local solvability of the equation

$$N^2 = b_1 M^4 + a M^2 e^2 + b_2 e^4, \tag{2.34}$$

and the (affine) equation

$$u^2 = b_1 v^4 + a v^2 + b_2 \tag{2.35}$$

associated with it. In general, the criterion for solvability over $\mathbb{R}$ (i.e. for $p = \infty$) of the equation $Y^2 = g(X)$ is very simple: the polynomial g must assume a non-negative value at some x. This will certainly be satisfied if g has real roots, and if g has no real roots, then the leading coefficient of g must be positive.

Concerning the solvability of Eq. (2.35) in $\mathbb{Q}_p$ (i.e. Eq. (2.34) modulo p^k for each $k \geq 1$), it is sufficient to consider only those prime numbers p for which $p \mid 2\Delta$. Namely, it can be shown that for all other values of p, the equations are solvable.

Equation (2.34) can be written in the form

$$N^2 = b_1 \left(M^2 + \frac{ae^2}{2b_1} \right)^2 - \frac{b'e^4}{4b_1},$$

from which we get the condition $(\frac{b_1}{p}) = 1$ for every odd prime divisor p of b'. This gives only a necessary but not a sufficient condition for solvability in $\mathbb{Q}_p$. The general algorithm uses Hensel's lemma, so that for a given solution modulo p^k it checks whether this solution can be "lifted" to a solution modulo p^{k+1}. If k is large enough ($k > v_p(\Delta)$), the algorithm certainly gives the answer to that question, and thus answers the question of solvability in $\mathbb{Q}_p$.

Suppose that in the above-described "descent via 2-isogeny" algorithm, we arrived at the equation

$$u^2 = b_1 v^4 + a v^2 + b_2, \quad b_1 b_2 = b, \tag{2.36}$$

which is locally solvable everywhere, but we have not been able to find a rational point on it. Then the "second descent" method can be applied, and it is sometimes possible to find a rational point (u, v) on (2.36) or prove that there are no rational points on it. The idea is that since (2.36) is locally solvable everywhere, then also the corresponding conic

$$u^2 = b_1 w^2 + a w + b_2, \tag{2.37}$$

is locally solvable everywhere. For such quadratic equations, the local-global principle of Hasse and Minkowski holds, which implies that (2.37) is also globally solvable, i.e. it has a rational point (u_0, w_0). We can assume that $w_0 \neq 0$, because if $w_0 = 0$, then b_2 is a square, so Eq. (2.36) certainly has a solution. All rational points on (2.37) can be obtained by the following parametric formulas:

$$w = \frac{w_0 t^2 - 2u_0 t + a + b_1 w_0}{t^2 - b_1},$$

$$u = \frac{-u_0 t^2 + (a + 2b_1 w_0)t - u_0 b_1}{t^2 - b_1}.$$

We want to find (or prove that it does not exist) a rational number t such that $w = \frac{f(t)}{g(t)}$ is the square of a rational number. This condition leads to new quartics. Now we repeat the procedure of examining local solvability using these quartics and searching for solutions with relatively small height (for details, see [73, Chapter 3.6.6]). Again, there is no guarantee that we will get an answer in each case.

The set of all $\widetilde{b}_1$ for which Eq. (2.30) has a local solution everywhere also forms a group (and analogously for b'_1). If the corresponding orders are 2^{f_1} and 2^{f_2}, then the number $s = f_1 + f_2 - 2$ is called the 2-*Selmer rank* of E (see [329, Chapter 10.4]). It is clear that $r \le s$. There is a conjecture that the congruence $r \equiv s$ (mod 2) always holds.

The described algorithm is implemented in the `mwrank` program by John Cremona, which is included in the SageMath software package. The algorithm and its implementation are explained in detail in [74]. Algorithms for rank computation also exist in the Magma software package [45] and, more recently, in PARI via the `ellrank` function. The output of `ellrank`$(E, \{eff\})$ is $[r, R, s, L]$ such that the rank of the elliptic curve E is between r and R, s gives information on the Tate-Shafarevich group (obtained by the Cassels-Tate pairing [55, 172]), and L is a list of independent, non-torsion rational points on the curve. The optional parameter *eff* is a measure of the effort done to find rational points before giving up with the search.

In the general case, when E does not need to have a point of order 2, again, the idea is to attach to the curve E a family of quartics. In this case, they have a more general form

$$H : \quad y^2 = g(x) = ax^4 + bx^3 + cx^2 + dx + e. \tag{2.38}$$

Here $a, b, c, d, e \in \mathbb{Q}$ are such that

$$12ae - 3bd + c^2 = \lambda^4 c_4, \quad 72ace + 9bcd - 27ad^2 - 27eb^2 - 2c^3 = 2\lambda^6 c_6$$

for some $\lambda \in \mathbb{Q}$. A detailed description of the algorithm can be found in [74, Chapter 3.6]. The core of the algorithm is given in the paper by Birch and Swinnerton-Dyer [39] from 1963. This algorithm is also contained in `mwrank` (although it works much less efficiently than the version for curves with a point of order 2).

If we choose an elliptic curve in a "random" way, it will most likely have the trivial torsion group and very small rank (0 or 1). A conjecture is that the average rank is 1/2 ("half" of curves have rank 0 and "half" have rank 1, and the number of curves of rank ≥ 2 is asymptotically negligible). From the results of Bhargava and Shankar [35], it follows that the average rank is strictly less than 1 (here we assume that elliptic curves $y^2 = x^3 + Ax + B$, $A, B \in \mathbb{Z}$, are ordered with respect to the size of $\max(4|A|^3, 27B^2)$).

We saw earlier how to ensure that the curve has a prescribed torsion subgroup. Now we will consider the methods for finding elliptic curves of relatively high rank (we cannot expect to get a very high rank bearing in mind that no elliptic curve is currently known with rank higher than 28). As we have already mentioned, although we are not aware of any examples of a curve with a very high rank, it has long been a widely accepted conjecture that the rank can be arbitrarily large. One theoretical result that gives some support to that conjecture is the result of Tate and Shafarevich, which says that the rank of elliptic curves over the field $\mathbb{F}_q(t)$ (field of functions of one variable over a finite field) is unbounded. However, a recent paper by Park, Poonen, Voight, and Wood [302] provides heuristic arguments suggesting that the rank might be bounded.

The general method for finding high-rank elliptic curves consists of the following three phases:

- *Construction of families:* We generate a family of elliptic curves over $\mathbb{Q}$ (e.g. a curve over the field of rational functions $\mathbb{Q}(t)$) which we believe (or know) to contain elliptic curves of high rank, for example, because the rank of the curve over $\mathbb{Q}(t)$ is relatively large. According to Silverman's specialization theorem [328, Theorem 11.4], for all but finitely many rational numbers t_0, the rank of the curve over $\mathbb{Q}$, which is obtained by inserting (specializing) $t = t_0$, will be greater than or equal to the generic rank over $\mathbb{Q}(t)$. Let us mention that the algorithm, whose authors are Ivica Gusić and Petra Tadić, for curves with at least one point of order 2 over $\mathbb{Q}(t)$ enables finding the corresponding injective specializations $t = t_0$ and calculating the generic rank of such curves (see [201, 202]).
- *Sieve:* We calculate some data for each curve in the observed family, which give us certain information on the rank (e.g. lower and upper bounds for the rank – maybe with the assumption that some of the generally accepted conjectures are valid). It is important here that this (although perhaps imprecise) information on the rank can be calculated much faster than the rank itself. Based on this information, we select ("sieve") in the observed family, a small subset of the best candidates for the high rank.
- *Rank computation:* For each curve from the (small) list of best candidates, we try to compute the exact rank or at least the best possible lower bound for the rank to confirm that this curve really has a high rank.

Most of the methods that are still used today in the first two phases were introduced by Jean-Francois Mestre in 1980s and 1990s.

We will describe one of his constructions (from [274]), by which he obtained in 1991 infinitely many elliptic curves of rank ≥ 11. That construction is called *Mestre's polynomial method*. The starting point in the construction is the following fact (by analogy with the "theorem of division with a remainder", we could call it the "lemma on rooting with a remainder."). We describe it here in the form presented in [315, Chapter 1.3].

Lemma 2.5.3 *Let K be a field of characteristic 0. If $A \in K[x]$ is a monic polynomial and $\deg A = mn$, where $m, n \geq 1$, then there exists a monic polynomial $B \in K[x]$ such that $\deg B = n$ and $\deg(A - B^m) < n(m - 1)$.*

Proof By mathematical induction on i, we will prove that for each $i \leq n$, there exists $B_i \in K[x]$ such that $\deg B_i = n$ and $\deg(A - B_i^m) < nm - i$. For $i = 0$, we can take $B_0 = x^n$. Assume that the statement holds for $i - 1$, where $0 < i \leq n$. Thus, there exists a polynomial B_{i-1} of degree n such that $\deg(A - B_{i-1}^m) < nm - i + 1$. We are looking for the polynomial B_i in the form $B_i = B_{i-1} + cx^{n-i}$. We have

$$B_i^m = B_{i-1}^m + mc B_{i-1}^{m-1} x^{n-i} + \binom{m}{2} c^2 B_{i-1}^{m-2} x^{2(n-i)} + \cdots ,$$

where the degrees of every summand, starting from the third one onwards, are $\leq n(m - 2) + 2(n - i) = nm - 2i < nm - i$. The degree of $B_{i-1}^{m-1} x^{n-i}$ is $nm - i$, so we can choose c so that coefficients of x^{nm-i} cancel, and we get $\deg(A - B_i^m) = \deg(A - B_{i-1}^m - mc B_{i-1}^{m-1} x^{n-i}) < nm - i$. Since the polynomial B_0 is monic, it follows from the construction that all polynomials B_i are monic. $\square$

Corollary 2.5.4 *Let $p(x) \in \mathbb{Q}[x]$ be a monic polynomial and $\deg p = 2n$. Then there exist unique polynomials $q(x), r(x) \in \mathbb{Q}[x]$ such that q is monic, $p = q^2 - r$ and $\deg r \leq n - 1$.*

Proof The existence of $q(x)$ and $r(x)$ follows from Lemma 2.5.3 for $m = 2$. To prove the uniqueness, assume that $q_1(x), r_1(x) \in \mathbb{Q}[x]$ have the same properties as $q(x), r(x)$. From $q^2 - r = q_1^2 - r_1$, we get

$$(q - q_1)(q + q_1) = r - r_1. \tag{2.39}$$

Assume that $q \neq q_1$. Then the polynomial on the left-hand side of (2.39) has the degree $\geq n$, while that of the right-hand side has the degree $\leq n - 1$. Thus, $q = q_1$ and $r = r_1$. $\square$

We can find the polynomial q from Corollary 2.5.4 by successively calculating its unknown coefficients or from the asymptotic expansion of $\sqrt{p}$.

Now suppose that $p(x) = \prod_{i=1}^{2n}(x - a_i)$, where $a_1, \ldots, a_{2n}$ are distinct rational numbers. Then on the curve

$$C : \quad y^2 = r(x)$$

we have points $(a_i, \pm q(a_i))$, $i = 1, \ldots, 2n$. If $\deg r = 3$ or 4, and $r(x)$ does not have multiple roots, then C represents an elliptic curve. For $\deg r = 3$ it is clear. If $\deg r = 4$, then we choose one rational point on C (e.g. $(a_1, q(a_1))$) for the point at infinity, and transform C in an elliptic curve (see Sect. 2.2).

For $n = 5$, almost all choices of a_i give $\deg r = 4$. Then C has 10 rational points of the form $(a_i, q(a_i))$, and we can expect to get an elliptic curve of rank ≥ 9. Mestre constructed a family of elliptic curves (i.e. an elliptic curve over the field of rational functions $\mathbb{Q}(t)$) of rank ≥ 11, by taking $n = 6$ and $a_i = b_i + t$, $i = 1, \ldots, 6$; $a_i = b_{i-6} - t$, $i = 7, \ldots, 12$. In this case, the polynomial $r(x)$ generally has degree 5. However, we can try to choose the numbers $b_1, \ldots, b_6$ so that the coefficient of x^5 is equal to 0. In the first Mestre example from 1991, he took $b_1 = -17$, $b_2 = -16$, $b_3 = 10$, $b_4 = 11$, $b_5 = 14$, $b_6 = 17$.

Later Mestre [275], Nagao [290], and Kihara [234], using similar constructions, improved this result and constructed curves over $\mathbb{Q}(t)$ of rank ≥ 14. In 2006, using significantly different methods that have their origins in algebraic geometry, Elkies constructed a curve over $\mathbb{Q}(t)$ of rank ≥ 18 (see [162], [320, Chapter 13.1]). All these curves have a trivial torsion group. Fermigier [166], Kulesz [243], Lecacheux [249–251], and Nagao [291] modified Mestre's method and obtained families of curves with (relatively) large rank and non-trivial torsion groups. As we will see later in Sect. 3.6, elliptic curves associated with Diophantine m-tuples can be used as well for the construction of elliptic curves (over $\mathbb{Q}$ and $\mathbb{Q}(t)$) of high rank for torsion groups $\mathbb{Z}/2\mathbb{Z} \times \mathbb{Z}/k\mathbb{Z}$, $k = 2, 4, 6, 8$.

In the second phase, "sieving", the rough idea is that it is more likely that a curve will have "a lot" of rational points (i.e. large rank) if there are many points under the reduction modulo p (i.e. if the number $N_p = |E(\mathbb{F}_p)|$ is large) for the "majority" of primes p. Note that according to Hasse's theorem, we have

$$p + 1 - 2\sqrt{p} \leq N_p \leq p + 1 + 2\sqrt{p}$$

(for a proof, see [239, Chapter 10]), so the fact that the number N_p is large actually means that it is close to the upper bound from Hasse's theorem.

A much more precise version of this rough idea is the famous *Birch and Swinnerton-Dyer (BSD) conjecture*, which states that

$$\prod_{p \leq X,\, p \nmid 2\Delta} \frac{N_p}{p} \sim const \cdot (\log X)^r,$$

where $r = \mathrm{rank}(E)$. The BSD conjecture is often expressed in terms of the L-function, which is defined by

$$L(E, s) = \prod_{p \nmid \Delta} \left(1 - a_p p^{-s} + p^{1-2s}\right)^{-1} \cdot \prod_{p \mid \Delta} \left(1 - a_p p^{-s}\right)^{-1},$$

where $a_p = p + 1 - N_p$ for primes p in which E has good reduction, $a_p = 0$ in the case of additive reduction, $a_p = 1$ in the case of multiplicative split reduction, and $a_p = -1$ in the case of multiplicative non-split reduction. We can understand this function as an analogue of the Riemann zeta function if we recall Euler's formula $\zeta(s) = \sum_{n=1}^{\infty} \frac{1}{n^s} = \prod_p (1 - p^{-s})^{-1}$ (see [116, Chapter 7.3]). The function

$L(E, s)$ has an analytic continuation to the entire complex plane $\mathbb{C}$ and satisfies the functional equation

$$\Lambda(s) = w_E \cdot \Lambda(2 - s),$$

where $w_E \in \{-1, 1\}$, while

$$\Lambda(s) = N^{s/2}(2\pi)^{-s}\Gamma(s)L(E, s),$$

where Γ is the gamma function, and N denotes the conductor of E.

Now the BSD conjecture can be expressed by stating that the vanishing order of $L(E, s)$ in $s = 1$ (the so-called *analytic rank* of E) is equal to the rank of r, i.e. that

$$L(E, s) = c \cdot (s - 1)^r + \text{terms with higher order in } (s - 1),$$

where $c \neq 0$ is a constant. It is known that the conjecture is valid if the analytic rank is equal to 0 or 1 [198, 241].

The parity conjecture states that the value of w_E determines the parity of the rank: $w_E = (-1)^r$, i.e. if $w_E = 1$, then the rank is even, and if $w_E = -1$, then the rank is odd (it allows us to conditionally determine the rank in cases when, by other methods, we conclude that $r \in \{r', r' + 1\}$ for some r'). The value of w_E can be calculated in PARI by the function `ellrootno`.

Although the Birch and Swinnerton-Dyer conjecture is very important for understanding the rank of elliptic curves, it is generally not suitable for direct calculation (at least conditionally) of the rank due to the slow convergence of the corresponding series. That is why in the "sieving" phase, some other variations of the rough idea mentioned above are usually used.

We can fix a finite set of primes $\mathcal{P}$, and for each $p \in \mathcal{P}$ find all values of the parameters modulo p that maximize N_p. If we observe the curve of the form $y^2 = x^3 + ax + b$, the parameters will be $(a, b) \in \mathbb{F}_p^2$, and the maximum N_p is $p + 1 + \lfloor 2\sqrt{p} \rfloor$. If we are looking for curves of high rank with the given torsion group, then we use the corresponding parametrizations given in Sect. 2.4, and the maximum N_p is $|E(\mathbb{Q})_{\text{tors}}| \cdot \left\lfloor \frac{p+1+\lfloor 2\sqrt{p} \rfloor}{|E(\mathbb{Q})_{\text{tors}}|} \right\rfloor$. After that, by the Chinese remainder theorem, we construct a list of parameters which maximize N_p for all $p \in \mathcal{P}$ (see [272]).

Mestre and Nagao (see [272, 289]) gave heuristic arguments (motivated by the BSD conjecture) which suggest that for curves of high rank certain sums should assume large values (the largest in the observed family). Some of these sums are

$$S_1(X) = \sum_{p \leq X} \frac{N_p + 1 - p}{N_p} \log p,$$

$$S_2(X) = \sum_{p \leq X} \frac{N_p + 1 - p}{N_p},$$

$$S_3(X) = \sum_{p \leq X} (N_p - p - 1) \log p$$

(see [163] for some optimizations, [236] for more precise connections between these sums and the BSD and Nagao's conjecture, and [229] for using deep convolutional neural networks in this context and comparison of several similar sums). In applications of this idea, several positive integers $X_1 < X_2 < \cdots < X_k$ are chosen, and $S_i(X_1), S_i(X_2), \ldots, S_i(X_k)$ are calculated, but so that in each step we discard, say, 80 % of the "worst" curves, i.e. those with the smallest values of the corresponding sums. We note that for an efficient implementation of this method, X_k should not be too big (say $X_k < 100000$) because we do not have a very efficient algorithm for calculating N_p for very large primes p. In PARI, the number a_p can be calculated by the function $\mathtt{ellap}(E, p)$, so N_p is obtained as $N_p = p + 1 - a_p$.

Let us see which heuristic argument connects the sum $S_1(N)$ and the BSD conjecture. The BSD conjecture implies that $L(E, s) = L(s) = (s - 1)^r \cdot g(s)$, where $g(1) \neq 0$. (In fact, a more precise version of the BSD conjecture predicts exactly the value of $g(1)$. In the description, among other quantities, there appear the order of the torsion and Tate-Shafarevich groups and the regulator.) The logarithmic derivative gives

$$\frac{L'(s)}{L(s)} = \frac{r}{s - 1} + \frac{g'(s)}{g(s)}.$$

So, when s tends to 1, we expect that the logarithmic derivative of L tends faster to infinity the higher the rank is. Now we look at the product

$$L(s, N) = \prod_{p \leq N} (1 - a_p p^{-s} + p^{1-2s})^{-1},$$

which differs from $L(s)$ in that we "cut" the product at N, and we ignored the difference in factors between prime numbers with good and bad reduction. Let us put $f(s, N) = \log L(s, N)$. Then

$$f'(s, N) = -\sum_{p \leq N} \frac{a_p p^{-s} - 2p^{1-2s}}{1 - a_p p^{-s} + p^{1-2s}} \log p.$$

It seems reasonable to assume that $\lim_{N \to \infty} f'(s, N)$ is a good approximation for $\frac{L'(s)}{L(s)}$. Therefore, the rate of divergence of the limit

$$\lim_{s \to 1} \lim_{N \to \infty} f'(s, N) \tag{2.40}$$

can be an indicator of the size of the rank. If we could change the order of limits in (2.40), we would get exactly

$$\lim_{N \to \infty} f'(1, N) = \lim_{N \to \infty} -\sum_{p \leq N} \frac{a_p - 2}{p + 1 - a_p} \log p = \lim_{N \to \infty} S_1(N).$$

Thus, we have shown the announced connection between the BSD conjecture and the sum $S_1(N)$.

We saw before that in the case when E has a rational point of order 2, we have a very simple upper bound for the rank: $r \leq \omega(b) + \omega(b') - 1$, and also $r \leq s$, where s is the 2-Selmer rank. In general, in the case where $E(\mathbb{Q})$ has a point of finite order, it is possible to give a simple upper bound for the rank, as we will now explain. If we are lucky, that upper bound matches the lower bound we get by searching for points with small heights, and in this way, we can get the exact value for the rank without the need to apply more demanding methods. The mentioned upper bound is called *Mazur's bound* [269]. Let E be an elliptic curve over $\mathbb{Q}$ given by its minimal Weierstrass equation, and assume that E has a rational point of odd prime order p. Then

$$r \leq m_p = b + a - m - 1,$$

where

- b is the number of primes with bad reduction;
- a is the number of primes with additive reduction;
- m is the number of primes q with multiplicative reduction for which it is additionally valid that p does not divide the exponent of q in Δ and that $q \not\equiv 1$ (mod p).

Example 2.5.5 (Dujella-Lecacheux, 2001) Let us compute the rank of the curve E given by the equation

$$y^2 + y = x^3 + x^2 - 1712371016075117860x + 8857879575356913895129401 64.$$

We have

$$E(\mathbb{Q})_{\text{tors}} = \{O, (888689186, 8116714362487),$$
$$(-139719349, -33500922231893), (-139719349, 33500922231892),$$
$$(888689186, -8116714362488)\} \cong \mathbb{Z}/5\mathbb{Z}.$$

Thus we calculate Mazur's bound m_5. The discriminant is

$$\Delta = -3^{15} \cdot 5^5 \cdot 7^5 \cdot 11^5 \cdot 19^5 \cdot 41^5 \cdot 127^5 \cdot 1409 \cdot 10864429,$$

so we have: $b = 9$, $a = 0$, $m = 2$ and we get that $r \leq m_5 = 6$.

By searching for points $P = (x, y)$ on E with integer coordinates and $|x| < 10^9$ (in PARI, this can be done with the function `ellratpoints`), we find the following six independent points modulo $E(\mathbb{Q})_{\text{tors}}$ (how to check the independence of points will be discussed in the next section):

$$(624069446, 7758948474007), (763273511, 4842863582287)$$

$$(680848091, 5960986525147), (294497588, 20175238652299)$$

$$(-206499124, 35079702960532), (676477901, 6080971505482),$$

by which we proved that $\text{rank}(E) = 6$ (this was, at the time of its discovery, the curve of the largest known rank with the torsion group $\mathbb{Z}/5\mathbb{Z}$, and it was obtained by specializing parameters in a family of elliptic curves from [249]).

If by the above methods, the obtained upper bound for the rank is greater than the lower bound by at least 2, so even with the information about conditional rank parity we cannot determine the rank, we can try to increase the lower bound or decrease the upper bound. To increase the lower bound, we can increase the limit up to which the algorithms search for points on the curve or the corresponding equations of fourth degree. In PARI, this can be done by using an additional parameter in the `ellrank` function, for example, `ellrank(e, 6)`, while in `mwrank`, this can be done using the option `-b`, for example `-b 12` (an additional useful option, necessary for curves with very large coefficients, is the option `-p` which increases the default precision).

To obtain an upper bound for the rank (or to decrease the previously obtained one), we can use *Mestre's conditional upper bound* [273], which gives an upper bound for the rank assuming that the Birch and Swinnerton-Dyer conjecture and the generalized Riemann conjecture hold:

$$\text{rank} \leq \frac{\pi^2}{8\lambda}\left(\log N - 2\sum_{p^m \leq e^\lambda} b(p^m)F_\lambda(m\log p)\frac{\log p}{p^m} - M_\lambda\right),$$

where N is the conductor, $b(p^m) = a_p^m$ if $p \mid N$, $b(p^m) = \alpha_p^m + \alpha'^m_p$ if $p \nmid N$, where α_p and α'_p are roots of $x^2 - a_px + p$,

$$M_\lambda = 2\left(\log 2\pi + \int_0^{+\infty}(F_\lambda(x)/(e^x - 1) - e^{-x}/x)dx\right),$$

$F_\lambda(x) = F(x/\lambda)$, while F is a function with specific properties, and we can take $F(x) = (1 - |x|)\cos(\pi x) + \sin(\pi|x|)/\pi$ for $x \in [-1, 1]$ and $F(x) = 0$ otherwise. For example, for the record Elkies curve of rank ≥ 28, Mestre's bound gives that the rank ≤ 30, while for the previous record curve of rank ≥ 24, it gives that the rank is exactly equal to 24 (assuming that BSD and GRH hold, see [42]).

Let T be one of the 15 possible torsion groups for an elliptic curve over $\mathbb{Q}$ (according to Mazur's theorem). We define

$$B(T) = \sup\{\text{rank}(E(\mathbb{Q})) \;:\; E(\mathbb{Q})_{\text{tors}} \cong T\}.$$

There was also a conjecture that $B(T)$ is unbounded for all possible torsion groups T, but as we have already mentioned, there are recent papers that suggest boundedness of the rank and thus also of the quantities $B(T)$. At present, we only know that $B(T) \geq 3$ for all T. The currently known best lower bounds for $B(T)$

Table 2.2 High rank elliptic curves with prescribed torsion

T	$B(T) \geq$	Authors
0	28	Elkies
$\mathbb{Z}/2\mathbb{Z}$	20	Elkies and Klagsbrun
$\mathbb{Z}/3\mathbb{Z}$	15	Elkies and Klagsbrun
$\mathbb{Z}/4\mathbb{Z}$	13	Elkies and Klagsbrun
$\mathbb{Z}/5\mathbb{Z}$	9	Klagsbrun
$\mathbb{Z}/6\mathbb{Z}$	9	Klagsbrun, Voznyy
$\mathbb{Z}/7\mathbb{Z}$	6	Klagsbrun
$\mathbb{Z}/8\mathbb{Z}$	6	Elkies, Dujella, MacLeod, and Peral, Voznyy
$\mathbb{Z}/9\mathbb{Z}$	4	Fisher, van Beek, Dujella and Petričević, Dujella, Petričević, and Rathbun
$\mathbb{Z}/10\mathbb{Z}$	4	Dujella, Elkies, Fisher
$\mathbb{Z}/12\mathbb{Z}$	4	Fisher
$\mathbb{Z}/2\mathbb{Z} \times \mathbb{Z}/2\mathbb{Z}$	15	Elkies
$\mathbb{Z}/2\mathbb{Z} \times \mathbb{Z}/4\mathbb{Z}$	**9**	Dujella and Peral, Klagsbrun
$\mathbb{Z}/2\mathbb{Z} \times \mathbb{Z}/6\mathbb{Z}$	**6**	Elkies, Dujella, Peral, and Tadić, Dujella and Peral
$\mathbb{Z}/2\mathbb{Z} \times \mathbb{Z}/8\mathbb{Z}$	**3**	Connell, Dujella, Campbell and Goins, Rathbun, Dujella and Rathbun, Flores, Jones, Rollick, Weigandt, and Rathbun, Fisher, AttarBashi, Rathbun, and Voznyy, AttarBashi, Fisher, Rathbun, and Voznyy, AttarBashi, Fisher, and Voznyy

are given in Table 2.2. Most of the results from this table were obtained by a combination of the methods described in this section, with several improvements described in the recent paper by Elkies and Klagsbrun [163]. The ranks that can be obtained with elliptic curves induced by rational Diophantine triples are marked in bold. Details about the record curves can be found on the web page [118].

We said that the first step in finding elliptic curves of high rank (and then also those with some additional property, such as given torsion group) is the construction of families of elliptic curves (most often, these are elliptic curves over the field of rational functions $\mathbb{Q}(t)$) of relatively large "generic" rank. Therefore, the following quantities are of interest:

$$G(T) = \sup\{\operatorname{rank}(E(\mathbb{Q}(t))) \; : \; E(\mathbb{Q}(t))_{\mathrm{tors}} \cong T\},$$

$$C(T) = \lim\sup\{\operatorname{rank}(E(\mathbb{Q})) \; : \; E(\mathbb{Q})_{\mathrm{tors}} \cong T\}.$$

In the case that in Tables 2.3 and 2.4, the bound for $C(T)$ is larger than the bound for $G(T)$, this means that the current record for $C(T)$ comes from the

Table 2.3 Elliptic curves over $\mathbb{Q}(t)$ with high rank and prescribed torsion

T	$G(T) \geq$	Authors
0	18	Elkies
$\mathbb{Z}/2\mathbb{Z}$	11	Elkies, Dujella and Peral
$\mathbb{Z}/3\mathbb{Z}$	7	Elkies, Eroshkin
$\mathbb{Z}/4\mathbb{Z}$	6	Dujella and Peral
$\mathbb{Z}/5\mathbb{Z}$	4	Eroshkin
$\mathbb{Z}/6\mathbb{Z}$	3	Lecacheux, Kihara, Eroshkin, Woo, Dujella and Peral, MacLeod, Voznyy
$\mathbb{Z}/7\mathbb{Z}$	1	Kulesz, Lecacheux, Rabarison, Harrache, MacLeod
$\mathbb{Z}/8\mathbb{Z}$	2	Dujella and Peral, MacLeod, Dujella, Kazalicki, and Peral
$\mathbb{Z}/9\mathbb{Z}$	0	Kubert
$\mathbb{Z}/10\mathbb{Z}$	0	Kubert
$\mathbb{Z}/12\mathbb{Z}$	0	Kubert
$\mathbb{Z}/2\mathbb{Z} \times \mathbb{Z}/2\mathbb{Z}$	7	Elkies
$\mathbb{Z}/2\mathbb{Z} \times \mathbb{Z}/4\mathbb{Z}$	**4**	Dujella and Peral
$\mathbb{Z}/2\mathbb{Z} \times \mathbb{Z}/6\mathbb{Z}$	**2**	Dujella and Peral, MacLeod, Dujella, Kazalicki, and Peral
$\mathbb{Z}/2\mathbb{Z} \times \mathbb{Z}/8\mathbb{Z}$	**0**	Kubert

family parameterized by rational points on some elliptic curve of positive rank. The references for the entries in the tables can be found on the web page [120]. In the table for $G(T)$, the ranks that come from the elliptic curves induced by rational Diophantine triples are again marked in bold.

There are also more precise predictions about how large a rank can be for elliptic curves with the given torsion group. It may be interesting to mention that in [302], Park, Poonen, Voight, and Wood stated a prediction that only finitely many elliptic curves with torsion groups $\mathbb{Z}/8\mathbb{Z}$ and $\mathbb{Z}/2\mathbb{Z} \times \mathbb{Z}/6\mathbb{Z}$ have rank ≥ 4. On the other hand, Dujella and Peral [148] proved that there are infinitely many elliptic curves with these torsion groups and rank ≥ 3 (for $\mathbb{Z}/2\mathbb{Z} \times \mathbb{Z}/6\mathbb{Z}$, see Sect. 3.6.4), so if the above prediction is correct, this result for these torsion groups would be the best possible. The same applies to the results on torsion groups $\mathbb{Z}/6\mathbb{Z}$ and $\mathbb{Z}/2\mathbb{Z} \times \mathbb{Z}/4\mathbb{Z}$ obtained by Eroshkin [165]. Those four "borderline" ranks are marked in bold in the table for $C(T)$. In general, the prediction is that only finitely many (non-isomorphic) curves with a given torsion group have rank larger than that specified in the PPVW column in Table 2.4. Let us also mention that in [138], Dujella, Kazalicki, and Peral gave additional infinite families of rank ≥ 3 and the torsion group $\mathbb{Z}/8\mathbb{Z}$ or $\mathbb{Z}/2\mathbb{Z} \times \mathbb{Z}/6\mathbb{Z}$ and performed experiments suggesting that there could be infinitely many curves with these torsion groups and rank ≥ 4. These results might indicate that the

Table 2.4 Infinite families of elliptic curves with high rank and prescribed torsion

T	$C(T) \geq$	PPVW	Authors
0	19	21	Elkies
$\mathbb{Z}/2\mathbb{Z}$	11	13	Elkies, Dujella and Peral
$\mathbb{Z}/3\mathbb{Z}$	8	9	Eroshkin
$\mathbb{Z}/4\mathbb{Z}$	6	7	Elkies, Dujella and Peral
$\mathbb{Z}/5\mathbb{Z}$	4	5	Eroshkin
$\mathbb{Z}/6\mathbb{Z}$	**5**	5	Eroshkin
$\mathbb{Z}/7\mathbb{Z}$	2	3	Lecacheux, Elkies, Rabarison, Harrache, Voznyy
$\mathbb{Z}/8\mathbb{Z}$	**3**	3	Dujella and Peral, Dujella, Kazalicki, and Peral
$\mathbb{Z}/9\mathbb{Z}$	1	2	Atkin and Morain, Kulesz, Rabarison, Gasull, Manosa, and Xarles
$\mathbb{Z}/10\mathbb{Z}$	1	2	Atkin and Morain, Kulesz, Rabarison, Voznyy
$\mathbb{Z}/12\mathbb{Z}$	1	2	Suyama, Kulesz, Rabarison, Halbeisen, Hungerbühler, Shamsi Zargar, and Voznyy
$\mathbb{Z}/2\mathbb{Z} \times \mathbb{Z}/2\mathbb{Z}$	8	9	Elkies
$\mathbb{Z}/2\mathbb{Z} \times \mathbb{Z}/4\mathbb{Z}$	**5**	5	Eroshkin
$\mathbb{Z}/2\mathbb{Z} \times \mathbb{Z}/6\mathbb{Z}$	**3**	3	Dujella and Peral, Dujella, Kazalicki, and Peral
$\mathbb{Z}/2\mathbb{Z} \times \mathbb{Z}/8\mathbb{Z}$	1	2	Atkin and Morain, Kulesz, Lecacheux, Campbell and Goins, Rabarison

heuristic in [302] needs some adjustments, at least in the case of curves with certain torsion groups.

Finding elliptic curves with positive rank and large torsion is not just a curiosity. Montgomery [281] and Atkin and Morain [15] proposed using elliptic curves over $\mathbb{Q}$ with a large torsion group and positive rank to increase the chance of success of Lenstra's algorithm for factorization of large integers [253]. In the background of that idea is Proposition 2.4.6, which implies that $|E(\mathbb{F}_p)|$ is divisible by the order of the torsion group of $E(\mathbb{Q})$. Hence, a larger torsion group of $E(\mathbb{Q})$ increases the likelihood that the order of $|E(\mathbb{F}_p)|$ has small prime factors, which is essential for the success of Lenstra's algorithm. Furthermore, it turns out that we can have an additional advantage obtained using elliptic curves with an even larger torsion group over some number field of a small degree. This raises the question of what torsion groups are possible, for example, over quadratic or cubic fields.

Let E be an elliptic curve over a quadratic field $\mathbb{K}$. Kenku and Momose [232] and Kamienny [226] proved that the torsion group of $E(\mathbb{K})$ is isomorphic to one of the following 26 groups:

$$\begin{aligned}
\mathbb{Z}/k\mathbb{Z}, &\quad 1 \le k \le 18,\ k \ne 17, \\
\mathbb{Z}/2\mathbb{Z} \times \mathbb{Z}/2k\mathbb{Z}, &\quad 1 \le k \le 6, \\
\mathbb{Z}/3\mathbb{Z} \times \mathbb{Z}/3k\mathbb{Z}, &\quad k = 1, 2,\ (\text{only if } \mathbb{K} = \mathbb{Q}(\sqrt{-3})), \\
\mathbb{Z}/4\mathbb{Z} \times \mathbb{Z}/4\mathbb{Z}, &\quad (\text{only if } \mathbb{K} = \mathbb{Q}(i)).
\end{aligned} \tag{2.41}$$

Note that if the torsion group over a number field $\mathbb{K}$ contains $\mathbb{Z}/k\mathbb{Z} \times \mathbb{Z}/k\mathbb{Z}$, then the k-th roots of unity lie in $\mathbb{K}$ (see [329, Chapter III.8]).

Similarly to Mazur's theorem, here also, each of these 26 groups appears as a torsion group over some quadratic field for infinitely many non-isomorphic elliptic curves. With cubic fields, the situation is different. Namely, Najman [295] proved that there exists a unique curve with torsion group $\mathbb{Z}/21\mathbb{Z}$ over some cubic field. For each of the 26 groups from (2.41), there is an elliptic curve over some quadratic field with that torsion group and positive rank (moreover, of rank ≥ 2). Current record ranks can be found on the web page [119].

2.6 Canonical Height and Mordell-Weil Basis

The two main steps in the proof of the Mordell-Weil theorem are:

- the proof that the index $[E(\mathbb{Q}) : 2E(\mathbb{Q})]$ is finite;
- properties of the *height* h, defined by $h(P) = \log H(x)$, where $P = (x, y)$ and $H(\frac{m}{n}) = \max(|m|, |n|)$, while $h(O) = 0$.

The function H is sometimes called the "naïve" height and h the logarithmic height. It is obvious that for any constant C, the set

$$\{P \in E(\mathbb{Q}) : h(P) \le C\}$$

is finite (it has no more than $2(2e^C + 1)^2$ elements).

We want to see the relationship between the heights of points P and $2P$ (roughly how many times the number of digits will increase in the decimal representation of the point $2P$ compared to the representation of the point P). Let the curve be given by the equation $y^2 = x^3 + ax + b$ and $P = (x, y) \in E(\mathbb{Q})$. Then the x-coordinate of the point $2P$ is

$$x(2P) = \frac{x^4 - 2ax^2 - 8bx + a^2}{4(x^3 + ax + b)}.$$

Since squaring a number roughly doubles its height and the highest power of x which appears in the expression of $x(2P)$ is x^4, we conclude that $h(2P) \approx 4h(P)$ (for the precise statement and its proof, see [356, Chapter 8.3]).

The *canonical height*, which is also known as the *Néron-Tate height*, is defined by

$$\hat{h}(P) = \lim_{n \to \infty} \frac{h(2^n P)}{4^n}, \tag{2.42}$$

and has important properties.

It can be shown that the canonical height satisfies $\hat{h}(2P) = 4\hat{h}(P)$, and, more generally, $\hat{h}(mP) = m^2 \hat{h}(P)$. Furthermore, the function $\hat{h}$ also has the following properties (for the proof, see [116, Chapter 15.4] or [356, Chapter 8.3]):

- For all $P \in E(\mathbb{Q})$, $\hat{h}(P) \geq 0$ holds. Here the equality $\hat{h}(P) = 0$ holds if and only if $P \in E(\mathbb{Q})_{\text{tors}}$.
- For any constant C, the set $\{P \in E(\mathbb{Q}) : \hat{h}(P) \leq C\}$ is finite.
- For all $P, Q \in E(\mathbb{Q})$, the "parallelogram law" holds:

$$\hat{h}(P + Q) + \hat{h}(P - Q) = 2\hat{h}(P) + 2\hat{h}(Q). \tag{2.43}$$

Let us now sketch the proof of the Mordell-Weil theorem (assuming that the index $[E(\mathbb{Q}) : 2E(\mathbb{Q})]$ is finite, about which we said something (at least in the special case of a curve with a point of order 2) in the previous section).

Proof Let $\{R_1, R_2, \ldots, R_n\}$ be a set of representatives from $E(\mathbb{Q})/2E(\mathbb{Q})$, i.e.

$$E(\mathbb{Q}) = (R_1 + 2E(\mathbb{Q})) \cup (R_2 + 2E(\mathbb{Q})) \cup \cdots \cup (R_n + 2E(\mathbb{Q})), \tag{2.44}$$

and $k = \max_i (\hat{h}(R_i))$. Let $Q_1, \ldots, Q_m$ be all the points in $E(\mathbb{Q})$ for which $\hat{h}(Q_i) \leq k$ (we know that the set of such points is finite). Let G be a subgroup of $E(\mathbb{Q})$ generated by

$$R_1, \ldots, R_n, Q_1, \ldots, Q_m.$$

We claim that $G = E(\mathbb{Q})$. Assume the opposite, i.e. that there exists $P \in E(\mathbb{Q})$ such that $P \notin G$. Since there are only finitely many points with smaller canonical height than $\hat{h}(P)$, without loss of generality, we can assume that P is the point with the smallest height for which $P \notin G$ holds. From (2.44), we have that for some index i and point $P_1 \in E(\mathbb{Q})$,

$$P = R_i + 2P_1.$$

From the properties of the canonical height mentioned above and due to the fact that $\hat{h}(P) > k$ (since $P \neq Q_i$), we have

$$4\hat{h}(P_1) = \hat{h}(2P_1) = \hat{h}(P - R_i) = 2\hat{h}(P) + 2\hat{h}(R_i) - \hat{h}(P + R_i)$$

$$\leq 2\hat{h}(P) + 2k < 2\hat{h}(P) + 2\hat{h}(P) = 4\hat{h}(P).$$

Therefore, $\hat{h}(P_1) < \hat{h}(P)$, and since P is the point with the smallest canonical height, which is not in G, it must be that $P_1 \in G$. However, then also $P = R_i + 2P_1 \in G$, so we got a contradiction. Therefore, $G = E(\mathbb{Q})$, which proves that the group $E(\mathbb{Q})$ is finitely generated. $\qquad\square$

Calculation of canonical height $\hat{h}(P)$ by definition (via limit) is problematic because it requires the calculation of very large integers that appear in the numerator and denominator of points $x(2^n P)$. We have to calculate them exactly because in the next step we need the values of the numerator and denominator after possible cancelling (and just from approximate values we cannot know which cancelling will occur). In 1988, Silverman [326] gave a more efficient algorithm for calculating the canonical height. The idea is to write the "global height" $\hat{h}(P)$ as the sum of the "local heights" $\hat{h}_p(P)$, where p is a prime number or $p = \infty$. In PARI, this algorithm is implemented in the function `ellheight`.

The parallelogram law (2.43) suggests that, analogously as in the case of a norm that satisfies the law of the same name, with the help of the canonical height, we will be able to define some variant of the "scalar product".

Definition 2.6.1 The *Néron-Tate pairing* of the heights of points $P, Q \in E(\mathbb{Q})$ is defined by

$$\langle P, Q \rangle = \frac{1}{2}(\hat{h}(P + Q) - \hat{h}(P) - \hat{h}(Q))$$

$$= \frac{1}{4}(\hat{h}(P + Q) - \hat{h}(P - Q)).$$

From $\hat{h}(2P) = 4\hat{h}(P)$, it follows that $\hat{h}(P) = \langle P, P \rangle$. The Néron-Tate height pairing is obviously symmetric, and it is not difficult to show (using the parallelogram law) that it is also bilinear.

Proposition 2.6.2 *If T is a torsion point and P is any point on $E(\mathbb{Q})$, then*

$$\hat{h}(P + T) = \hat{h}(P) \quad \text{and} \quad \langle P, T \rangle = 0.$$

Proof Let T be a point of order m. Then

$$\hat{h}(P + T) = \frac{\hat{h}(m(P + T))}{m^2} = \frac{\hat{h}(mP)}{m^2} = \hat{h}(P),$$

so

$$\langle P, T \rangle = \frac{1}{2}(\hat{h}(P) - \hat{h}(P) - 0) = 0.$$

$\square$

We now define an analogue of Gram's determinant of scalar products.

Definition 2.6.3 The *height determinant* of points $P_1, \ldots, P_m$ is $\det((\langle P_i, P_j \rangle))$.

In the software package PARI, the height determinant can be calculated by the function `ellheightmatrix`, followed by the function `matdet`.

It holds that

$$\hat{h}(n_1 P_1 + \cdots + n_m P_m) = N((\langle P_i, P_j \rangle))N^{\tau},$$

where $N = (n_1, \ldots, n_m)$.

Furthermore, the points $P_1, \ldots, P_m$ are dependent mod $E(\mathbb{Q})_{\mathrm{tors}}$, i.e. there are integers $n_1, \ldots, n_m$, that are not all equal to 0, such that $n_1 P_1 + \cdots + n_m P_m \in E(\mathbb{Q})_{\mathrm{tors}}$, if and only if $\det((\langle P_i, P_j \rangle)) = 0$.

A particularly important case of independent points is the Mordell-Weil basis. A *Mordell-Weil basis* $Q_1, \ldots, Q_r$ for $E(\mathbb{Q})$ is a $\mathbb{Z}$-basis for $E(\mathbb{Q})$ mod $E(\mathbb{Q})_{\mathrm{tors}}$, i.e. each $P \in E(\mathbb{Q})$ can be expressed in a unique way as

$$P = n_1 Q_1 + \cdots + n_r Q_r + T, \quad n_i \in \mathbb{Z}, \quad T \in E(\mathbb{Q})_{\mathrm{tors}}.$$

The *regulator* of E is $\mathrm{Reg}(E) = \det((\langle Q_i, Q_j \rangle))$, where $Q_1, \ldots, Q_r$ is a Mordell-Weil basis. If $r = 0$, then it is defined that $\mathrm{Reg}(E) = 1$. From the independence of points in a basis, it follows that $\mathrm{Reg}(E) \neq 0$. Moreover, it can be shown that $\mathrm{Reg}(E) > 0$ (see [73, Chapter 3.5]).

Let $P_1, \ldots, P_r$ be r independent points mod $E(\mathbb{Q})_{\mathrm{tors}}$, and suppose that

$$\{P_1, \ldots, P_r\} \cup E(\mathbb{Q})_{\mathrm{tors}}$$

generates a subgroup G of $E(\mathbb{Q})$ of index q. Then

$$\det((\langle P_i, P_j \rangle)) = q^2 \mathrm{Reg}(E).$$

Thus, $\mathrm{Reg}(E)$ is the smallest height determinant of r independent points on $E(\mathbb{Q})$. Furthermore, r independent points form a basis if and only if their height determinant is equal to the regulator.

If a Mordell-Weil basis $P_1, \ldots, P_r$ (or at least some independent set of points) of an elliptic curve is known, it is often of interest to find a basis (or a set of points that generates the same subgroup as the initial set of points) with elements with smaller (canonical) heights. For this purpose, the LLL algorithm can be used, with the role of the Gram matrix being played by the height matrix $(\langle P_i, P_j \rangle)$. This idea was presented and implemented by Rathbun [306] in 2003.

First, let us say something about the LLL algorithm, which is related to the problem of finding the shortest non-zero vector in a lattice.

Let n be a positive integer and let $b_1, \ldots, b_n$ be linearly independent vectors in $\mathbb{R}^n$. The *lattice* ($\mathbb{Z}$-module) L spanned by these vectors is the set of all their integer linear combinations

$$L = \left\{ \sum_{i=1}^{n} x_i b_i \ : \ x_i \in \mathbb{Z} \right\}.$$

We call $B = \{b_1, \ldots, b_n\}$ a *lattice basis* of L. For example, in $\mathbb{R}^2$, if $b_1 = (1, 0)$, $b_2 = (0, 1)$, then L is the lattice of all points in the plane with integer coordinates.

A lattice typically has many different bases, so we may ask if we can choose a basis that would have some additional good properties. It is clear that B represents the basis of the vector space $\mathbb{R}^n$. We know that we can obtain an orthogonal basis for the same vector space by the Gram-Schmidt process ($b_i^* = b_i - \sum_{j=1}^{i-1} \mu_{ij} b_j^*$, $i = 1, \ldots, n$, where $\mu_{ij} = \frac{\langle b_i, b_j^* \rangle}{\langle b_j^*, b_j^* \rangle}$). However, the new basis does not have to span the same lattice as the initial basis B because the coefficients μ_{ij} do not have to be integers. In general, a lattice does not need to have an orthogonal basis. In 1982, A. K. Lenstra, H. W. Lenstra, and L. Lovász [252] introduced the notion of an *LLL-reduced basis* with the following properties:

(1) $|\mu_{i,j}| \leq \frac{1}{2}, 1 \leq j < i \leq n$;
(2) $\|b_i^*\|^2 \geq (\frac{3}{4} - \mu_{i,i-1}^2)\|b_{i-1}^*\|^2$.

The first condition can be interpreted as saying that an LLL-reduced basis is "almost orthogonal", while the second condition says that the sequence of norms of the vectors $\|b_i^*\|$ is "almost increasing". An additional important property of an LLL-reduced basis is that the first vector in that basis is very short, i.e. it has a small norm. It can be proved that $\|b_1\| \leq 2^{(n-1)/2}\|x\|$ for all non-zero vectors $x \in L$ (see [332, Chapter V.3]), but in applications, it very often happens that b_1 is exactly the shortest non-zero vector from L.

In their paper from 1982, A. K. Lenstra, H. W. Lenstra, and L. Lovász [252] presented a polynomial time algorithm for constructing an LLL-reduced basis from an arbitrary lattice basis (named after them the *LLL algorithm*). The algorithm soon found numerous applications, for example, in the factorization of polynomials with rational coefficients, cryptanalysis of the RSA cryptosystem with a small public or private exponent, the knapsack problem, Diophantine approximations (see Sect. 4.1.5) and Diophantine equations (see Sects. 4.4 and 4.6).

In 1989, de Weger [78] proposed a variant of the LLL algorithm that uses only integer arithmetic (if the input data are integers) and avoids problems related to the numerical stability of the algorithm.

In the software package PARI, the LLL algorithm (the L^2 variant from [296]) is implemented by the function $\mathtt{qflll}(x)$, which returns as a result the transformation matrix T such that xT is an LLL-reduced basis of the lattice generated by the columns of the matrix x. There is also the function $\mathtt{qflllgram}(x)$, which does the same thing as $\mathtt{qflll}$, except that here x is the Gram matrix of scalar products of lattice vectors and not the vector coordinates matrix. The result is again the transformation matrix T whose columns provide the connection between the initial vectors and the vectors in the reduced basis. Programs for finding an LLL-reduced basis also exist in other software packages (e.g. $\mathtt{lattice}$ in Maple, $\mathtt{LatticeReduce}$ in Mathematica, $\mathtt{LLL}$ in Magma).

Let us now present the previously mentioned Rathbun's algorithm.

Reduced Mordell-Weil basis

```
ellLreduce (e, plist) =
e = ellinit (e);
n = length (plist);
u = qflllgram (ellheightmatrix (e, plist));
newplist = vector (n, j, [0]);
for (i = 1, n,  for (j = 1, n,
newplist[i] = elladd (e, newplist[i],  ellpow (e, plist[j], u[j, i])))));
```

Example 2.6.4 Consider the curve

$$y^2 + xy = x^3 - 39139760676569376374592499967383835x$$
$$+ 8061422259431089866408009166162570044567355791 3297.$$

It is one of the curves with the highest known rank among the curves with the torsion group $\mathbb{Z}/10\mathbb{Z}$, and the curve with the smallest conductor of all known curves whose Mordell-Weil group is isomorphic to $\mathbb{Z}/10\mathbb{Z} \times \mathbb{Z}^4$ (found by Dujella in 2005).

By the program $\mathtt{mwrank}$, we get that its rank is equal to 4 and that the Mordell-Weil group is generated by the points

$$P_1 = \left(\frac{6302726293975449488626684139017006}{13379318255014009}, \right.$$
$$\left. \frac{136233789132437251836981528851741590439605 5887491}{154757240337717 0172063027} \right),$$
$$P_2 = \left(\frac{1010862761896550838303235 0174}{590486201761}, \right.$$
$$\left. -\frac{1958345587631673357656809634618006468198497}{453747902505406991} \right),$$

$$P_3 = \left(\frac{2747445167847502233647380243466 86}{4890306578748529}, \right.$$

$$\left. \frac{21095091791152839218465210600937926397829067896 39}{34198269430476738315058 3} \right),$$

$$P_4 = \left(\frac{50839337272548006001}{64}, \frac{36139644164828072797955276737 1}{512} \right).$$

The canonical heights of points P_1, P_2, P_3, P_4 are 34.02261806, 26.72628169, 45.34998979, 21.98466653, respectively.

If we apply Rathbun's algorithm, we get a new Mordell-Weil basis:

$$Q_1 = \left(-\frac{343612010825901006209}{6724}, \frac{668899306736487700573297621576 9}{551368} \right),$$

$$Q_2 = \left(-\frac{10216528923584657172449}{145924}, \right.$$

$$\left. \frac{18867039044714009240612294658973 9}{55742968} \right),$$

$$Q_3 = \left(-71051466385703906, -13442883241925418821620 7 \right),$$

$$Q_4 = \left(\frac{3127754920096993023081873 4}{515244601}, \right.$$

$$\left. \frac{9552822287995333042843125194339637846 7}{11695537198099} \right),$$

with heights 10.27648431, 12.33469261, 15.949425, 24.802228. The transformation matrix is

$$u = \begin{pmatrix} -1 & -1 & 0 & -1 \\ 1 & 1 & -1 & 2 \\ 0 & 0 & 1 & 0 \\ 0 & 1 & 0 & 1 \end{pmatrix}.$$

This means that $Q_1 = -P_1 + P_2$, $Q_2 = -P_1 + P_2 + P_4$, $Q_3 = -P_2 + P_3$, $Q_4 = -P_1 + 2P_2 + P_4$.

So far, we have not used torsion points. We know from Proposition 2.6.2 that they cannot change the canonical heights of the points. However, they can "simplify" the basis in the sense that they reduce the naïve height of points or that points with rational coordinates are replaced by points with integer coordinates. Since, in our example, the torsion group is large ($\mathbb{Z}/10\mathbb{Z}$), at least some simplification will likely appear. Finally, by adding appropriate torsion points to Q_1, Q_2, Q_3, Q_4, we get the following Mordell-Weil basis:

$$R_1 = (8185642345602334, 700887231585412212447883 3),$$

$$R_2 = (-12386639730434786, -1127806570763234366872948 7),$$

$$R_3 = (-71051466385703906, -13442883241925418821620 7),$$

$$R_4 = \left(\frac{145126829591796568531936}{528529}, \right.$$

$$\left. \frac{5394312322844185507936647126243388 9}{384240583} \right).$$

Suppose that we calculated the rank r of an elliptic curve E by one of the previously described methods. Those methods will often give us r points $P_1, \ldots, P_r$ on E, which are independent modulo $E(\mathbb{Q})_{\text{tors}}$. However, that does not mean that $\{P_1, \ldots, P_r\} \cup E(\mathbb{Q})_{\text{tors}}$ will necessarily generate the whole group $E(\mathbb{Q})$, but maybe only a subgroup of finite index. We denote that subgroup by H. We would like, if possible, to find the generators of the whole Mordell-Weil group, i.e. a Mordell-Weil basis $Q_1, \ldots, Q_r$ (so that every point $P \in E(\mathbb{Q})$ can be, in a unique way, represented in the form $P = n_1 Q_1 + \cdots + n_r Q_r + T, n_i \in \mathbb{Z}, T \in E(\mathbb{Q})_{\text{tors}}$).

By following [73, Chapter 3.6], we will first look at a simple case when $r = 1$ (so the Mordell-Weil basis consists of one point, called a free generator) and $\Delta > 0$. Then E has an equation of the form $y^2 = (x - e_1)(x - e_2)(x - e_3)$, where $e_i \in \mathbb{R}$. Let $e_1 < e_2 < e_3$. Then $E^0(\mathbb{Q}) = \{(x, y) \in E(\mathbb{Q}) : x \geq e_3\} \cup \{O\}$ is a subgroup of $E(\mathbb{Q})$ which is called even or neutral component, while $E^{gg}(\mathbb{Q}) = \{(x, y) \in E(\mathbb{Q}) : e_1 \leq x \leq e_2\}$ is called odd component ("egg"). The odd component can be empty, and if it is non-empty, then $E^0(\mathbb{Q})$ has index 2 in $E(\mathbb{Q})$. Note that in the lattice $\mathbb{C}/L$ we may assume that ω_1 is real and ω_2 purely imaginary, so the points on $E^0(\mathbb{R})$ correspond to the parameters $z \in \mathbb{R}$, while the points on $E^{gg}(\mathbb{R})$ correspond to the parameters z (from the fundamental rectangle) for which $z - \omega_2/2 \in \mathbb{R}$.

Proposition 2.6.5 *Let an elliptic curve E with integer coefficients satisfy the following conditions:*

(i) rank$(E(\mathbb{Q})) = 1$;

(ii) $E(\mathbb{Q})$ *has a point P of infinite order such that $P + T$ has integer coordinates for all $T \in E(\mathbb{Q})_{\text{tors}}$;*

(iii) $\Delta > 0$;

(iv) *the odd component is non-empty.*

Then one free generator Q is one of the finitely many points with integer coordinates in the odd component.

Proof Let Q be a free generator. Then $nQ = P + T$ for some $n \in \mathbb{Z}$ and some torsion point T. By assumption (ii), the point nQ has integer coordinates. Then $Q = (x, y)$ also has integer coordinates.

This follows from a fact that is also used in the proof of the Lutz-Nagell theorem. Namely, suppose that $v_p(x) < 0$ for some prime number p. Then $v_p(x) = -2k$, $v_p(y) = -3k$ for some $k \in \mathbb{N}$. The fact that we need here is that

$$E(p^k) := \{(x, y) \in E(\mathbb{Q}) : v_p(x) \leq -2k, \ v_p(y) \leq -3k\} \cup \{O\}$$

is a subgroup of $E(\mathbb{Q})$ (see [356, Chapter 8.1]). Therefore, $Q \in E(p^k)$ implies that $nQ \in E(p^k)$, which is in contradiction with nQ having integer coordinates.

For each $T' \in E(\mathbb{Q})_{\text{tors}}$, the point $Q + T'$ is a free generator, so according to what has just been proved, it has integer coordinates. We claim that at least one of these points is located in the odd component. Let us assume the opposite. Then Q is in the even component, and all $T' \in E(\mathbb{Q})_{\text{tors}}$ are in the even component. However, $E(\mathbb{Q})$ is generated by Q and $E(\mathbb{Q})_{\text{tors}}$, so it would be contained in its even component, which contradicts assumption (iv).

There are obviously only finitely many points with integer coordinates in the odd component because their x-coordinates are in the segment $[e_1, e_2]$. $\square$

Example 2.6.6 Let us find a free generator of the curve

$$y^2 = (x + 1)(3x + 1)(8x + 1).$$

As in Example 2.5.2, we will work with the isomorphic curve

$$E : y^2 = x^3 + 11x^2 - 80x.$$

In Example 2.5.2, we saw that the rank of E is equal to 1. The algorithm also gave us several points of infinite order: $P_1 = (-10, 30)$, $P_2 = (-2, 14)$, $P_3 = (8, 24)$, $P_4 = (40, 240)$. By Example 2.4.7, we know that the only non-trivial torsion points on E are points of order 2: $T_1 = (-16, 0)$, $T_2 = (0, 0)$, $T_3 = (5, 0)$. Let us take the point $P_1 = (-10, 30)$. The points P_1, $P_1 + T_1 = -P_4$, $P_1 + T_2 = P_3$, $P_1 + T_3 = -P_2$ have integer coordinates. Therefore, all assumptions of Proposition 2.6.5 are satisfied. It is easy to see that the only points with integer coordinates with x-coordinate from the segment $[-16, 0]$ are the points $\pm P_1 = (-10, \pm 30)$, $\pm P_2 = (-2, \pm 14)$, T_1 and T_2. We conclude that P_1 is a free generator of $E(\mathbb{Q})$ (the other free generators are $-P_1$, $\pm P_2$, $\pm P_3$ and $\pm P_4$). Now the free generators of the initial curve are obtained by the transformation $x \mapsto \frac{x-8}{24}$. These are the points: $(-\frac{3}{4}, \pm\frac{5}{4})$, $(-\frac{5}{12}, \frac{7}{12})$, $(0, \pm 1)$, $(\frac{4}{3}, \pm\frac{35}{3})$. So, the obvious point $(0, 1)$ (corresponding to the trivial extension of the Diophantine triple with 0) is one of the free generators.

In the general case, we have r independent points on $E(\mathbb{Q})$, and they, together with $E(\mathbb{Q})_{\text{tors}}$, generate a subgroup H of $E(\mathbb{Q})$. We could get those independent points as a byproduct of the 2-descent algorithm or by searching for points of small heights on $E(\mathbb{Q})$. We would like to "enlarge" the subgroup H to the entire group $E(\mathbb{Q})$ (or to prove that $H = E(\mathbb{Q})$).

For this purpose, explicit estimates are used for the difference between naïve and canonical height. We will quote a general result of this type, which was proved in 1990 by Silverman [327]. Let us note that it is often possible to get significantly better estimates for concrete curves. We use the notation $\log^+(x) = \log\max(1, |x|)$ for $x \in \mathbb{R}$.

Proposition 2.6.7 *Let E be an elliptic curve given by a Weierstrass equation with integer coefficients, and let Δ be its discriminant, and j its j-invariant. Let us put $\eta = 2$ if $b_2 \neq 0$, and $\eta = 1$ if $b_2 = 0$, and define*

$$\mu(E) = \frac{1}{6}(\log|\Delta| + \log^+(j)) + \log^+(b_2/12) + \log(\eta).$$

Then, for every $P \in E(\mathbb{Q})$,

$$-\frac{1}{12}h(j) - \mu(E) - 1.922 \leq \hat{h}(P) - h(P) \leq \mu(E) + 2.14.$$

The proposition is easy to apply if the curve has rank 1 and we know one point P of infinite order. If P is not a free generator, we have $P = kQ + T$ for a generator Q and $k \geq 2$. Therefore $\hat{h}(Q) \leq \frac{1}{4}\hat{h}(P)$, so Proposition 2.6.7 gives us an upper bound B for the naïve height of Q. If we do not find any rational point with $h(Q) \leq B$, then we know that P is a generator, and otherwise, a generator is one of the found points.

Notice that Proposition 2.6.7 gives us an additional method for finding torsion points. Namely, since torsion points have the canonical height equal to 0, their naïve height is bounded by $\exp\left(\frac{1}{12}h(j) + \mu(E) + 1.922\right)$.

2.7 Exercises

1. Show that the curve

$$y^2 = x^3 + 3x^2 - 4$$

   is singular. Determine its singular point, and find its rational parametrization.
2. Transform the cubic curve given by the equation

$$u^3 + v^3 - 6 = 0$$

   into the short Weierstrass form. Find three solutions to the equation $u^3 + v^3 = 6$ in rational numbers u, v, where $u < v$.

3. Write the elliptic curve given by the equation

$$y^2 + xy + y = x^3 - x^2 - 4x$$

in the short Weierstrass form.

4. Find three distinct positive rational numbers x_1, x_2, x_3 with the property that $(x_i + 1)(3x_i + 1)(8x_i + 1)(120x_i + 1)$ is the square of a rational number for $i = 1, 2, 3$.

5. Determine the j-invariant of the elliptic curve given by the equation

$$y^2 = x(x + (t^2 - 1)^2)\left(x + \left(\frac{4t^2}{t^2 - 1}\right)^2\right),$$

where $t \in \mathbb{Q} \setminus \{-1, 0, 1\}$. Find a value of the parameter t for which this curve is isomorphic to the curve $y^2 = (-\frac{145}{408}x + 1)(\frac{408}{145}x + 1)(\frac{145439}{59160}x + 1)$.

6. An elliptic curve over $\mathbb{Q}$ is given by the equation

$$E : \quad y^2 = x^3 + 625x + 359375.$$

Determine its minimal Weierstrass equation. What type of reduction (good or bad; additive or multiplicative; split or non-split) does this curve have at $p = 13$?

7. Let E be an elliptic curve with the equation

$$y^2 = x^3 + ax + b.$$

Let $P = (x_1, y_1)$ be a point on E, and let $2P = (x_2, y_2)$. Prove that

$$y_1^2(4x_2(3x_1^2 + 4a) - 3x_1^3 + 5ax_1 + 27b) = 4a^3 + 27b^2.$$

8. Find points of finite order and determine the structure of the torsion group for the elliptic curve

$$y^2 = \left(\frac{12}{7}x + 1\right)\left(-\frac{15}{28}x + 1\right)\left(\frac{7}{4}x + 1\right)$$

induced by the rational Diophantine triple $\{\frac{12}{7}, -\frac{15}{28}, \frac{7}{4}\}$. Determine all prime numbers p for which $|E(\mathbb{Q})_{\text{tors}}| = |E(\mathbb{F}_p)|$.

9. Compute the rank of the elliptic curve over $\mathbb{Q}$ given by the equation

$$y^2 = x^3 - 82x.$$

Which of the fourth degree equations appearing in the descent via 2-isogeny algorithm have solutions?

10. Check that the elliptic curves over $\mathbb{Q}$ given by the equations

$$y^2 = x^3 - 10081x,$$

$$y^2 = x^3 + 1630368370x^2 + 134972837533033073x,$$

$$y^2 = x^3 + 4510328029x^2 + 622726581362777216x,$$

are of maximal rank, i.e. in (2.33) the equality holds.

11. Find elliptic curves of maximal rank with torsion group $\mathbb{Z}/2\mathbb{Z} \times \mathbb{Z}/2\mathbb{Z}$ and rank r, for $r = 2, 3, \ldots, 10$.

12. Show that there are infinitely many triples of rational numbers (a, b, c) such that

$$a + b + c = 8, \quad abc = 10.$$

13. Let E be a genus 1 curve over $\mathbb{Q}$ given by the equation

$$Y^2 = aX^4 + bX^2 + c.$$

Assume that E has a rational point (p, q) and regard E as an elliptic curve so that the point (p, q) corresponds to the point at infinity. Prove that then we have $2(-p, -q) = O$.

14. Let $a_1, a_2, a_3, a_4, a_5, a_6$ be distinct positive integers. Let $p(x) = (x^2 - a_1^2)(x^2 - a_2^2)(x^2 - a_3^2)(x^2 - a_4^2)(x^2 - a_5^2)(x^2 - a_6^2)$, and let $q(x)$ and $r(x)$ be polynomials with rational coefficients such that $p = q^2 - r$ and $\deg r \le 4$. Find positive integers $a_1, a_2, a_3, a_4, a_5, a_6$ such that:

- $\deg r = 4$,
- r has no multiple roots,
- the leading coefficient of r is the square of a rational number,
- the elliptic curve E equivalent to the curve $y^2 = r(x)$ has rank ≥ 6.

Note that the polynomial $r(x)$ is even, so the elliptic curve E has a point of order 2.

15. Find an elliptic curve over $\mathbb{Q}$ such that

$$E(\mathbb{Q}) \cong \mathbb{Z}/2\mathbb{Z} \times \mathbb{Z}/8\mathbb{Z} \times \mathbb{Z}^2.$$

16. Prove that the $D(-1)$-triple $\{1, 5, 10\}$ cannot be extended to a $D(-1)$-quadruple.

Chapter 3
Elliptic Curves Induced by Diophantine Triples

In the introduction, we have mentioned in several places connections between Diophantine m-tuples and elliptic curves. In this chapter, we will systematically study these connections. We will see how some classical results, like extensions of quadruples to rational quintuples, can be interpreted in terms of elliptic curves. Moreover, elliptic curves play a crucial role in the construction of infinite families of rational Diophantine sextuples. The question of the existence of such families was an open problem from the time when Euler showed that there are infinitely many rational Diophantine quintuples. Four different constructions of families of sextuples are known, and they all use elliptic curves (of different forms and used in different ways). Another fruitful connection between Diophantine m-tuples and elliptic curves is in constructing elliptic curves with large rank and given torsion group (here, four types of torsion groups can appear). Apart from the mentioned topics, we will also cover some other problems where connections between Diophantine m-tuples (integer and rational) and elliptic curves naturally appear.

3.1 Obvious Rational Points and Regular m-Tuples

We will now describe in more detail the connection between rational Diophantine m-tuples and elliptic curves. Let $\{a, b, c\}$ be a rational Diophantine triple, i.e. let

$$ab + 1 = r^2, \quad ac + 1 = s^2, \quad bc + 1 = t^2$$

for non-negative rational numbers r, s, t. In order to extend this triple to a rational Diophantine quadruple, we need to find a rational number x with the property that

A. Dujella, *Diophantine m-tuples and Elliptic Curves*,
Developments in Mathematics 79, https://doi.org/10.1007/978-3-031-56724-7_3

$ax + 1$, $bx + 1$ and $cx + 1$ are squares of rational numbers. If we multiply these three conditions, we get

$$y^2 = (ax + 1)(bx + 1)(cx + 1),$$

which is the equation of an elliptic curve. We will soon explain which points on the curve satisfy the initial system of three equations and give extensions to rational Diophantine quadruples.

Let us denote the curve $y^2 = (ax + 1)(bx + 1)(cx + 1)$ by $\mathcal{E}$. We say that $\mathcal{E}$ is *induced by the Diophantine triple* $\{a, b, c\}$. On curve $\mathcal{E}$, we have three obvious rational points of order 2; these are $A = (-\frac{1}{a}, 0)$, $B = (-\frac{1}{b}, 0)$, $C = (-\frac{1}{c}, 0)$. We also have two additional obvious rational points

$$P = (0, 1), \quad S = \left(\frac{1}{abc}, \frac{rst}{abc} \right).$$

It is easy to check that the x-coordinate of the point $P - S$ is exactly the number d_+ from the definition of the regular quadruple (while the x-coordinate of the point $P + S$ is equal to d_-; recall that $d_\pm = a + b + c + 2abc \pm 2rst$). To do this, by transformations $x \mapsto x/abc$, $y \mapsto y/abc$, from the curve $\mathcal{E}$ we get the curve

$$\mathcal{E}' : \quad y^2 = (x + bc)(x + ac)(x + ab)$$

with a monic cubic polynomial on the right-hand side of the equation. The points A, B, C, P and S on $\mathcal{E}$ become $A' = (-bc, 0)$, $B' = (-ac, 0)$, $C' = (-ab, 0)$, $P' = (0, abc)$ and $S' = (1, rst)$ on $\mathcal{E}'$. Now we calculate the x-coordinate of the point $P' - S'$ on $\mathcal{E}'$ (note that $\lambda = (y(-S') - y(P'))/(x(S') - x(P')) = -(rst + abc)$):

$$(rst + abc)^2 - (ab + ac + bc) - 0 - 1$$

$$= (ab + 1)(ac + 1)(bc + 1) + 2abcrst + a^2b^2c^2 - ab - ac - ab - 1$$

$$= abc(a + b + c + 2abc + 2rst),$$

from where, by dividing by abc, we get that the x-coordinate of the point $P - S$ on $\mathcal{E}$ is equal to d_+, as claimed. It can be proved analogously that the x-coordinate of the point $P + S$ on $\mathcal{E}$ is equal to d_-.

We can generally expect P and S to be independent points of infinite order. Nevertheless, the important question, with significant consequences, is whether these points can be of finite order and which orders are possible. In particular, we will see in Sect. 3.2 that one of the constructions of parametric families of rational Diophantine sextuples is closely connected to triples for which the point S has order 3. Note that if the triple $\{a, b, c\}$ is regular, then P and S are not independent, since then we have $2P = -S$. Indeed, in Sect. 1.4, we showed that the triple $\{a, b, c\}$ is regular if and only if $d_- = 0$. And we just saw that d_- is the x-coordinate of the point $P + S$. Since the x-coordinate of the point P is equal to 0, we conclude that

$P = P + S$ or $P = -P - S$. The first possibility is ruled out because $S \neq O$, and the second possibility gives $2P = -S$, as claimed.

Now we can answer which points on $\mathcal{E}$ provide extensions of the starting triple to a rational Diophantine quadruple. Namely, the x-coordinate of the point $T \in \mathcal{E}(\mathbb{Q})$ satisfies the starting three conditions that $ax + 1$, $bx + 1$, and $cx + 1$ are squares of rational numbers if and only if $T - P \in 2\mathcal{E}(\mathbb{Q})$. This follows from Theorems 2.4.9 and 2.4.10.

Corollary 3.1.1 *Let $T \in \mathcal{E}(\mathbb{Q})$ and $x = x(T)$. Then $ax + 1$, $bx + 1$, and $cx + 1$ are squares of rational numbers if and only if $T - P \in 2\mathcal{E}(\mathbb{Q})$.*

Proof For an arbitrary point $X = (x, y) \in \mathcal{E}(\mathbb{Q})$, we will denote by $X' = (xabc, yabc) = (x', y')$ the corresponding point in $\mathcal{E}'(\mathbb{Q})$. According to Theorem 2.4.10, the function $\varphi_a : \mathcal{E}'(\mathbb{Q}) \to \mathbb{Q}^*/\mathbb{Q}^{*2}$ defined by

$$\varphi_a(X') = \begin{cases} (x' + bc)\mathbb{Q}^{*2} & \text{if } X' = (x', y') \neq O, A', \\ (ac - bc)(ab - bc)\mathbb{Q}^{*2} & \text{if } X = A', \\ \mathbb{Q}^{*2} & \text{if } X = O, \end{cases}$$

is a group homomorphism. The same applies to analogously defined functions φ_b and φ_c. The point P satisfies the given condition because $x(P) = 0$. We have: $\varphi_a(P') = bc\mathbb{Q}^{*2}$, $\varphi_b(P') = ac\mathbb{Q}^{*2}$, $\varphi_c(P') = ab\mathbb{Q}^{*2}$. Now $x = x(T)$ satisfies the given condition that $ax + 1$, $bx + 1$ and $cx + 1$ are squares if and only if $x' = xabc$ satisfies that $x' + bc = bc\square$, $x' + ac = ac\square$ and $x' + ab = ab\square$, where $\square$ denotes a rational square, i.e. if and only if

$$\varphi_a(T') = \varphi_a(P'), \quad \varphi_b(T') = \varphi_b(P'), \quad \varphi_c(T') = \varphi_c(P').$$

Since $\varphi_a, \varphi_b, \varphi_c$ are homomorphisms, this is equivalent to

$$\varphi_a(T' - P') = \varphi_b(T' - P') = \varphi_c(T' - P') = \mathbb{Q}^{*2}.$$

By Theorem 2.4.9, this is equivalent to $T' - P' \in 2\mathcal{E}'(\mathbb{Q})$. $\qquad \square$

It follows from Theorem 2.4.9 that $S' = (1, rst) \in 2\mathcal{E}'(\mathbb{Q})$. Namely, $1 + bc$, $1 + ac$ and $1 + ab$ are squares, so $S' \in 2\mathcal{E}'(\mathbb{Q})$. It is not difficult to check this directly. Indeed, we have $S' = 2R'$, where

$$R' = (rs + rt + st + 1, (r + s)(r + t)(s + t)).$$

This, together with Corollary 3.1.1, implies that if $x(T)$ satisfies the condition that $ax + 1$, $bx + 1$ and $cx + 1$ are squares, then the numbers $x(T \pm S)$ also satisfy it.

It is interesting that $x(T)x(T + S) + 1$ is always the square of a rational number. Indeed, this statement is a consequence of the following more general fact.

Proposition 3.1.2 *Let Q, T and $(0, q)$ be three rational points on an elliptic curve E over $\mathbb{Q}$ given by the equation $y^2 = f(x)$, where f is a monic polynomial of degree 3. Assume that $O \notin \{Q, T, Q + T\}$. Then $x(Q)x(T)x(Q + T) + q^2$ is the square of a rational number.*

Proof Consider the curve

$$y^2 = f(x) - (x - x(Q))(x - x(T))(x - x(Q + T)), \tag{3.1}$$

which represents a conic (curve of the second degree) containing three collinear points Q, T, $-(Q + T)$; therefore, it must be the union of two rational lines, i.e. $y^2 = (kx + \ell)^2$. If we insert $x = 0$ in (3.1), we get $x(Q)x(T)x(Q + T) + q^2 = \ell^2$, as claimed. $\qquad\square$

To apply Proposition 3.1.2 for the construction of rational Diophantine quintuples (and later also sextuples), recall that by transformations $x \mapsto x/abc$, $y \mapsto y/abc$, from the curve $\mathcal{E}$, we got the curve

$$\mathcal{E}' : \quad y^2 = (x + ab)(x + ac)(x + bc)$$

with a monic cubic polynomial on the right-hand side of the equation, and that points P and S on $\mathcal{E}$ became $P' = (0, abc)$ and $S' = (1, rst)$ on $\mathcal{E}'$. We apply Proposition 3.1.2 with $Q = \pm S'$. Since the first coordinate of the point S' is equal to 1, we conclude that $x(T)x(T \pm S) + 1$ is a perfect square (after dividing the equality $x(T')x(T' \pm S') + a^2b^2c^2 = \square$ by $a^2b^2c^2$).

In this way, we can extend an arbitrary rational Diophantine quadruple $\{a, b, c, d\}$ to the rational Diophantine quintuple $\{a, b, c, d, e\}$, so that for T we take the point on the curve $\mathcal{E}$ whose first coordinate is equal to d, and for e we take the x-coordinate of the point $T + S$ (or $T - S$) on the same curve. This construction gives a rational Diophantine quintuple provided all the elements of the quintuple are different from zero, they are mutually distinct, and the point $T + S$ (respectively $T - S$) is not the point O. From here, we can obtain an explicit formula for the extension of a rational Diophantine quadruple to a quintuple (which generalizes Euler's formula for quintuples from Sect. 1.3).

Theorem 3.1.3 *Let x_1, x_2, x_3, x_4 be rational numbers such that $x_i x_j + 1 = y_{ij}^2$, $y_{ij} \in \mathbb{Q}$, for all $1 \le i < j \le 4$. Assume that $x_1 x_2 x_3 x_4 \ne 1$. Then the rational number $x_5 = A/B$, where*

$$A = \pm 2 y_{12} y_{13} y_{14} y_{23} y_{24} y_{34} + x_1 x_2 x_3 x_4 (x_1 + x_2 + x_3 + x_4)$$

$$+ 2(x_1 x_2 x_3 + x_1 x_2 x_4 + x_1 x_3 x_4 + x_2 x_3 x_4) + (x_1 + x_2 + x_3 + x_4),$$

$$B = (x_1 x_2 x_3 x_4 - 1)^2,$$

has the property that $x_i x_5 + 1$ are squares of rational numbers for $i = 1, 2, 3, 4$. More precisely, for $i \in \{1, 2, 3, 4\}$ the following holds:

$$x_i x_5 + 1 = \left(\frac{x_i y_{jk} y_{jl} y_{kl} \pm y_{ij} y_{ik} y_{il}}{x_1 x_2 x_3 x_4 - 1} \right)^2 ,$$

where $\{i, j, k, l\} = \{1, 2, 3, 4\}$.

Remark 3.1.4 Let us note that if $x_1 x_2 x_3 x_4 = 1$, then $x_1 x_2$, $x_1 x_3$, $x_1 x_4$, $x_2 x_3$, $x_2 x_4$, $x_3 x_4$ are all perfect squares (see Proposition 5.4.8), and $\{x_1, x_2, x_3, x_4\}$ can be extended to a rational Diophantine quintuple by $x_5 = \frac{(x_1 + x_2 - x_3 - x_4)^2 - 4(x_1 x_2 + 1)(x_3 x_4 + 1)}{4 x_4 (x_1 x_2 + 1)(x_1 x_3 + 1)(x_2 x_3 + 1)}$. Here, the point T corresponding to $x_4 = \frac{1}{x_1 x_2 x_3}$ has the same first coordinate as the point S on the elliptic curve induced by the triple $\{x_1, x_2, x_3\}$. Thus, $T = \pm S$, so one of the points $T \pm S$ is the point at infinity O.

The quintuples obtained using the formula from Theorem 3.1.3, i.e. quintuples of the form $\{a, b, c, x(T), x(T \pm S)\}$, are called *regular rational Diophantine quintuples*.

If the initial quadruple, as in Euler's construction of quintuples, is of the form $\{a, b, a + b + 2r, 4r(a + r)(b + r)\}$, then $T = P - S$ and $2P = -S$, so $T = 3P$ and $T - S = 5P$. Since $T + S = P$, the described construction gives only one non-trivial extension to a quintuple since $x(T + S) = x(P) = 0$. The same happens with the extension of an arbitrary triple to quintuple due to Arkin, Hoggatt, and Strauss [13], because it also uses the extension of the triple to regular quadruple, so again we have $T = P - S$ and $T + S = P$. However, when extending a quadruple via Theorem 3.1.3 (which first appeared in [92]), we generally have two different extensions to quintuple. We could say that the situation is analogous to extensions of integer triples to quadruples, where in general, we have two extensions: with d_+ and d_-, and in the case when the starting triple is regular, one of those two extensions is discarded because it gives the trivial extension with zero.

A significant difference is, however, that in the integer case, it is conjectured that there are no other possibilities for extension, while in the rational case, there are examples of quadruples that can be extended to quintuples in more than two ways. For example, in 2016, Gibbs [196] found an example of a quadruple

$$\left\{ \frac{81}{1400}, \frac{5696}{4725}, \frac{2875}{168}, \frac{4928}{3} \right\},$$

which can be extended to a quintuple in at least six different ways, i.e. with any of these six rational numbers:

$$\frac{98}{27}, \frac{104}{525}, \frac{96849}{350}, \frac{1549429}{1376646}, \frac{3714303488}{6103383075}, \frac{76943337252154322}{18574246299984075}.$$

3.2   Rational Diophantine Sextuples via Points of Order 3

In this section, we will present the construction of infinitely many rational Diophantine sextuples according to the paper [137] by Dujella, Kazalicki, Mikić, and Szikszai from 2017. At the end of the previous section, we presented the construction by which a rational Diophantine quadruple extends to (regular) quintuples in two different ways. Therefore, the union of those two quintuples,

$$\{a, b, c, x(T - S), x(T), x(T + S)\},$$

is "almost" a rational Diophantine sextuple.

If we assume that $T, T \pm S \notin \{O, \pm P\}$, the only missing condition is that

$$x(T - S)x(T + S) + 1$$

be a square. To find examples that satisfy this last condition, we will apply Proposition 3.1.2 for $Q = 2S'$. That will give us the desired conclusion if the condition $x(2S') = 1$ is met. Now from $x(2S') = x(S')$ we get the condition $2S' = -S'$ (since $2S' = S'$ is not possible), i.e. $3S' = O$. In this way, we connected the problem of the construction of rational Diophantine sextuples and elliptic curves with the torsion group $\mathbb{Z}/2\mathbb{Z} \times \mathbb{Z}/6\mathbb{Z}$. Using the formula for the duplication of points on an elliptic curve, the condition $x(2S') = x(S')$ can be written explicitly in the form

$$3 + 4(ab + ac + bc) + 6abc(a + b + c) + 12(abc)^2$$
$$- (abc)^2(a^2 + b^2 + c^2 - 2ab - 2ac - 2bc) = 0. \qquad (3.2)$$

This is a symmetric equation, so by introducing elementary symmetric polynomials $\sigma_1 = a + b + c$, $\sigma_2 = ab + ac + bc$, $\sigma_3 = abc$, we get a simpler equation, which we can solve in the variable σ_2:

$$\sigma_2 = (\sigma_1^2\sigma_3^2 - 12\sigma_3^2 - 6\sigma_1\sigma_3 - 3)/(4 + 4\sigma_3^2). \qquad (3.3)$$

If we insert (3.3) in $(ab + 1)(ac + 1)(bc + 1) = (rst)^2$, we get $(2\sigma_3^2 + \sigma_1\sigma_3 - 1)^2/(4 + 4\sigma_3^2) = (rst)^2$, i.e. we get the condition that $1 + \sigma_3^2$ is a square, so we can put $\sigma_3 = \frac{t^2-1}{2t}$ to satisfy this condition.

The polynomial

$$X^3 - \sigma_1 X^2 + \sigma_2 X - \sigma_3 \qquad (3.4)$$

should have rational roots, so its discriminant has to be a perfect square. From this condition, we get the quartic in σ_1:

$$(t^6 - 2t^4 + t^2)\sigma_1^4 + (23t^3 - 23t^5 + t^7 - t)\sigma_1^3$$

$$+ (-45t^2 - 45t^6 + 126t^4)\sigma_1^2 + (162t^5 - 162t^3 - 54t^7 + 54t)\sigma_1$$

$$- 27t^8 + 108t^6 - 54t^4 + 108t^2 - 27 = w^2.$$

Since the quartic has a rational point (the point at infinity), it can therefore be birationally transformed (using Proposition 2.2.2) into the equation of an elliptic curve over $\mathbb{Q}(t)$. We obtain the following elliptic curve

$$y^2 = x^3 + (3t^4 - 21t^2 + 3)x^2$$

$$+ (3t^8 + 12t^6 + 18t^4 + 12t^2 + 3)x + (t^2 + 1)^6. \tag{3.5}$$

This curve has positive rank, because it contains the point $R = (0, (t^2 + 1)^3)$ of infinite order.

If we take the points mR, transfer them back to the quartic, and calculate the corresponding triples $\{a, b, c\}$, we can expect that we will get infinitely many parametric families of rational triples for which the corresponding point S' on E' satisfies $3S' = O$.

Since the condition $1 + \sigma_3^2 = \square$ implies $rst \in \mathbb{Q}$, and $S' = -2S' \in 2\mathcal{E}'(\mathbb{Q})$, Theorem 2.4.9 applied to the curve $\mathcal{E}'$ implies that $ab + 1$, $ac + 1$, $bc + 1$ are squares, and $\{a, b, c\}$ obtained by this construction is indeed a rational Diophantine triple.

In particular, we may take the point $2R$, which has x-coordinate $-\frac{3}{4}(t^2 - 6t + 1)(t^2 + 6t + 1)$. Transferring it back to the quartic, we get

$$\sigma_1 = \frac{t^8 + 130t^6 - 390t^4 + 130t^2 + 1}{3(t - 1)(t + 1)(t^2 - 6t + 1)(t^2 + 6t + 1)t}.$$

If we insert the obtained values of $\sigma_1, \sigma_2, \sigma_3$ in $X^3 - \sigma_1 X^2 + \sigma_2 X - \sigma_3 = 0$, we get three rational solutions for X. Thus, we find the following family of Diophantine triples $\{a, b, c\}$ with the desired property:

$$a = \frac{18t(t - 1)(t + 1)}{(t^2 - 6t + 1)(t^2 + 6t + 1)},$$

$$b = \frac{(t - 1)(t^2 + 6t + 1)^2}{6t(t + 1)(t^2 - 6t + 1)},$$

$$c = \frac{(t + 1)(t^2 - 6t + 1)^2}{6t(t - 1)(t^2 + 6t + 1)}.$$

Let us consider the elliptic curve over $\mathbb{Q}(t)$ induced by the triple $\{a, b, c\}$. It has positive rank because the point $P = (0, 1)$ is of infinite order. Therefore, the described construction gives infinitely many rational Diophantine sextuples which contain the triple $\{a, b, c\}$. One such sextuple $\{a, b, c, d, e, f\}$ is obtained from the x-coordinates of the points $3P$, $3P + S$, $3P - S$ (more generally, we can take the points $(2n + 1)P$, $(2n + 1)P + S$, $(2n + 1)P - S$, for $n \in \mathbb{N}$).

We get $d = d_1/d_2$, $e = e_1/e_2$, $f = f_1/f_2$, where

$$d_1 = 6(t + 1)(t - 1)(t^2 + 6t + 1)(t^2 - 6t + 1)(8t^6 + 27t^5 + 24t^4 - 54t^3 + 24t^2 + 27t + 8)$$
$$\times\ (8t^6 - 27t^5 + 24t^4 + 54t^3 + 24t^2 - 27t + 8)(t^8 + 22t^6 - 174t^4 + 22t^2 + 1),$$

$$d_2 = t(37t^{12} - 885t^{10} + 9735t^8 - 13678t^6 + 9735t^4 - 885t^2 + 37)^2,$$

$$e_1 = -2t(4t^6 - 111t^4 + 18t^2 + 25)(3t^7 + 14t^6 - 42t^5 + 30t^4 + 51t^3 + 18t^2 - 12t + 2)$$
$$\times\ (3t^7 - 14t^6 - 42t^5 - 30t^4 + 51t^3 - 18t^2 - 12t - 2)(t^2 + 3t - 2)(t^2 - 3t - 2)$$
$$\times\ (2t^2 + 3t - 1)(2t^2 - 3t - 1)(t^2 + 7)(7t^2 + 1),$$

$$e_2 = 3(t + 1)(t^2 - 6t + 1)(t - 1)(t^2 + 6t + 1)$$
$$\times\ (16t^{14} + 141t^{12} - 1500t^{10} + 7586t^8 - 2724t^6 + 165t^4 + 424t^2 - 12)^2,$$

$$f_1 = 2t(25t^6 + 18t^4 - 111t^2 + 4)(2t^7 - 12t^6 + 18t^5 + 51t^4 + 30t^3 - 42t^2 + 14t + 3)$$
$$\times\ (2t^7 + 12t^6 + 18t^5 - 51t^4 + 30t^3 + 42t^2 + 14t - 3)(2t^2 + 3t - 1)(2t^2 - 3t - 1)$$
$$\times\ (t^2 - 3t - 2)(t^2 + 3t - 2)(t^2 + 7)(7t^2 + 1),$$

$$f_2 = 3(t + 1)(t^2 - 6t + 1)(t - 1)(t^2 + 6t + 1)$$
$$\times\ (12t^{14} - 424t^{12} - 165t^{10} + 2724t^8 - 7586t^6 + 1500t^4 - 141t^2 - 16)^2.$$

These formulas give infinitely many rational Diophantine sextuples. Furthermore, by choosing the rational parameter t from the appropriate intervals, we get infinitely many sextuples for any combination of signs. For example, for $5.83 < t < 6.86$, all elements are positive. As a particular example, let us take $t = 6$, so we get the sextuple with positive elements:

$$\left\{ \frac{3780}{73},\ \frac{26645}{252},\ \frac{7}{13140},\ \frac{79136175260255068460}{18278930922345566928 01}, \right.$$

$$\left. \frac{95104852709815809228981184}{35104191165465133563326695 5},\ \frac{3210891270762333567521084544}{21712719223923581005355} \right\}.$$

The construction of the above parametric family of rational Diophantine sextuples is based on the fact that the cubic polynomial (3.4) corresponding to $2R$ has rational roots. It can be shown that every multiple mR of R has the same property. For small multiples of R, this can be checked by direct computation. However, the proof for general multiples is much more involved. In [137], two proofs were presented, and here we will very roughly sketch one of the proofs.

Since the discriminant of the corresponding cubic polynomial is a perfect square, it is sufficient to prove that the cubic polynomial has at least one rational root, because then the remaining two roots will necessarily be rational. However, this property does not have to be satisfied for other points on the curve (3.5) (in the case that the rank is > 1). For example, for $t = 31$ (when the rank of (3.5) is equal to 2) and the point $(x, y) = (-150072, 682327360)$ (which is not a multiple of R), the polynomial $X^3 - \sigma_1 X^2 + \sigma_2 X - \sigma_3$ has no rational roots.

To each multiple $mR = (x, y)$ of the point R (with $m > 1$, so that $x \neq 0$), we attach the elliptic curve

$$E' : \quad Y^2 = X^3 + \sigma_2 X^2 + \sigma_1 \sigma_3 X + \sigma_3^2,$$

where $\sigma_3 = \frac{t^2-1}{2t}, t \notin \{-1, 0, 1\}$,

$$\sigma_1 = \frac{-t^4 + 4t^2 - 1 - x^{-1}(t^2 + 1)^4}{(t^2 - 1)t},$$

while σ_2 is given by (3.3).

With appropriate coordinate transformations, the curve E' becomes

$$E'' : \quad Y^2 = X^3 + \frac{((t^2 + 1)^2 x^{-1} + 1)^2}{4} X^2 \tag{3.6}$$
$$+ \frac{t^2((t^2 + 1)^2 x^{-2} + x^{-1})}{2} X + \frac{t^4 x^{-2}}{4}.$$

We want to show that E'' has a rational point of order 2.

In [137], this problem is related to the type of reduction of E'' at certain primes. Namely, it is shown that if E'' has no rational 2-torsion points, then it should have additive reduction for primes from an explicitly described set. On the other hand, by analyzing various cases (depending on $v_p(x)$ and $v_p(y)$, where v_p denotes the standard p-adic valuation), the necessary conditions on p in the case of additive reduction are described. In that way, the following result is obtained. Let $t \in \mathbb{Z}$ and (x, y) be such that E'' has good or multiplicative reduction (the cubic polynomial considered modulo p has either simple roots or one double root) for all $p \,|\, t(t^2 + 1)$ and let $v_3(y) \leq 0$. Then E'' has three rational points of order 2.

This result is combined with various properties of $v_p(x(mR))$ and $v_p(y(mR))$ to show that if $t \neq 1$ is a positive integer such that the number $t^2 + 1$ is square-free, then the elliptic curve E'', which corresponds to the multiple mR, $m > 1$, has three rational points of order 2; that is, the corresponding numbers a, b, c are rational (see [137] for details).

An effective version of Hilbert's irreducibility theorem (for an irreducible polynomial $f(x, t) \in \mathbb{Z}[x, t]$ there are "many" specializations $t = \tau \in \mathbb{Z}$ such that the polynomial $f(x, \tau)$ is irreducible in $\mathbb{Z}[x]$; see, e.g. [85] for a precise statement) now implies that the same statement holds for every rational number $t \notin \{-1, 0, 1\}$.

Note that in the previously mentioned example of rank 2, for $t = 31$ and $(x, y) = (-150072, 682327360)$, the curve E'' has additive reduction at 13, 31 and 37.

As we mentioned in Sect. 1.4, it is an open question whether there exist a rational Diophantine septuple. Note that by just applying Proposition 3.1.2, we cannot expect to find a candidate for a septuple. Namely, a natural candidate for the extension of the sextuple $\{a, b, c, x(T - S), x(T), x(T + S)\}$ would be $x(T + 2S)$, but it is already contained in the sextuple because $x(T + 2S) = x(T - S)$ since the point S has order 3.

3.3 Rational Diophantine Sextuples via Regularity Conditions

Rational Diophantine sextuples constructed in the previous section contain two regular quintuples. In this section, we will present an alternative construction of infinitely many rational Diophantine sextuples (based on the paper [140] by Dujella, Kazalicki, and Petričević from 2021) which contain two regular quadruples and one regular quintuple.

In 2016, Gibbs [196] sorted about a thousand known examples of rational Diophantine sextuples according to their structure, especially including data on regular triples, quadruples and quintuples contained in them. Petričević expanded the collection of examples with sextuples with relatively small numerators and denominators (and included sextuples with mixed signs—Gibbs observed only sextuples with positive elements). Many of these examples of sextuples contained two regular quadruples and one regular quintuple, which motivated the attempt to prove that there are infinitely many such sextuples.

Theorem 3.3.1 *There are infinitely many rational Diophantine sextuples that contain two regular Diophantine quadruples and one regular Diophantine quintuple.*

In the construction, we start with a parametrization of rational Diophantine triples. Let $\{a_1, a_2, a_3\}$ be a rational Diophantine triple and let $a_2 = (r^2 - 1)/a_1$ and $a_3 = (s^2 - 1)/a_1$ for rational numbers r and s. If we put $a_2 a_3 + 1 = (a_2 s^2 - a_2 + a_1)/a_1 = (1 + (s - 1)u)^2$, we get

$$a_3 = \frac{-4u(u - 1)(a_1 u - a_2)}{(-a_2 + a_1 u^2)^2}.$$

This parametrization was used in [113] to construct some examples of rational Diophantine sextuples with mixed signs. We will use here an equivalent parametrization found by Lasić (it is published in the appendix of [228]), which is cyclically symmetric in three parameters:

$$a_1 = \frac{2t_1(1 + t_1 t_2(1 + t_2 t_3))}{(-1 + t_1 t_2 t_3)(1 + t_1 t_2 t_3)},$$

$$a_2 = \frac{2t_2(1 + t_2 t_3(1 + t_3 t_1))}{(-1 + t_1 t_2 t_3)(1 + t_1 t_2 t_3)}, \tag{3.7}$$

$$a_3 = \frac{2t_3(1 + t_3 t_1(1 + t_1 t_2))}{(-1 + t_1 t_2 t_3)(1 + t_1 t_2 t_3)}.$$

The connection between the two mentioned parametrizations is given by

$$t_1 = \frac{a_1}{r - 1}, \quad t_2 = \frac{-(1 - r^2 + a_1^2 u^2)}{2(u - 1)a_1}, \quad t_3 = \frac{2a_1 u(u - 1)}{1 - r^2 + a_1^2 u^2},$$

$$a_1 = \frac{2t_1(1 + t_1 t_2(1 + t_2 t_3))}{(-1 + t_1 t_2 t_3)(1 + t_1 t_2 t_3)}, \quad r = \frac{1 + 2t_1 t_2 + 2t_1 t_2^2 t_3 + t_2^2 t_3^2 t_1^2}{(-1 + t_1 t_2 t_3)(1 + t_1 t_2 t_3)}, \quad u = -t_2 t_3.$$

Let now $\{a_1, a_2, a_3, a_4\}$ and $\{a_1, a_2, a_3, a_5\}$ be regular rational Diophantine quadruples, i.e. a_4 and a_5 are solutions of the quadratic equation

$$(a_1 + a_2 - a_3 - x)^2 - 4(a_1 a_2 + 1)(a_3 x + 1) = 0.$$

We obtain that

$$a_4 = \frac{-2(1 - t_3 + t_2 t_3)(t_3 t_1 + 1 - t_1)(-t_2 + 1 + t_1 t_2)(-1 + t_1 t_2 t_3)}{(1 + t_1 t_2 t_3)^3},$$

$$a_5 = \frac{2(t_3 + t_2 t_3 + 1)(t_3 t_1 + t_1 + 1)(1 + t_2 + t_1 t_2)(1 + t_1 t_2 t_3)}{(-1 + t_1 t_2 t_3)^3}.$$

For $\{a_1, a_2, a_3, a_4, a_5\}$ to be a rational Diophantine quintuple, it remains to satisfy the condition that $a_4 a_5 + 1$ is a perfect square. In this way, we obtain the condition that

$$p(t_1, t_2, t_3) = (-3t_2^4 t_3^4 + 4t_2^4 t_3^2 - 8t_2^3 t_3^3 + 8t_2^3 t_3 + 4t_2^2 t_3^4 - 8t_2^2 t_3^2 + 4t_2^2)t_1^4 \tag{3.8}$$

$$+ (-8t_2^4 t_3^3 - 8t_2^3 t_3^4 - 8t_2^3 t_3^2 - 8t_2^2 t_3^3 + 8t_2^2 t_3$$

$$+ 8t_2 t_3^4 - 16t_2 t_3^2 + 8t_2)t_1^3$$

$$+ (4t_2^4 t_3^4 - 8t_2^4 t_3^2 - 8t_2^3 t_3^3 - 8t_2^2 t_3^4 - 18t_2^2 t_3^2 - 16t_2^3 t_3 - 8t_2^2 + 8t_2 t_3^3$$

$$- 8t_2 t_3 + 4t_3^4 - 8t_3^2 + 4)t_1^2$$

$$+ (8t_2^4 t_3^3 - 16t_2^2 t_3^3 + 8t_2^3 t_3^2 - 8t_2^2 t_3 - 8t_2 t_3^2 - 8t_2 + 8t_3^3 - 8t_3)t_1$$

$$+ 4t_2^4 t_3^2 + 8t_2^3 t_3 - 8t_3^2 t_2^2 + 4t_2^2 - 8t_2 t_3 + 4t_3^2 - 3$$

is a perfect square. If we compute the discriminant of the polynomial p with respect to the variable t_1 and factor it, we get that one of the factors is

$$p_1(t_2, t_3) = 3 + 10t_2t_3 - 3t_3^2 + 3t_3^2 t_2^2$$

(the other factors are either of a much higher degree or correspond to quintuples having an element equal to 0). The condition $p_1(t_2, t_3) = 0$ (which ensures that the polynomial p with respect to t_1 has a double root, and hence a square factor) leads to the condition that $9t_3^2 + 16$ is a perfect square, say $9t_3^2 + 16 = (3t_3 + u)^2$. From here, we get

$$t_3 = \frac{16 - u^2}{6u},$$

$$t_2 = \frac{u^2 + 10u + 16}{(u - 4)(u + 4)}.$$

If we insert this into (3.8), we get that $a_4a_5 + 1$ is a perfect square. Thus, we obtained a two-parameter family (in parameters t_1 and u) of rational Diophantine quintuples containing two regular quadruples (such quintuples appeared in the construction of elliptic curves of high rank with torsion group $\mathbb{Z}/2\mathbb{Z} \times \mathbb{Z}/2\mathbb{Z}$ in [99]).

Let us now extend the irregular quadruple $\{a_1, a_3, a_4, a_5\}$ (or any other irregular quadruple contained in the quintuple $\{a_1, a_2, a_3, a_4, a_5\}$) to regular quintuples $\{a_1, a_3, a_4, a_5, a_6\}$ and $\{a_1, a_3, a_4, a_5, a_7\}$. By Theorem 3.1.3, this means that a_6 and a_7 are the solutions of the quadratic equation

$$(a_1a_3a_4a_5x + 2a_1a_3a_4 + a_1 + a_3 + a_4 - a_5 - x)^2$$
$$- 4(a_1a_3 + 1)(a_1a_4 + 1)(a_3a_4 + 1)(a_5x + 1) = 0. \tag{3.9}$$

In the further construction, we will not use a_7, so we only list the value of a_6:

$$a_6 = 6(u + 4)(u + 8)(u + 2)(u - 4)(2t_1u^2 + 3u^2 + 20t_1u + 12u + 32t_1)$$
$$\times (t_1u^2 + 10t_1u + 16t_1 - 6u)(t_1u^2 + 10t_1u + 16t_1 + 6u)(t_1u^2 + 10t_1u + 16t_1 - 24 - 6u)$$
$$\times (4096t_1^2 + 15360t_1^2u + 15168t_1^2u^2 + 5920t_1^2u^3 + 948t_1^2u^4 + 60t_1^2u^5 + t_1^2u^6 - 12288t_1u$$
$$- 7680t_1u^2 + 480t_1u^4 + 48t_1u^5 - 5184u^2 - 2592u^3 - 324u^4)^{-2}.$$

The only remaining condition in order to $\{a_1, a_2, a_3, a_4, a_5, a_6\}$ be a rational Diophantine sextuple is that $a_2a_6 + 1$ is a perfect square. This condition leads to a quartic in the variable t_1 over $\mathbb{Q}(u)$:

$$(u^{12} + 120u^{11} + 5496u^{10} + 125600u^9 + 1639440u^8 + 13075200u^7 + 65656320u^6$$
$$+ 209203200u^5 + 419696640u^4 + 514457600u^3 + 360185856u^2$$
$$+ 125829120u + 16777216)t_1^4$$

$$+ (24u^{12} + 1296u^{11} + 32256u^{10} + 446208u^9 + 3461760u^8 + 13047552u^7 - 208760832u^5$$

$$- 886210560u^4 - 1827667968u^3 - 2113929216u^2 - 1358954496u - 402653184)t_1^3$$

$$+ (36u^{12} + 1296u^{11} + 18072u^{10} + 48096u^9 - 1681632u^8 - 22516992u^7 - 127051776u^6$$

$$- 360271872u^5 - 430497792u^4 + 197001216u^3 + 1184366592u^2$$

$$+ 1358954496u + 603979776)t_1^2$$

$$+ (-432u^{11} - 15552u^{10} - 259200u^9 - 2267136u^8 - 9116928u^7 + 145870848u^5$$

$$+ 580386816u^4 + 1061683200u^3 + 1019215872u^2 + 452984832u)t_1$$

$$+ 1296u^{10} + 41472u^9 + 670032u^8 + 6054912u^7 + 31643136u^6 + 96878592u^5 + 171528192u^4$$

$$+ 169869312u^3 + 84934656u^2 = z^2.$$

Since this quartic has the $\mathbb{Q}(u)$-rational point at infinity (because the term with t_1^4 is a square: $(u^2+40u+16)^2(u+8)^4(u+2)^4$), it can be transformed into an elliptic curve over $\mathbb{Q}(u)$, as described in Sect. 2.2. Here, the singular point at infinity on the quartic corresponds to the point at infinity and an additional point P_1 on the elliptic curve. The quartic also has a $\mathbb{Q}(u)$-rational point, denoted by P_2, which corresponds to $t_1 = \frac{-3(u+4)u}{2(u^2+10u+16)}$. That point gives $a_6 = 0$, so we do not get a rational Diophantine sextuple from it. However, the point $2P_2$ on the elliptic curve corresponds to the point on the quartic for which

$$t_1 = \frac{3(3u^4 + 40u^3 + 368u^2 + 1280u + 1024)}{4(u^2 + 10u + 16)(u + 20)u},$$

so by inserting this value, we get a parametric family of rational Diophantine sextuples

$$\left\{ \frac{-12u(u + 4)(3u^4 + 8u^3 + 224u^2 + 576u + 512)(3u^3 + 28u^2 + 256u + 256)}{(u + 8)(u + 2)(u - 4)(3u^3 + 8u^2 + 144u + 128)(3u^4 + 48u^3 + 528u^2 + 1280u + 1024)}, \right.$$

$$\frac{8u(u + 20)(3u^5 + 8u^4 + 64u^3 - 640u^2 - 2304u - 2048)(u + 8)(u + 2)}{3(u + 4)(u - 4)(3u^3 + 8u^2 + 144u + 128)(3u^4 + 48u^3 + 528u^2 + 1280u + 1024)},$$

$$\frac{2(u + 4)(u - 4)(39u^7 + 776u^6 + 8096u^5 + 48640u^4 + 226048u^3 + 587776u^2 + 770048u + 393216)}{3(u + 8)(u + 2)(3u^3 + 8u^2 + 144u + 128)(3u^4 + 48u^3 + 528u^2 + 1280u + 1024)},$$

$$\frac{-8(u^2 + 4u + 32)(3u^3 + 14u^2 - 40u - 64)(9u^3 + 8u^2 + 112u + 384)(3u^4 + 48u^3 + 528u^2 + 1280u + 1024)}{3(u + 8)(u + 4)(u + 2)(u - 4)(3u^3 + 8u^2 + 144u + 128)^3},$$

$$\frac{4u(u+2)(17u^2+48u+48)(3u^5+8u^4-176u^3-2944u^2-9216u-8192)(3u^3+8u^2+144u+128)(u+8)^2}{3(u + 4)(u - 4)(3u^4 + 48u^3 + 528u^2 + 1280u + 1024)^3},$$

$$\left. \frac{12(u+2)(u-4)(5u+8)(u+4)(3u^2+8u+64)(3u^3+8u^2+144u+128)(3u^4+48u^3+528u^2+1280u+1024)}{(u + 8)(16384 + 69632u + 64768u^2 + 22272u^3 + 3680u^4 + 576u^5 + 9u^6)^2} \right\},$$

which satisfies the properties from Theorem 3.3.1.

For example, for $u = -1$, we get the rational Diophantine sextuple

$$\left\{ \frac{27900}{17479}, \frac{471352}{112365}, \frac{261770}{17479}, \frac{185535272}{419265}, \frac{63737828}{526368735}, \frac{79554420}{408480247} \right\}.$$

By taking other linear combinations of points P_1 and P_2, we can get new (but more complicated) families of rational Diophantine sextuples.

3.4 Rational Diophantine Sextuples via Edwards Curves

In this section, we will describe another construction of infinite families of rational Diophantine sextuples. It was presented in 2017 by Dujella and Kazalicki [135] and used the idea of Piezas [305]. This construction also uses genus 1 curves, but in the Edwards form, which can be transformed into elliptic curves in the Weierstrass form, we have seen in Sect. 2.2. The construction does not start from a triple but from a quadruple $\{a, b, c, d\}$.

Let $\{a, b, c, d\}$ be a rational Diophantine quadruple and let

$$ab + 1 = t_{12}^2, \quad ac + 1 = t_{13}^2, \quad ad + 1 = t_{14}^2,$$
$$bc + 1 = t_{23}^2, \quad bd + 1 = t_{24}^2, \quad cd + 1 = t_{34}^2.$$

Set $m = abcd$, and consider the curve

$$D: \quad (x^2 - 1)(y^2 - 1) = m. \tag{3.10}$$

With substitutions $x = 1/u$, $y = 1/v$, we obtain the curve in the Edwards form

$$u^2 + v^2 = 1 + (1 - m)u^2 v^2.$$

Since $(t_{12}^2 - 1)(t_{34}^2 - 1) = (t_{13}^2 - 1)(t_{24}^2 - 1) = (t_{14}^2 - 1)(t_{23}^2 - 1) = abcd$, we have the following three points on D: $Q_1 = (t_{12}, t_{34})$, $Q_2 = (t_{13}, t_{24})$, $Q_3 = (t_{14}, t_{23})$. The elements of the quadruple corresponding to these three points are distinct if and only if no two of these points can be transformed from one to another by changing signs and switching coordinates, e.g. for the points (t_{12}, t_{34}), $(-t_{34}, t_{12})$ and (t_{14}, t_{23}), we have that $a = d$.

The following proposition gives a criterion for extending quadruples to sextuples.

Theorem 3.4.1 (Piezas [305]) *Let $\{a, b, c, d\}$ be a rational Diophantine quadruple, and e and f the roots of the equation*

$$(abcdx + 2abc + a + b + c - d - x)^2 = 4(ab + 1)(ac + 1)(bc + 1)(dx + 1). \tag{3.11}$$

If $ef \neq 0$ and

$$(abcd - 3)^2 = 4(ab + cd + 3), \tag{3.12}$$

then $\{a, b, c, d, e, f\}$ is a rational Diophantine sextuple. Furthermore,

$$(a + b - e - f)^2 = 4(ab + 1)(ef + 1) \tag{3.13}$$

$$(c + d - e - f)^2 = 4(cd + 1)(ef + 1). \tag{3.14}$$

Proof From (3.11), Vieta's formulas give that

$$ef + 1 = \frac{a^2b^2c^2d^2 - 6abcd + a^2 - 2ab - 2ac - 2ad + b^2 - 2bc - 2bd + c^2 - 2cd + d^2 - 3}{(abcd - 1)^2}.$$

$$\tag{3.15}$$

By (3.12), the numerator of (3.15) is equal to

$$a^2 + 2ab - 2ac - 2ad + b^2 - 2bc - 2bd + c^2 + 2cd + d^2 = (a + b - c - d)^2.$$

Hence, we proved that

$$ef + 1 = \left(\frac{a + b - c - d}{abcd - 1} \right)^2.$$

Relations (3.13) and (3.14) can be checked similarly by replacing $e + f$ and ef with the expressions obtained from Vieta's formulas and using (3.12). $\qquad\square$

Note that e and f are the extensions of a rational Diophantine quadruple given in Theorem 3.1.3, i.e. $\{a, b, c, d, e\}$ and $\{a, b, c, d, f\}$ are regular rational Diophantine quintuples, while (3.13) and (3.14) state that $\{a, b, e, f\}$ and $\{c, d, e, f\}$ are regular rational Diophantine quadruples. An example of a sextuple with such configuration is $\left\{ \frac{96}{847}, \frac{287}{484}, \frac{129}{28}, \frac{847}{4}, \frac{1209}{700}, \frac{147}{25} \right\}$ (see [196]).

Let us write condition (3.12) in terms of t_{12} and t_{34}. We have

$$((t_{12}^2 - 1)(t_{34}^2 - 1) - 3)^2 - 4(t_{12}^2 - 1) - 4(t_{34}^2 - 1) - 12 =$$

$$(t_{12}t_{34} + t_{12} + t_{34})(t_{12}t_{34} - t_{12} - t_{34})(t_{12}t_{34} - t_{12} + t_{34})(t_{12}t_{34} + t_{12} - t_{34}) = 0.$$

We see that condition (3.12) is equivalent to $t_{12}t_{34} = \pm t_{12} \pm t_{34}$, i.e. $t_{34} = \pm \frac{t_{12}}{t_{12} \pm 1}$, with some choice of signs.

Thus, we set $t_{12} = t$, $t_{34} = \frac{t}{t-1}$ and

$$m = (t_{12}^2 - 1)(t_{34}^2 - 1) = \frac{2t^2 + t - 1}{t - 1}.$$

The curve

$$D: \quad (x^2 - 1)(y^2 - 1) = \frac{2t^2 + t - 1}{t - 1}$$

is birationally equivalent to the elliptic curve

$$E: \quad S^2 = T^3 - 2 \cdot \frac{2t^2 - t + 1}{t - 1} T^2 + \frac{(2t - 1)^2(t + 1)^2}{(t - 1)^2} T$$

over $\mathbb{Q}(t)$. The map is given by $T = 2(x^2 - 1)y + 2x^2 - (2 - m)$, and $S = 2Tx$. Rational specializations of E are non-singular for $t \neq 0, \pm 1, \frac{1}{2}$.

Let us denote by

$$P = \left(\frac{(2t - 1)^2(t + 1)}{t - 1}, \frac{2t(2t - 1)^2(t + 1)}{t - 1} \right)$$

a point of infinite order on E, and by

$$R = \left(\frac{(t + 1)(2t - 1)}{t - 1}, \frac{2(t + 1)(2t - 1)}{t - 1} \right)$$

a point of order 4. The point $(t_{12}, t_{34}) = \left(t, \frac{t}{t-1} \right) \in D(\mathbb{Q}(t))$ corresponds to the point $P \in E(\mathbb{Q}(t))$.

Proposition 3.4.2 *The Mordell-Weil group of $E(\mathbb{Q}(t))$ is generated by P and R.*

Proof We use the injectivity criterion by Gusić and Tadić from [202, Theorem 1.3]. It states that given an elliptic curve $y^2 = x^3 + A(t)x^2 + B(t)x$, where $A, B \in \mathbb{Z}[t]$, with exactly one non-trivial 2-torsion point over $\mathbb{Q}(t)$, the specialization homomorphism at $t_0 \in \mathbb{Q}$ is injective if the following condition is satisfied: for every non-constant square-free divisor $h(t) \in \mathbb{Z}[t]$ of $B(t)$ or $A(t)^2 - 4B(t)$, the rational number $h(t_0)$ is not a square in $\mathbb{Q}$.

After clearing out the denominators in the defining equation of E, we can write it in the form $y^2 = x^3 + A(t)x^2 + B(t)x$, where $A(t) = -2(2t^2 - t + 1)(t - 1)$, $B(t) = (2t - 1)^2(t + 1)^2(t - 1)^2$, $A(t)^2 - 4B(t) = -32t^2(t - 1)^3$. Thus, we see that the injectivity criterion is satisfied for $t_0 = 6$. Furthermore, it is easy to check (e.g. by `mwrank`) that the specializations of points P and R generate the Mordell-Weil group of $E_{t_0}(\mathbb{Q})$ (the torsion group of the specialized curve is $\mathbb{Z}/4\mathbb{Z}$, the rank is equal to 1, and the specialization of P is a free generator). $\qquad\square$

If $Q \in E$ is the point that corresponds to the point $(x, y) \in D$, then the points $-Q$ and $Q + R$ correspond to the points $(-x, y)$ and $(y, -x)$, respectively. Hence the triple $(Q_1, Q_2, Q_3) \in E(\mathbb{Q}(t))^3$ corresponds to the quadruple whose elements are not distinct if and only if there are two points, say Q_i and Q_j, such that $Q_i = \pm Q_j + kR$, where $k \in \{0, 1, 2, 3\}$.

We now fix $Q_1 = P$ and state conditions under which the point (Q_1, Q_2, Q_3) $\in E(\mathbb{Q}(t))^3$ corresponds to the degenerate Diophantine sextuple (i.e. $ef = 0$). We call such triple degenerate. It can be shown that the triple $(Q_1, Q_2, Q_3) \in E(\mathbb{Q}(t))^3$ is degenerate if and only if $\pm Q_1 \pm Q_2 \pm Q_3 = R$ for some choice of the signs.

It remains to investigate the question when for a triple (Q_1, Q_2, Q_3), where $Q_1 = P$, the corresponding Diophantine quadruple $\{a, b, c, d\}$, obtained from $a^2 = (t_{12}^2 - 1)(t_{13}^2 - 1)/(t_{23}^2 - 1)$, $b^2 = (t_{12}^2 - 1)(t_{23}^2 - 1)/(t_{13}^2 - 1)$, $c^2 = (t_{13}^2 - 1)(t_{23}^2 - 1)/(t_{12}^2 - 1)$, $d^2 = (t_{14}^2 - 1)(t_{24}^2 - 1)/(t_{12}^2 - 1)$, is rational. Note that if one element of the quadruple is rational, then all elements are rational.

For a point $(T, S) \in E(\mathbb{Q}(t))$, we have

$$x^2 - 1 = \left(\frac{S}{2T}\right)^2 - 1$$

$$= \frac{1}{4T^2}\left(T^3 - \frac{2(2t^2 - t + 1)}{t - 1}T^2 + \frac{(2t - 1)^2(t + 1)^2}{(t - 1)^2}T - 4T^2\right)$$

$$= \frac{1}{4T}\left(T - \frac{(2t - 1)(t + 1)}{t - 1}\right)^2 =: f(T).$$

Since

$$a^2 = \frac{(t_{12}^2 - 1)(t_{13}^2 - 1)(t_{14}^2 - 1)}{m} = \frac{f(x(Q_1))f(x(Q_2))f(x(Q_3))}{m}$$

$$\equiv x(Q_1)x(Q_2)x(Q_3)m \equiv (2t - 1)x(Q_2)x(Q_3)$$

$$\equiv x(-P + R)x(Q_2)x(Q_3) \,(\mathrm{mod}\ \mathbb{Q}(t)^{*2})$$

(where we used that $x(Q_1) = x(P) = \frac{(2t-1)^2(t+1)}{t-1}$, $m = \frac{(2t-1)(t+1)}{t-1}$, $x(-P + R) = \frac{(2t-1)(t+1)^2}{(t-1)^2}$), for the rationality of a, it is sufficient to prove that $x(-P + R)x(Q_2)x(Q_3)$ is a square in $\mathbb{Q}(t)$.

Since the point $(0, 0) \in E(\mathbb{Q}(t))$ is a point of order 2, the 2-descent homomorphism $E(\mathbb{Q}(t)) \to \mathbb{Q}(t)^*/\mathbb{Q}(t)^{*2}$, defined as in Theorem 2.4.10, together with the above-given coordinates of the points P, R and $-P + R$, lead us to the following result:

Proposition 3.4.3 *Let $Q_2, Q_3 \in E(\mathbb{Q}(t))$.*

(a) If $Q_2 + Q_3 \equiv O$ (mod $2E(\mathbb{Q}(t))$), then $a^2 \equiv 2t - 1$ (mod $\mathbb{Q}(t)^{*2}$).
(b) If $Q_2 + Q_3 \equiv R$ (mod $2E(\mathbb{Q}(t))$), then $a^2 \equiv (t - 1)(t + 1)$ (mod $\mathbb{Q}(t)^{*2}$).
(c) If $Q_2 + Q_3 \equiv P$ (mod $2E(\mathbb{Q}(t))$), then $a^2 \equiv (t - 1)(t + 1)(2t - 1)$ (mod $\mathbb{Q}(t)^{*2}$).
(d) If $Q_2 + Q_3 \equiv P + R$ (mod $2E(\mathbb{Q}(t))$), then $a^2 \equiv 1$ (mod $\mathbb{Q}(t)^{*2}$).

The proposition covers all the possibilities since, by Proposition 3.4.2, the Mordell-Weil group of $E(\mathbb{Q}(t))$ is generated by P and R.

Remark 3.4.4 In cases (a) and (b), we can still obtain parametric families of Diophantine sextuples if we specialize to those t for which $2t - 1$ or $(t - 1)(t + 1)$ are squares (e.g. if we replace t by $\frac{t^2+1}{2}$ and $\frac{t^2+1}{2t}$, respectively). Concerning the case (c), notice that the elliptic curve $y^2 = (x - 1)(x + 1)(2x - 1)$ has the Mordell-Weil group isomorphic to $\mathbb{Z}/2\mathbb{Z} \times \mathbb{Z}/4\mathbb{Z}$, so we do not get a parametric family of sextuples in this case.

For an illustration, we calculate a parametric family $\{a, b, c, d, e, f\}$ of rational Diophantine sextuples that corresponds to the triple $(P, 2P, 4P)$. It is easily seen that the triple is not degenerate. Since $2P + 4P \equiv O \pmod{2E(\mathbb{Q}(t))}$, by part a) of Proposition 3.4.3, the rationality of the sextuple will follow if we replace t by $\frac{t^2+1}{2}$. The corresponding Diophantine quadruple is given by

$$a = \frac{(t^8 - 8t^6 - 14t^4 + 32t^2 - 27)(t^8 + 26t^4 - 40t^2 - 3)}{64(t - 1)t(t + 1)(t^4 - 2t^2 + 5)(t^4 + 6t^2 - 3)},$$

$$b = \frac{16t(t - 1)^2(t + 1)^2(t^2 + 3)(t^4 - 2t^2 + 5)(t^4 + 6t^2 - 3)}{(t^8 - 8t^6 - 14t^4 + 32t^2 - 27)(t^8 + 26t^4 - 40t^2 - 3)},$$

$$c = \frac{t(t^8 - 8t^6 - 14t^4 + 32t^2 - 27)(t^8 + 26t^4 - 40t^2 - 3)}{(t - 1)(t + 1)(t^2 - 3)^2(t^2 + 1)^2(t^4 - 2t^2 + 5)(t^4 + 6t^2 - 3)},$$

$$d = \frac{4t(t^2 - 3)^2(t^2 + 1)^2(t^4 - 2t^2 + 5)(t^4 + 6t^2 - 3)}{(t - 1)(t + 1)(t^8 - 8t^6 - 14t^4 + 32t^2 - 27)(t^8 + 26t^4 - 40t^2 - 3)}.$$

Using Theorem 3.4.1, we find that $e = e_1/e_2$ and $f = f_1/f_2$ are given by

$$e_1 = (t + 1)(t^2 - 2t + 3)(t^2 + 2t - 1)(t^6 - 2t^5 + t^4 + 12t^3 + 7t^2 - 2t - 9)$$

$$\times (t^6 + 2t^5 - 3t^4 + 4t^3 - 17t^2 + 18t + 3)$$

$$\times (t^{12} - 4t^{11} + 6t^{10} + 20t^9 - t^8 + 24t^7 - 12t^6 - 88t^5 - 177t^4 + 364t^3 - 90t^2 - 60t + 81)$$

$$\times (t^{12} + 4t^{11} - 2t^{10} - 4t^9 - 41t^8 + 40t^7 + 100t^6 - 72t^5 + 63t^4 + 212t^3 - 66t^2 - 180t + 9),$$

$$e_2 = 64(t - 1)t(t^2 - 3)^2(t^2 + 1)^4(t^4 - 2t^2 + 5)(t^4 + 6t^2 - 3)(t^8 - 8t^6 - 14t^4 + 32t^2 - 27)$$

$$\times (t^8 + 26t^4 - 40t^2 - 3),$$

$$f_1 = (t - 1)(t^2 - 2t - 1)(t^2 + 2t + 3)(t^6 - 2t^5 - 3t^4 - 4t^3 - 17t^2 - 18t + 3)$$

$$\times (t^6 + 2t^5 + t^4 - 12t^3 + 7t^2 + 2t - 9)$$

$$\times (t^{12} - 4t^{11} - 2t^{10} + 4t^9 - 41t^8 - 40t^7 + 100t^6 + 72t^5 + 63t^4 - 212t^3 - 66t^2 + 180t + 9)$$

$$\times (t^{12} + 4t^{11} + 6t^{10} - 20t^9 - t^8 - 24t^7 - 12t^6 + 88t^5 - 177t^4 - 364t^3 - 90t^2 + 60t + 81),$$

$$f_2 = 64t(t + 1)(t^2 - 3)^2(t^2 + 1)^4(t^4 - 2t^2 + 5)(t^4 + 6t^2 - 3)(t^8 - 8t^6 - 14t^4 + 32t^2 - 27)$$

$$\times (t^8 + 26t^4 - 40t^2 - 3).$$

If we replace t by $u^2 + 1$, the elliptic curve E will have another point of infinite order (independent of P), $L = \left(\frac{(2+u^2)^2}{u^2}, \frac{(2+u^2)^2(1+u^2)}{u^3} \right)$. Now the triple $(P, 2P +$

$L, R + L - P)$ is not degenerate and satisfies the condition of Proposition 3.4.3.d). The construction gives the following family of rational Diophantine sextuples

$$a = \frac{(u^3 + 3u^2 + u + 1)(u^3 + u^2 + 3u + 1)(2u - 1)}{2(u - 2)(u - 1)(u^2 + u + 1)(u + 1)},$$

$$b = \frac{2(u - 1)(u^2 + u + 1)(u + 1)(u^2 + 2)(u - 2)u^2}{(u^3 + 3u^2 + u + 1)(u^3 + u^2 + 3u + 1)(2u - 1)},$$

$$c = \frac{(u^3 + 3u^2 + u + 1)(u^3 + u^2 + 3u + 1)(u - 2)}{2(2u - 1)(u - 1)(u^2 + u + 1)(u + 1)u^2},$$

$$d = \frac{2(2u^2 + 1)(2u - 1)(u - 1)(u^2 + u + 1)(u + 1)}{u^2(u^3 + 3u^2 + u + 1)(u^3 + u^2 + 3u + 1)(u - 2)},$$

$$e = \frac{8u^2(u - 1)(2u + 1)(u + 2)(u + 1)(u^2 + 1)}{(u - 2)(2u - 1)(u^2 + u + 1)(u^3 + u^2 + 3u + 1)(u^3 + 3u^2 + u + 1)},$$

$$f = \frac{3(3u^2 + 2u + 1)(u^2 + 2u + 3)(u^4 + 1)(u^4 + 4u^2 + 1)}{2(u - 1)(u - 2)(2u - 1)(u + 1)(u^2 + u + 1)(u^3 + u^2 + 3u + 1)(u^3 + 3u^2 + u + 1)}.$$

If we further require $2(u^2 + 1)$ to be a square, then the resulting parametrization $u \mapsto \frac{v^2 - 2v - 1}{v^2 + 2v - 1}$ yields the point K on E,

$$K = \left(\frac{(3v^2 - 2v + 1)(v^2 + 2v + 3)(3v^2 + 2v + 1)^2}{(v^2 - 2v - 1)^2(v^2 + 2v - 1)^2}, \right.$$

$$\left. \frac{4(3v^2 - 2v + 1)(v^2 + 1)(v^2 + 2v + 3)(3v^2 + 2v + 1)^2}{(v^2 + 2v - 1)^2(v^2 - 2v - v)^3} \right),$$

with the property that $2K = S$. When we apply our construction to the triple $(P, K, -2K + R)$, we obtain a relatively simple (in fact, the simplest known) family of sextuples (also found by Piezas [305]):

$$a = \frac{(v^2 - 2v - 1)(v^2 + 2v + 3)(3v^2 - 2v + 1)}{4v(v^2 - 1)(v^2 + 2v - 1)},$$

$$b = \frac{4v(v^2 - 1)(v^2 - 2v - 1)}{(v^2 + 2v - 1)^3},$$

$$c = \frac{4v(v^2 - 1)(v^2 + 2v - 1)}{(v^2 - 2v - 1)^3},$$

$$d = \frac{(v^2 + 2v - 1)(v^2 - 2v + 3)(3v^2 + 2v + 1)}{4v(v^2 - 1)(v^2 - 2v - 1)},$$

$$e = \frac{-v(v^2 + 4v + 1)(v^2 - 4v + 1)}{(v - 1)(v + 1)(v^2 + 2v - 1)(v^2 - 2v - 1)},$$

$$f = \frac{(v - 1)(v + 1)(3v^2 - 1)(v^2 - 3)}{4v(v^2 + 2v - 1)(v^2 - 2v - 1)}.$$

$$(3.16)$$

Note that with this parametrization we have $x(-P+R) \equiv (3v^2+2v+1)(v^2-2v+3)$ $(\mathrm{mod}\ \mathbb{Q}(t)^{*2})$, $x(K) \equiv (3v^2-2v+1)(v^2+2v+3)$ $(\mathrm{mod}\ \mathbb{Q}(t)^{*2})$, $x(-2K+R) \equiv (3v^2+2v+1)(v^2-2v+3)(3v^2-2v+1)(v^2+2v+3)$ $(\mathrm{mod}\ \mathbb{Q}(t)^{*2})$, and thus the rationality condition is satisfied.

Proposition 3.4.5 *There are infinitely many rational Diophantine sextuples in which one element is a positive integer.*

Proof It suffices to choose in the parametric family (3.16), the rational parameter $v = w/z$ such that $w^2 + 2wz - z^2 = \pm 1$ (then b will be an integer) or $w^2 - 2wz - z^2 = \pm 1$ (then c will be an integer). The solutions of $(w+z)^2 - 2z^2 = \pm 1$ are given in terms of the Pell numbers P_k, defined by $P_0 = 0$, $P_1 = 1$, $P_k = 2P_{k-1} + P_{k-2}$, which satisfy $(P_k + P_{k-1})^2 - 2P_k^2 = \pm 1$. So we can take $v = \pm \frac{P_{k-1}}{P_k}$ or $v = \pm \frac{P_{k+1}}{P_k}$.
$\square$

As a concrete example illustrating Proposition 3.4.5, we may take $v = 5/2$. We get the sextuple

$$\left\{ \frac{1121}{11480}, \frac{840}{68921}, 34440, \frac{23001}{280}, \frac{2530}{287}, \frac{19383}{1640} \right\},$$

in which the third element is a positive integer.

In Sect. 3.2, we showed that there exist infinitely many triples, each of which can be extended to rational Diophantine sextuples in infinitely many ways. We cannot expect the analogous result for quadruples to hold since we know that any rational Diophantine quadruple can be extended in at most finitely many ways to a rational Diophantine quintuple (see [214]). However, the construction presented in this section gives a partial analogue for quadruples; namely it shows that there are infinitely many sextuples $\{a, b, c, d, e, f\}$ with fixed product $abcd$ (more precisely, with fixed products ab and cd).

Dražić and Kazalicki [87] generalized some of the results from this section to rational $D(q)$-quadruples. By considering curves of the form $(x^2 - q)(y^2 - q) = m$ (which are equivalent to the so-called twisted Edwards curves), they characterized rationals m with the property that there exists a $D(q)$-quadruple with the product of elements equal to m.

3.5 Rational Diophantine Sextuples with Square Denominators

In this section, we study the arithmetic properties of rational Diophantine sextuples. In particular, we prove the following theorem due to Dujella, Kazalicki, and Petričević [139].

Theorem 3.5.1 *There are infinitely many rational Diophantine sextuples such that denominators of all the elements (in the lowest terms) in the sextuples are perfect squares.*

Using a variant of the construction of sextuples presented in the previous section, we will obtain the following family of Diophantine sextuples, which we denote by $\mathcal{F}(t)$:

$$\left\{ -\frac{9(t^2+1)}{8(t-1)(t+1)}, \; \frac{(t^2-7)(t^2+1)(7t^2-1)}{8(t-1)^3(t+1)^3}, \right.$$

$$\frac{8(t-1)^3(t+1)^3}{9(t^2+1)^3}, \; -\frac{2(t^2+5)(5t^2+1)}{9(t-1)(t+1)(t^2+1)},$$

$$-\frac{2t(t^2-4t-3)(3t^2-4t-1)(t^3+8t^2+5t+4)(4t^3-5t^2+8t-1)}{(t-1)(t+1)(t^2+1)(t^4+34t^2+1)^2},$$

$$\left. \frac{2t(t^2+4t-3)(3t^2+4t-1)(t^3-8t^2+5t-4)(4t^3+5t^2+8t+1)}{(t-1)(t+1)(t^2+1)(t^4+34t^2+1)^2} \right\}.$$

Note that this construction always produces sextuples with mixed signs since the product of the first and the third element is negative.

Our starting point is an example of a Diophantine sextuple with square denominators

$$\mathcal{L} = \left\{ \frac{75}{8^2}, \; -\frac{3325}{64^2}, \; -\frac{12288}{125^2}, \; \frac{123}{10^2}, \; \frac{3498523}{2260^2}, \; \frac{698523}{2260^2} \right\},$$

which has been discovered by a numerical search.

In order to describe a family of Diophantine sextuples containing $\mathcal{L}$, we will use the construction from the previous section. Thus, we consider the Edwards curve D given by the equation $(x^2-1)(y^2-1) = \frac{2t^2+t-1}{t-1}$, which is birationally equivalent to the elliptic curve F, and we know that points P (of infinite order) and R (of order 4) generate the Mordell-Weil group $E(\mathbb{Q}(t))$ (see Proposition 3.4.2). The construction from the previous section is based on rational Diophantine quadruples $\{a, b, c, d\}$ that satisfy the condition

$$(abcd - 3)^2 = 4(ab + cd + 3), \tag{3.17}$$

and this condition allows the extension of the quadruple to a sextuple (see Theorem 3.4.1). To specify the family of Diophantine quadruples which satisfies (3.17) and whose product is $\frac{2t^2+t-1}{t-1}$, we need to specify two points Q_2 and Q_3 in $E(\mathbb{Q}(t))$ (in the previous section, we gave sufficient conditions for three points P, Q_2 and Q_3 in $E(\mathbb{Q}(t))$ to define the quadruple whose elements are rational, distinct and non-zero).

If we go back to our example, we can observe that the first four elements of $\mathcal{L}$ satisfy condition (3.17), and their product is equal to $\frac{2t_0^2+t_0-1}{t_0-1}$ where $t_0 = 113/625$. Inspired by the fact that $\frac{1-t_0}{2} = \frac{256}{625}$ is a square, we restrict to the subfamily obtained by replacing t with $1-2t^2$. In fact, after this substitution, $E(\mathbb{Q}(t))$ has torsion group

$\mathbb{Z}/2\mathbb{Z} \times \mathbb{Z}/4\mathbb{Z}$. If we further replace t with $\frac{3t^2+3}{4t^2-4}$ (and denote the resulting elliptic curve also by E), then we obtain another point of infinite order in $E(\mathbb{Q}(t))$,

$$Q = \left(-\frac{\left(t^2 - 7\right)^2 \left(7t^2 - 1\right)^2}{36(t-1)^2(t+1)^2 \left(t^2+1\right)^2}, \frac{t\left(t^2 - 7\right)^2 \left(7t^2 - 1\right)^2}{9(t-1)^2(t+1)^2 \left(t^2+1\right)^3} \right).$$

It is easy to check that the family $\mathcal{F}(t)$ corresponds to the triple $(P, Q, P + R + T_1) \in E(\mathbb{Q}(t))^3$, where $T_1 = \left(-\frac{(t^2+5)^2(7t^2-1)^2}{36(t-1)^2(t+1)^2(t^2+1)^2}, 0 \right)$ is a point of order 2. We obtain $\mathcal{L}$ if we specialize to $t = -1/7$, i.e. $\mathcal{L} = \mathcal{F}(-1/7)$.

We can notice that some square factors are already present in the denominators of elements of $\mathcal{F}(t)$. The remaining factors involve $(t - 1)$, $(t + 1)$, $(t^2 + 1)$ and powers of 2 and 3. Thus, we are interested in rational parameters t such that $t^4 - 1 = (t-1)(t+1)(t^2+1)$ is "almost" a square. Let $C : t^4 - 1 = -6s^2$ be a genus 1 curve defined over $\mathbb{Q}$. It is birationally equivalent to the elliptic curve $E' : y^2 = x^3 + 9x$ via the map $h : E' \to C, h(x, y) = \left(\frac{3-x}{3+x}, \frac{-2y}{(x+3)^2} \right)$. This elliptic curve has the 2-torsion point $(0, 0)$ and the rank equal to 1 with a free generator $T = (4, 10)$ (corresponding to $\left(-\frac{1}{7}, \frac{120}{49} \right)$ on C). Set $(t_k, s_k) := h(kT) \in C$.

Proposition 3.5.2 *If k is a positive integer such that $k \equiv 1, 2 \pmod 3$, then $\mathcal{F}(t_k)$ is a rational Diophantine sextuple such that denominators of all the elements in the sextuple are perfect squares.*

We recall that for a prime p and $q = a/b \in \mathbb{Q}$, $\gcd(a, b) = 1$, the p-adic valuation of q is $v_p(q) = r$, where $a/b = p^r a_1/b_1$ and $p \nmid a_1 b_1$.

If $(x, y) \in E'(\mathbb{Q})$ and $v_p(x) < 0$, then there exists a positive integer r such that $v_p(x) = -2r$ and $v_p(y) = -3r$. Let $E'^{(r)}(p) = \{(x, y) \in E'(\mathbb{Q}) : v_p(x) \leq -2r, \ v_p(y) \leq -3r\} \cup \{O\}$. Then $E'^{(r)}(p)$ is a subgroup of $E'(\mathbb{Q})$ (see, e.g. [356, Chapter 8.1]).

Lemma 3.5.3 *Let k be a positive integer. If $k \equiv 1, 2 \pmod 3$, then $v_3(t_k - 5) \geq 2$ and $v_3(s_k) = 0$.*

Proof Define $(x_k, y_k) := kT$. Direct computation shows that $v_3(x_1) = 0$, $v_3(x_2) = 0$ and $v_3(x_3) = -2$. Since $E'^{(1)}(3)$ is a subgroup of $E'(\mathbb{Q})$, we conclude that $v_3(x_k) < 0$ if and only if $k \equiv 0 \pmod 3$. Assume that $v_3(x_k) \geq 1$. Then $v_3(y_k) \geq 2$ and $(x_{2k}, y_{2k}) \in E'^{(1)}(3)$, which is a contradiction since $k \not\equiv 0 \pmod 3$. Therefore, if $k \equiv 1, 2 \pmod 3$, then $v_3(x_k) = 0$. Since $y_k^2 = x_k^3 + 9x_k$, it follows $v_3(x_k - 1) \geq 1$. Now, $t_k - 5 = \frac{3-x_k}{3+x_k} - 5 = -\frac{6(2+x_k)}{3+x_k}$ implies that $v_3(t_k - 5) \geq 2$. Finally, from $t_k^4 - 1 = -6s_k^2$, it follows that $v_3(s_k) = 0$. $\qquad \square$

Proof (Sketch of the Proof of Proposition 3.5.2) We have to analyze the denominators of all the elements in $\mathcal{F}(t) = \{a_1(t), a_2(t), \ldots, a_6(t)\}$.

Since $a_1(t) = -\dfrac{9(t^2+1)}{8(t-1)(t+1)} = -\dfrac{9(t^2+1)^2}{8(t^4-1)}$, we have that $a_1(t_k) = \dfrac{3(t_k^2+1)^2}{16s_k^2}$.

Similarly, $a_3(t) = \dfrac{8(t-1)^3(t+1)^3}{9(t^2+1)^3} = \dfrac{8(t^4-1)^3}{9(t^2+1)^6}$, and $a_3(t_k) = \dfrac{-3 \cdot 2^6 s_k^6}{(t_k^2+1)^6}$. The claim for $a_1(t)$ and $a_3(t)$ now follows from Lemma 3.5.3 since for $k \equiv 1, 2 \pmod 3$ we have that $v_3(s_k) = 0$.

We have that $a_2(t) = \dfrac{(t^2-7)(t^2+1)(7t^2-1)}{8(t-1)^3(t+1)^3} = \dfrac{(t^2-7)(t^2+1)^4(7t^2-1)}{8(t^4-1)^3}$. The only primes that can divide the denominator of $a_2(t_k) = \dfrac{-(t_k^2-7)(t_k^2+1)^4(7t_k^2-1)}{3 \cdot 24^2 s_k^6}$ are the primes that divide s_k or $1/t_k$. Assume that p is a prime different from 2 and 3. Since $t_k^4 - 1 = -6s_k^2$, if $v_p(t_k) < 0$, then $v_p(s_k) = 2v_p(t_k)$. Hence, for such p, $v_p(a_2(t_k)) = 1$ if $p = 7$ and $v_p(a_2(t_k)) = 0$ otherwise. If $v_p(s_k) > 0$, then $v_p(a_2(t_k))$ is even unless $v_p((t_k^2-7)(7t_k^2-1)) > 0$. This cannot happen because, by the Euclidean algorithm, we find polynomials $f_1(t)$, $f_2(t) \in \mathbb{Z}[t]$ such that $f_1(t)(t^4-1) + f_2(t)(t^2-7)(7t^2-1) = 1152 = 2^7 3^2$, which is divisible only by 2 and 3. To rule out $p = 3$ case, we note that $v_3\big((t_k^2-7)(7t_k^2-1)\big) \geq 3$ since by Lemma 3.5.3, we have that $v_3(t_k - 5) \geq 2$. Hence $v_3(a_2(t_k)) \geq 0$. If $v_2(s_k) > 0$, then $v_2(t_k) = 0$ and $v_2((t_k^2-7)(7t_k^2-1)) = 2$ is even. Finally, if $v_2(t_k) < 0$, then $4v_2(t_k) = 2v_2(s_k) + 1$, which is not possible.

The case of $a_4(t)$ can be handled similarly, while that cases of $a_5(t)$ and $a_6(t)$ are slightly more involved, but again the claims follow from Lemma 3.5.3. The details can be found in [139]. $\qquad\square$

3.6 Elliptic Curves of High Rank with Prescribed Torsion Group

In Sect. 2.5, we listed the tables with the highest known ranks over $\mathbb{Q}$ and $\mathbb{Q}(t)$ for each of the 15 torsion groups from Mazur's theorem and specifically indicated the record ranks associated with elliptic curves induced by rational Diophantine triples. Curves induced by rational Diophantine triples can have one of the following four torsion groups: $\mathbb{Z}/2\mathbb{Z} \times \mathbb{Z}/2\mathbb{Z}$, $\mathbb{Z}/2\mathbb{Z} \times \mathbb{Z}/4\mathbb{Z}$, $\mathbb{Z}/2\mathbb{Z} \times \mathbb{Z}/6\mathbb{Z}$, and $\mathbb{Z}/2\mathbb{Z} \times \mathbb{Z}/8\mathbb{Z}$, so we will now take a closer look at these four cases.

3.6.1 Torsion Group $\mathbb{Z}/2\mathbb{Z} \times \mathbb{Z}/2\mathbb{Z}$ over $\mathbb{Q}(t)$

Elliptic curves induced by rational Diophantine triples were used in constructing curves with a large rank for the first time by Dujella [99] in 2000. It was shown that there exist elliptic curves with torsion group $\mathbb{Z}/2\mathbb{Z} \times \mathbb{Z}/2\mathbb{Z}$ and rank 4 over $\mathbb{Q}(t)$ and curves with the same torsion and rank 7 over $\mathbb{Q}$. The idea was to use a rational Diophantine quintuple $\{a, b, c, d, e\}$, which contains two regular quadruples $\{a, b, c, d\}$ and $\{a, b, c, e\}$. In [99], a one-parameter family of such quintuples was found, and in Sect. 3.3, we saw how such two-parameter family can be obtained. Now, for example, if we consider the curve induced by $\{b, d, e\}$, i.e.

$$y^2 = (bx + 1)(dx + 1)(ex + 1),$$

then we may expect that it will have at least four independent points of infinite order, namely the points with x-coordinates

$$0, \quad \frac{1}{bde}, \quad a, \quad c.$$

Indeed, it was shown in [99] that for the mentioned one-parameter family of curves, these points are independent, so the rank of the curve over $\mathbb{Q}(t)$ is ≥ 4. By computing the rank (with a version of `mwrank` available at that time) for small values of the parameter t, an example of a rank 7 curve was obtained, which at the time was the record for the torsion group $\mathbb{Z}/2\mathbb{Z} \times \mathbb{Z}/2\mathbb{Z}$.

A similar idea can be applied to obtain families of curves with a torsion group $\mathbb{Z}/2\mathbb{Z} \times \mathbb{Z}/2\mathbb{Z}$ and rank ≥ 5 over $\mathbb{Q}(t)$, if a parametric family of rational Diophantine sextuples $\{a, b, c, d, e, f\}$ is taken as a starting point, for example, the one we got in Sect. 3.2.

Consider the curve

$$E(t): \quad y^2 = (dx + 1)(ex + 1)(fx + 1).$$

It, in addition to three obvious points of order two, has rational points with x-coordinates

$$0, \quad \frac{1}{def}, \quad a, \quad b, \quad c.$$

We claim that these five points are independent points of infinite order over $\mathbb{Q}(t)$. Since the specialization map which, for $t_0 \in \mathbb{Q}$, maps the curve $E(t)$ over $\mathbb{Q}(t)$ to the curve E_{t_0} over $\mathbb{Q}$, is a homomorphism, it is sufficient to find one parameter t_0 for which these five points are independent points of infinite order on E_{t_0}. It is not difficult to verify that $t_0 = 2$ has the required property (by checking that the determinant of the corresponding height matrix is different from zero). Therefore, the rank of the curve $E(t)$ over $\mathbb{Q}(t)$ is indeed ≥ 5.

To obtain a curve over $\mathbb{Q}(t)$ of rank ≥ 6, we will start from a "simple" family of rank ≥ 3 and increase the rank in several steps so that we try to satisfy the condition that an additional point (which will be suggested by the 2-descent method) be on the curve. We will follow the construction from [149, 150], which is an improved version of the constructions from [6].

We start from a two-parameter family of irregular Diophantine quadruples, which contains two regular triples (like Diophantus' example with which we started our discussion in Sect. 1.1). If we then take as a starting point one of the irregular subtriples of that quadruple, we can expect to get a curve of rank ≥ 3 (apart from the points P and S (i.e. R so that $2R = S$), which are independent due to the irregularity of the triple, we also have a point that comes from the fourth element from the quadruple, and for which we can expect that it is independent of P and S due to the irregularity of the quadruple).

We follow the construction and notation from [91]. Let us put: $a_1 = a$, $a_2 = b$, $ab + q = x^2$, $a_3 = a + b + 2x$, $a_4 = a + 4b + 4x$. Then $\{a_1, a_2, a_3\}$ and $\{a_2, a_3, a_4\}$ are regular $D(q)$-triples. In order to obtain a $D(q)$-quadruple, it remains to satisfy the condition that $a_1 a_4 + q$ is a perfect square, i.e.

$$a(a + 4b + 4x) + q = a^2 + 4ax + 4(x^2 - q) + q = y^2. \tag{3.18}$$

Equation (3.18) can be written as

$$3q = (a + 2x)^2 - y^2 = (a + 2x - y)(a + 2x + y),$$

and we can try to satisfy it by putting

$$a + 2x - y = 3, \quad a + 2x + y = q.$$

By adding, we get $q = 2a + 4x - 3$. If we insert this into $ab + q = x^2$, we get

$$a(b + 2) = (x - 1)(x - 3).$$

We put here $x = ak + 1$, so we get $b = ak^2 - 2k - 2$ and $q = 2a(2k + 1) + 1$. Thus, we obtained the $D(2a(2k + 1) + 1)$-quadruple

$$\{a,\ ak^2 - 2k - 2,\ a(k + 1)^2 - 2k,\ a(2k + 1)^2 - 8k - 4\}$$

with the required property of (ir)regularity. In order to get a rational $D(1)$-quadruple from it, let us first put $2a(2k + 1) + 1 = n^2$, i.e. $k = \frac{n^2 - 2a - 1}{4a}$, and then divide the elements of the $D(n^2)$-quadruple by n. Thus, after rearranging the order of elements according to [6, 149, 150], we obtain the following two-parameter family of rational $D(1)$-quadruples:

$$c_1(a, n) = \frac{a}{n},$$

$$c_2(a, n) = \frac{((n - 3)(n - 1) + 2a)((n + 1)(n + 3) + 2a)}{16an},$$

$$c_3(a, n) = \frac{(n - 3)(n - 1)(n + 1)(n + 3)}{4an},$$

$$c_4(a, n) = \frac{((n - 3)(n - 1) - 2a)((n + 1)(n + 3) - 2a)}{16an}.$$

$$\tag{3.19}$$

By construction, this quadruple is not regular, but it contains two regular triples $\{c_1, c_2, c_4\}$ and $\{c_2, c_3, c_4\}$.

Let us now consider the elliptic curve induced by the triple $\{c_1, c_2, c_3\}$, i.e.

$$y^2 = (c_1(a, n)x + 1)(c_2(a, n)x + 1)(c_3(a, n)x + 1). \tag{3.20}$$

Notice that we have chosen an irregular subtriple so that the three known points on the curve, which correspond to x-coordinates

$$0, \quad c_4(a, n), \quad \frac{t_{1,2}\, t_{1,3} + t_{1,2}\, t_{2,3} + t_{1,3}\, t_{2,3} + 1}{c_1(a, n)c_2(a, n)c_3(a, n)},$$

where $t_{i,j} = t_{i,j}(a, n) = \sqrt{c_i(a, n)c_j(a, n) + 1}$, $1 \le i < j \le 3$, would be independent. In terms of a and n, these three points have coordinates

$$P_1 = (0, 1),$$

$$P_2 = \Big(\frac{(n^2 + 4n - 2a + 3)(n^2 - 4n - 2a + 3)}{16an},$$

$$- \frac{(n^2 - 2a + 3)(n^4 - 10n^2 - 4a^2 + 9)(n^4 - 2an^2 - 10n^2 - 6a + 9)}{512a^2n^3} \Big),$$

$$P_3 = \Big(\frac{6n}{(n - 3)(n + 3)}, \frac{(n^2 + 6a - 9)(3n^2 + 2a - 3)}{4a(n - 3)(n + 3)} \Big).$$

Curve (3.20) has torsion subgroup $\mathbb{Z}/2\mathbb{Z} \times \mathbb{Z}/2\mathbb{Z}$ and rank ≥ 3 over $\mathbb{Q}(n, a)$. The points P_1, P_2 and P_3 are independent points of infinite order.

Indeed, it suffices to find a specialization for which the points corresponding to P_1, P_2 and P_3 are independent points of infinite order. For example, for $a = 2$ and $n = 5$, the specialized points $(0, 1)$, $(11/10, -1173/125)$, $(15/8, 133/8)$ have this property.

Let us now look for a condition on a and n so that the curve has an additional rational point. In the standard way, we transform the equation of the curve to the form $y^2 = x^3 + Ax^2 + Bx$.

First, we get the equation

$$y^2 = x(x + c_1(a, n)c_3(a, n) - c_1(a, n)c_2(a, n))(x + c_2(a, n)c_3(a, n)$$

$$- c_1(a, n)c_2(a, n)),$$

and then we get rid of the denominators so that we get the form

$$y^2 = x^3 + A(a, n)x^2 + B(a, n)x \tag{3.21}$$

in which $A(a, n)$ and $B(a, n)$ are polynomials with integer coefficients:

$$A(a, n) = 81 + 108a + 108a^2 - 96a^3 - 32a^4 - 180n^2 - 84an^2 - 120a^2n^2$$
$$- 32a^3n^2 + 118n^4 - 28an^4 + 12a^2n^4 - 20n^6 + 4an^6 + n^8,$$
$$B(a, n) = 4a^2(9 + 2a - n^2)(3 + 2a - 4n + n^2)(3 + 2a + 4n + n^2)$$
$$\times (-3 + 2a + 3n^2)(-9 + 4a^2 + 10n^2 - n^4).$$

After these transformations, the x-coordinates of the three known independent points of infinite order become

$$x_1 = 4a^2(3 + 2a - 4n + n^2)(3 + 2a + 4n + n^2),$$

$$x_2 = \frac{(3 + 2a - 4n + n^2)(3 + 2a + 4n + n^2)(9 - 6a - 10n^2 - 2an^2 + n^4)^2}{16n^2},$$

$$x_3 = 2a(3 + 2a - 4n + n^2)(3 + 2a + 4n + n^2)(-3 + 2a + 3n^2).$$

Motivated by the 2-descent method, we look at the factors of B (these are quantities b_1 from the 2-descent method) as candidates for an additional point, and among them, we are looking for those leading to simple conditions for obtaining an extra point. For example, the condition that $(3+2a-4n+n^2)(-3+2a+3n^2)(-9+4a^2 + 10n^2 - n^4)$ is the x-coordinate of an additional point is that

$$2(9 + 6a + 8a^2 - 18n - 4an + 8n^2 - 2an^2 + 2n^3 - n^4) \tag{3.22}$$

is a square. One of the roots of the discriminant of this polynomial (in a) is $n = 7/3$, so we can satisfy this condition by putting $n = 7/3$ since (3.22) becomes $\frac{16}{81}(9a - 8)^2$. By inserting $n = 7/3$, the coefficients of the curve become

$$A(a) = -2(-51200 + 109440a + 38880a^2 + 55404a^3 + 6561a^4),$$

$$B(a) = 243a^2(20 + 3a)(-4 + 9a)(16 + 9a)(80 + 9a)(320 + 81a^2),$$

while the x-coordinates of the previous three points and of the new fourth point are

$$x_1 = 81a^2(-4 + 9a)(80 + 9a),$$
$$x_2 = 27a(20 + 3a)(-4 + 9a)(80 + 9a),$$
$$x_3 = \frac{1}{441}(-4 + 9a)(80 + 9a)(160 + 171a)^2, \tag{3.23}$$
$$x_4 = 3(20 + 3a)(-4 + 9a)(320 + 81a^2).$$

Similarly as before, through appropriate specialization, it can be shown that these four points are independent, so we obtained a curve of rank ≥ 4 over $\mathbb{Q}(a)$.

Using a slight modification of the above-described method of testing the factors of polynomial $B(a)$ as candidates for additional points (so that we also look at the factors of $B'(a) = A(a)^2 - 4B(a)$), it is possible to find a large number of curves of rank ≥ 5. We will give details for one of them, for which we can additionally increase the rank to 6.

For $9a(-20 + 9a)(16 + 9a)(80 + 9a)$ to be the x-coordinate of a new point on the curve of rank 4, the condition that

$$10(-20 + 9a)(-2 + 9a)$$

is a square should be satisfied. This condition corresponds to a curve of genus 0, which has an obvious point corresponding to $a = 2/9$, so it has a parametric solution

$$a = \frac{2(-1 + 10w)(1 + 10w)}{9(-1 + 10w^2)}. \tag{3.24}$$

The corresponding curve of rank ≥ 5 is $Y^2 = X^3 + A(w)X^2 + B(w)X$, where

$$A(w) = -2(-169 - 12020w^2 + 678000w^4 - 12680000w^6 + 80000000w^8),$$

$$B(w) = (-1 + 10w)^2(1 + 10w)^2(-1 + 20w^2)(1 + 80w^2)(-31 + 400w^2)$$

$$\times (-41 + 500w^2)(9 - 200w^2 + 2000w^4).$$

Five independent points of infinite order have x-coordinates

$$x_1 = \frac{1}{9}(-1 + 10w)^2(1 + 10w)^2(1 + 80w^2)(-41 + 500w^2),$$

$$x_2 = \frac{1}{9}(-1 + 10w)(1 + 10w)(1 + 80w^2)(-31 + 400w^2)(-41 + 500w^2),$$

$$x_3 = \frac{1}{49}(1 + 80w^2)(-11 + 300w^2)^2(-41 + 500w^2),$$

$$x_4 = (1 + 80w^2)(-31 + 400w^2)(9 - 200w^2 + 2000w^4),$$

$$x_5 = 9(-1 + 10w)(1 + 10w)(-1 + 20w^2)(-41 + 500w^2).$$

$$\tag{3.25}$$

Finally, to obtain a curve of rank ≥ 6, we ask that

$$-9(-1 + 10w)^2(1 + 10w)^2(-1 + 20w^2)(-41 + 500w^2)$$

be the x-coordinate of a new point on the previous curve of rank 5. This gives us the condition that

$$- (-2 + 7w)(2 + 7w)$$

is a square, which again corresponds to a curve of genus 0, so we get a parametric solution

$$w = \frac{2(-1 + v)(1 + v)}{7(1 + v^2)}. \tag{3.26}$$

The resulting curve of rank ≥ 6 has the equation $Y^2 = X^3 + A(v)X^2 + B(v)X$, where

$$A(v) = -2(130752711 - 35202346632v^2 + 260292593988v^4 - 1337869740984v^6$$
$$+ 1975889131370v^8 - 1337869740984v^{10} + 260292593988v^{12}$$
$$- 35202346632v^{14} + 130752711v^{16}),$$

$$B(v) = -(9 - 80v + 9v^2)(9 + 80v + 9v^2)(-27 + 13v^2)^2(-13 + 27v^2)^2$$
$$\times (9 + 8018v^2 + 9v^4)(31 - 258v^2 + 31v^4)(369 - 542v^2 + 369v^4)$$
$$\times (14409 - 41564v^2 + 400054v^4 - 41564v^6 + 14409v^8).$$

Six independent points of infinite order have x-coordinates

$$x_1 = -\frac{1}{9}(-27 + 13v^2)^2(-13 + 27v^2)^2(9 + 8018v^2 + 9v^4)(369 - 542v^2 + 369v^4),$$

$$x_2 = -\frac{1}{9}(9 - 80v + 9v^2)(9 + 80v + 9v^2)(-27 + 13v^2)$$
$$\times (-13 + 27v^2)(9 + 8018v^2 + 9v^4)(369 - 542v^2 + 369v^4),$$

$$x_3 = -\frac{1}{49}(9 + 8018v^2 + 9v^4)(369 - 542v^2 + 369v^4)(661 - 3478v^2 + 661v^4)^2,$$

$$x_4 = (9 - 80v + 9v^2)(9 + 80v + 9v^2)(369 - 542v^2 + 369v^4)$$
$$\times (14409 - 41564v^2 + 400054v^4 - 41564v^6 + 14409v^8),$$

$$x_5 = -441(1 + v^2)^2(-27 + 13v^2)(-13 + 27v^2)(9 + 8018v^2 + 9v^4)$$
$$\times (31 - 258v^2 + 31v^4),$$

$$x_6 = -9(-27 + 13v^2)^2(-13 + 27v^2)^2(9 + 8018v^2 + 9v^4)(31 - 258v^2 + 31v^4),$$

and the corresponding quadruple $(\{c_1(a, n), c_2(a, n), c_3(a, n), c_4(a, n)\}$ after performing the substitutions) is $\{q_1, q_2, q_3, q_4\}$, where

$$q_1 = \frac{2(-27 + 13v^2)(-13 + 27v^2)}{21(9 + 178v^2 + 9v^4)},$$

$$q_2 = -\frac{(9 + 8018v^2 + 9v^4)(369 - 542v^2 + 369v^4)}{42(-27 + 13v^2)(-13 + 27v^2)(9 + 178v^2 + 9v^4)},$$

$$q_3 = -\frac{160(9 + 178v^2 + 9v^4)}{21(-27 + 13v^2)(-13 + 27v^2)},$$

$$q_4 = \frac{3(111 - 418v^2 + 111v^4)(237 + 2074v^2 + 237v^4)}{14(-27 + 13v^2)(-13 + 27v^2)(9 + 178v^2 + 9v^4)}.$$

By taking the appropriate specialization, we can easily check that the given six points are independent and that the rank over $\mathbb{Q}(v)$ is ≥ 6. However, using the algorithm developed by Gusić and Tadić [202], we can also show that the rank over $\mathbb{Q}(v)$ is exactly equal to 6. To do this, we need to find a specialization for which the specialization homomorphism is injective (by Silverman's specialization theorem [328, Theorem 11.4], we know that this is true for all but finitely many specializations, but the Gusić-Tadić algorithm allows us to find specializations that are certainly injective) and for which the rank is equal to 6. One such specialization is $v = 5$ for which the hypothesis of the following theorem from [202] is fulfilled.

Theorem 3.6.1 (Gusić-Tadić) *Let E be an elliptic curve over $\mathbb{Q}(t)$ (which is not isomorphic to a curve over $\mathbb{Q}$), given by the equation*

$$E : \quad y^2 = (x - e_1)(x - e_2)(x - e_3), \quad e_1, e_2, e_3 \in \mathbb{Z}[t].$$

Assume that $t_0 \in \mathbb{Q}$ satisfies the condition that for every non-constant square-free divisor h in $\mathbb{Z}[t]$ of $(e_1 - e_2)(e_1 - e_3)$, $(e_2 - e_1)(e_2 - e_3)$ and $(e_3 - e_1)(e_3 - e_2)$, the rational number $h(t_0)$ is not a square in $\mathbb{Q}$. Then the specialization homomorphism $\sigma : E(\mathbb{Q}(t)) \to E(t_0)(\mathbb{Q})$ is injective.

In our case, we have

$$(e_1 - e_2)(e_1 - e_3) =$$

$$- (9 - 80v + 9v^2)(9 + 80v + 9v^2)(-27 + 13v^2)^2(-13 + 27v^2)^2$$

$$\times (9 + 8018v^2 + 9v^4)(31 - 258v^2 + 31v^4)(369 - 542v^2 + 369v^4)$$

$$\times (14409 - 41564v^2 + 400054v^4 - 41564v^6 + 14409v^8).$$

Inserting $v = 5$ in the irreducible factors in this expression and by factoring the obtained integers, it is easy to see that none of the products gives a square. The conditions for $(e_2 - e_1)(e_2 - e_3)$ and $(e_3 - e_1)(e_3 - e_2)$ can be checked analogously.

That the rank of the curve obtained by specialization $v = 5$,

$$y^2 = x^3 + 2873456813402026684928x^2 + 1243229530387248524833275121109500959 1296x,$$

is indeed equal to 6, can be checked, for example, with the program `mwrank` or `ellrank` in PARI.

Combining the substitutions (3.24) and (3.26), we see that from the curve of rank 4 given by the equation $y^2 = x^3 + A(a)x^2 + B(a)x$ we can obtain a curve of rank 6 by the substitution

$$a = -\frac{2(-27 + 13v^2)(-13 + 27v^2)}{9(9 + 178v^2 + 9v^4)}.$$

In the papers [149, 150], four more substitutions are presented, which also give curves of rank 6:

$$a = -\frac{64(831744 - 40128v + 4288v^2 - 44v^3 + v^4)}{9(-1520 + 88v + v^2)(-2736 - 264v + 5v^2)},$$

$$a = \frac{10732176 + 628992v + 19192v^2 + 576v^3 + 9v^4}{36v(27 + v)(364 + 9v)},$$

$$a = \frac{5(-10 + 6v + v^2)(-18 - 18v + 5v^2)}{9(12 - 2v + v^2)(3 - v + v^2)},$$

$$a = \frac{5(584820 + 135432v - 18288v^2 + 396v^3 + 5v^4)}{9(684 - 66v + v^2)(171 - 33v + v^2)}.$$

The question arises whether some of these families of rank 6 have an infinite intersection and whether we can thus obtain an infinite family of curves of rank ≥ 7. To answer that question, we can compare their j-invariants j_i and j_k, factorize their difference $j_i(v_i) - j_k(v_k)$, and look whether a factor corresponding to a curve of genus ≤ 1 appears. Comparing the second and third of the above five substitutions, we find two such factors:

$$v_2^2 v_3^2 + 72v_2^2 v_3 + 88v_2 v_3^2 + 1820v_2^2 - 1520v_3^2$$
$$- 96096v_2 - 65664v_3 - 995904 = 0,$$

$$5v_2^2 v_3^2 + 216v_2^2 v_3 - 264v_2 v_3^2 + 3276v_2^2 - 2736v_3^2$$
$$+ 288288v_2 - 196992v_3 - 4979520 = 0.$$

Solving these equations with respect to v_2, we obtain in both cases the condition

$$54v_3^4 + 2736v_3^3 + 66592v_3^2 + 2987712v_3 + 64393056 = \square.$$

This quartic can be transformed in the standard way into the elliptic curve

$$y^2 = x^3 + x^2 - 28174550x + 45644288448,$$

which has rank 3, and therefore the elliptic curve and the quartic have infinitely many rational points. It can be shown (details are given in [150]) that in this way, we obtain infinitely many elliptic curves of rank ≥ 7 that are induced by rational Diophantine triples (the independence of the 7 points is checked again using the appropriate specialization, in this case $(v_2, v_3) = (-\frac{76}{3}, 26)$).

3.6.2   Torsion Group $\mathbb{Z}/2\mathbb{Z} \times \mathbb{Z}/2\mathbb{Z}$ over $\mathbb{Q}$

In this subsection, we will present the construction from [150] of an elliptic curve of rank 12 induced by a rational Diophantine triple, the current record for such curves. Previous records were 7 (already mentioned construction which uses rational Diophantine quintuples [99]), 9 (found among the curves induced by irregular subtriples of the quadruple $\{k - 1, k + 1, 4k, 16k^3 - 4k\}$ [110]) and 11 (found within families of rank ≥ 5 like those from the previous subsection [6, 149]).

We consider curves of the form

$$E' : \quad y^2 = (x + ab)(x + ac)(x + bc), \tag{3.27}$$

where $\{a, b, c\}$ is a rational Diophantine triple. We use the parametrization mentioned in Sect. 3.3:

$$a = \frac{2t_1(1 + t_1 t_2(1 + t_2 t_3))}{(-1 + t_1 t_2 t_3)(1 + t_1 t_2 t_3)},$$

$$b = \frac{2t_2(1 + t_2 t_3(1 + t_3 t_1))}{(-1 + t_1 t_2 t_3)(1 + t_1 t_2 t_3)},$$

$$c = \frac{2t_3(1 + t_3 t_1(1 + t_1 t_2))}{(-1 + t_1 t_2 t_3)(1 + t_1 t_2 t_3)}.$$

As we have already mentioned, we can expect such a curve to have rank ≥ 2 because we have points P and S on it. We observed experimentally that the rank is ≥ 3 if $t_3(t_3 - t_2)$ is a perfect square (and, cyclically, if $t_1(t_1 - t_3)$ or $t_2(t_2 - t_1)$ is a perfect square). Indeed, if we insert

$$x = -\frac{4(t_2^2 t_3 - t_3 + t_2)(t_3 t_1^2 t_2 + 1 + t_3 t_1)(t_2 t_3 + t_2 t_3^2 t_1 + 1)}{t_3(-1 + t_1 t_2 t_3)^2(1 + t_1 t_2 t_3)^2}$$

(note that $x + ab = \frac{b(c-b)}{t_2 t_3}$) in (3.27), we get

$$y^2 = 64(1 + t_3 t_1)^2 (t_1 t_2 t_3 - t_2 - t_2^2 t_3 + t_3)^2 (t_2 t_3 + t_2 t_3^2 t_1 + 1)^2 (1 + t_2 t_3)^2$$
$$\times (t_3 t_1^2 t_2 + 1 + t_3 t_1)^2 (t_3 - t_2) t_3^{-3} (-1 + t_1 t_2 t_3)^{-6} (1 + t_1 t_2 t_3)^{-6},$$

which gives exactly the condition that $t_3(t_3 - t_2)$ is a perfect square.

So, if we find a triple of rational numbers (t_1, t_2, t_3) such that

$$t_3(t_3 - t_2), \quad t_1(t_1 - t_3), \quad t_2(t_2 - t_1) \tag{3.28}$$

are all perfect squares, we may expect that the rank will jump at least by 3 and that we will get a curve with the rank ≥ 5.

One way to satisfy conditions (3.28) is by the so-called *almost perfect cuboids* (a *perfect cuboid* is one for which the lengths of all edges, face diagonals and space diagonal are positive integers – it is an open problem whether such a cuboid exists, and an almost perfect cuboid is one in which we drop one of the seven conditions). Indeed, if we put

$$t_3 = s_3^2, \ \ t_1 = -s_1^2, \ \ t_2 = s_2^2, \ \ s_3^2 - s_2^2 = s_4^2,$$

we have

$$s_1^2 + s_2^2 = \square, \ \ s_2^2 + s_4^2 = \square, \ \ s_1^2 + s_2^2 + s_4^2 = \square, \tag{3.29}$$

so we got an almost perfect cuboid (only one face diagonal is not an integer). Parametric solutions of system (3.29) are known (see, e.g. [350]):

$$s_1 = 2(m^2 + m + 1)(m^2 - 1)^2(m^2 + 1 + 4m),$$
$$s_2 = 4(m^2 + m + 1)(2m + 1)(m^2 - 1)(2m + m^2),$$
$$s_4 = (2m + 1)(2m + m^2)(3m^2 + 2m + 1)(m^2 + 2m + 3),$$

which gives us

$$t_1 = -4(m^2 + m + 1)^2(m^2 - 1)^4(m^2 + 1 + 4m)^2,$$
$$t_2 = 16(m^2 + m + 1)^2(2m + 1)^2(m^2 - 1)^2(2m + m^2)^2,$$
$$t_3 = m^2(2m + 1)^2(m + 2)^2(5m^2 + 8m + 5)^2(m^2 + 1)^2.$$

We will now present another way to obtain a two-parameter solution of (3.28), which will be more suitable for searching for higher-rank specializations. We can satisfy the first two conditions by putting

$$t_3(t_3 - t_2) = (t_3 + u)^2, \quad t_1(t_1 - t_3) = (t_1 + v)^2$$

so we get

$$t_2 = -\frac{u(2t_3 + u)}{t_3}, \qquad t_3 = -\frac{v(2t_1 + v)}{t_1}.$$

By inserting this in the third condition $t_2(t_2 - t_1) = \square$, we obtain

$$(8uv^2 - 2u^2v)t_1^3 + (u^4 - 8u^3v + 15u^2v^2 + 8uv^3)t_1^2$$
$$+ (-4u^3v^2 + 16u^2v^3 + 2uv^4)t_1 + 4u^2v^4 = \square. \tag{3.30}$$

Equation (3.30) can be understood as an elliptic curve over $\mathbb{Q}(u, v)$, with an obvious point $P_1 = (0, 2uv^2)$. If we take the point $2P_1$, we get

$$t_1 = \frac{v^2(-v + 16u)}{8u(-4v + u)},$$

which gives the triple

$$a = -\frac{v^2(-v + 16u)(16v^2 - 64u^2 - v^4 + 16uv^3 - 4v^5u + v^4u^2)}{u(v + 2)(v^2 - 2v + 4)(v - 2)(v^2 + 2v + 4)(2u - v)(2u + v)(-4v + u)},$$

$$b = \frac{16u(-4v + u)v(4v - 64u + 16uv^2 - 4u^2v - v^5 + 4u^2v^3)}{(v + 2)(v^2 - 2v + 4)(v - 2)(v^2 + 2v + 4)(2u - v)(2u + v)(-v + 16u)},$$

$$c = \frac{4(256uv - 64u^2 - 16v^4 + 64u^2v^2 + v^6 - 16v^5u)(2u - v)(2u + v)}{u(v + 2)(v^2 - 2v + 4)(v - 2)(v^2 + 2v + 4)(-v + 16u)(-4v + u)}.$$

Thus, we obtained an elliptic curve of rank ≥ 5 over $\mathbb{Q}(u, v)$. Indeed, if we write the equation of the curve in the form $y^2 = x^3 + Ax^2 + Bx$, where

$$A = v(256v^{13} - 32v^{15} + v^{17} + 140288v^9u^2 + 741888v^7u^4 - 4096v^{10}u - 1167360v^8u^3$$
$$- 21258240v^6u^5 - 7936v^{12}u + 664832v^{10}u^3 + 11440128v^8u^5 + 32192v^{11}u^2$$
$$- 2785824v^9u^4 - 32380416v^7u^6 + 28747776v^5u^6 + 6463488v^6u^7 + 71860224u^7v^4$$
$$- 2205696u^8v^5 + 1536v^{14}u - 24192v^{13}u^2 - 22528v^{12}u^3 + 591360v^{11}u^4$$
$$- 3244800u^5v^{10} - 128483328v^3u^8 - 12979200v^8u^7 + 7816v^{15}u^2 - 36160v^{14}u^3$$
$$- 8616v^{13}u^4 + 100992v^{12}u^5 - 128v^{16}u - 2023776v^{11}u^6 + 4v^{18}u - 449v^{17}u^2$$
$$+ 7824v^{16}u^3 - 31368v^{15}u^4 + 2860032v^{10}u^7 + 70176v^{14}u^5 + 112296v^{13}u^6$$
$$+ 9461760v^7u^8 - 2785824v^9u^8 - 332160v^{12}u^7 + 128188416v^2u^9 - 37027840v^4u^9$$
$$- 1441792v^6u^9 + 2659328v^8u^9 + 46368v^{11}u^8 - 6193152u^{10}v^5 + 515072u^{10}v^7$$
$$- 291840u^9v^{10} + 16818240v^9u^6 - 29425664vu^{10} + 32014336u^{10}v^3 + 140288v^9u^{10}$$

$$- 2097152u^{11}v^2 + 1572864u^{11}v^4 - 507904u^{11}v^6 - 16384u^{11}v^8 + 65536u^{12}v^5$$

$$+ 65536u^{12}v - 131072u^{12}v^3 + 1048576u^{11}),$$

$$B = 4(8vu^2 - 8u^2 + 16vu - v^2u + v^2 + 2v^3)(8vu^2 + 8u^2 - 16vu - v^2u - v^2 + 2v^3)$$

$$\times (-16v^2 + 64u^2 + v^4 - 16v^3u)(4v - 64u + 16v^2u - 4vu^2 - v^5 + 4v^3u^2)$$

$$\times (2vu^2 - 16u^2 + 2vu + 8v^2u - 4v^2 - v^3)(16vu - 4u^2 - v^4 + 4v^2u^2)$$

$$\times (16v^2 - 64u^2 - v^4 + 16v^3u - 4v^5u + v^4u^2)$$

$$\times (2vu^2 + 16u^2 - 2vu + 8v^2u + 4v^2 - v^3)(-v + 16u)^2(-4v + u)^2u^2v^3,$$

then the x-coordinates of five independent points of infinite order are:

$$x(P) = -4(4v - 64u + 16v^2u - 4vu^2 - v^5 + 4v^3u^2)(-4v + u)^2(-v + 16u)^2$$

$$\times (16v^2 - 64u^2 - v^4 + 16v^3u - 4v^5u + v^4u^2)u^2v^3,$$

$$x(R) = 4(16vu - 4u^2 - v^4 + 4v^2u^2)(16v^2 - 64u^2 - v^4 + 16v^3u - 4v^5u + v^4u^2)$$

$$\times (8vu^2 - 8u^2 + 16vu - v^2u + v^2 + 2v^3)(8vu^2 + 8u^2 - 16vu - v^2u - v^2 + 2v^3)$$

$$\times (-v + 16u)(-4v + u)vu,$$

$$x(T_1) = 16(16vu - 4u^2 - v^4 + 4v^2u^2)(2vu^2 - 16u^2 + 2vu + 8v^2u - 4v^2 - v^3)$$

$$\times (2vu^2 + 16u^2 - 2vu + 8v^2u + 4v^2 - v^3)(4v - 64u + 16v^2u - 4vu^2 - v^5 + 4v^3u^2)$$

$$\times (-v + 16u)(-4v + u)u,$$

$$x(T_2) = -4(8vu^2 - 8u^2 + 16vu - v^2u + v^2 + 2v^3)(-16v^2 + 64u^2 + v^4 - 16v^3u)$$

$$\times (16vu - 4u^2 - v^4 + 4v^2u^2)(8vu^2 + 8u^2 - 16vu - v^2u - v^2 + 2v^3)$$

$$\times (16v^2 - 64u^2 - v^4 + 16v^3u - 4v^5u + v^4u^2)(-4v + u)u/v^2,$$

$$x(T_3) = (4v - 64u + 16v^2u - 4vu^2 - v^5 + 4v^3u^2)(-16v^2 + 64u^2 + v^4 - 16v^3u)$$

$$\times (16v^2 - 64u^2 - v^4 + 16v^3u - 4v^5u + v^4u^2)(2u^2 + 8vu - v^2)^2(-v + 16u).$$

Here the point P corresponds to the point $(0, abc)$ on $y^2 = (x+ab)(x+ac)(x+bc)$, the point R satisfies $2R = S$, where S corresponds to the point $(1, rst)$ on $y^2 = (x + ab)(x + ac)(x + bc)$, the point T_1 corresponds to the condition $t_3(t_3 - t_2) = \square$, the point T_2 corresponds to the condition $t_1(t_1 - t_3) = \square$, while the point T_3 corresponds to the condition $t_2(t_2 - t_1) = \square$. To check the independence, it suffices to find a specialization (u_0, v_0) for which the points P, R, T_1, T_2 and T_3 are independent on the specialized curve over $\mathbb{Q}$. It can be checked that this is true, for example, for $(u_0, v_0) = (2, 1)$, since $(-1218332736, 2549361250008000)$, $(33504115024, -425318435221680)$, $(20402783424, -4098511134213120)$, $(-22588281936, -157553266503600)$, $(67893188319, -20235210553147590)$ are independent points on the curve $y^2 = x^3 + 21361758597x^2 - 28803989016278714304x$.

Now we are looking for specializations (u, v) for which the rank is significantly higher than 5. We combine some of the methods for the construction of elliptic curves of high rank, which we described in Sect. 2.5. We are looking for curves with a relatively large Mestre-Nagao sum

$$S(N, E) = \sum_{p=2}^{N} \frac{-a_p + 2}{p + 1 - a_p} \log p,$$

where $a_p = a_p(E) = p + 1 - |E(\mathbb{F}_p)|$, and with large 2-Selmer rank (we use mwrank with the -s option). In the search for curves of rank 12, we can additionally use the condition that $w_E = 1$ (via the PARI function ellrootno), because the parity conjecture implies that then the rank is even. In addition to the two-parameter family, we also search within certain one-parameter subfamilies (for example, for $u = v$). For curves that pass all tests, we can try to calculate the rank using the programs mwrank or ellrank.

We find curves of rank 11 for the following values of the parameters: $(u, v) =$

$$\left(\frac{11}{24}, \frac{5}{9}\right), \left(-\frac{145}{6}, \frac{29}{12}\right), \left(\frac{136}{19}, \frac{68}{5}\right), \left(-\frac{16}{77}, \frac{4}{21}\right), \left(\frac{473}{705}, \frac{43}{47}\right), \left(-\frac{89}{135}, \frac{89}{45}\right),$$

$$\left(-\frac{62}{43}, \frac{93}{43}\right), \left(\frac{71}{273}, \frac{142}{91}\right), \left(\frac{224}{67}, \frac{7}{2}\right), \left(-\frac{1032}{923}, \frac{172}{71}\right), \left(-\frac{1501}{87}, \frac{158}{87}\right),$$

$$\left(\frac{1358}{1007}, \frac{194}{53}\right), \left(-\frac{2072}{1819}, \frac{148}{107}\right), \left(\frac{454}{481}, \frac{227}{37}\right), \left(\frac{77}{173}, \frac{77}{173}\right), \left(\frac{163}{137}, \frac{163}{137}\right).$$

Let us mention that the curve corresponding to the parameters $(u, v) = (-\frac{62}{43}, \frac{93}{43})$, i.e. to the triple

$$\{a, b, c\} = \left\{\frac{21409906185}{74591676404}, -\frac{31580198976}{18647919101}, -\frac{10309975195}{18647919101}\right\},$$

with minimal Weierstrass equation

$$y^2 + xy = x^3 - x^2 - 21252276640652798739707819217x$$

$$+ 938627524108684110053910801619511357084941,$$

is the curve with the smallest discriminant among all known curves of rank 11 and torsion group $\mathbb{Z}/2\mathbb{Z} \times \mathbb{Z}/2\mathbb{Z}$.

Finally, we find one curve of rank 12, and that is for $(u, v) = (-\frac{95}{33}, \frac{50}{57})$, i.e.

$$\{a, b, c\} = \left\{\frac{6125241375}{11907531272}, \frac{5535371271425}{14277129995128}, -\frac{273138178560}{153430695649}\right\},$$

with minimal Weierstrass equation

$$y^2 + xy + y = x^3 - x^2 - 144449170752859135685608918646049119571126895 0880x$$
$$+ 5599215837796254212486835849395617624562242901704374615558514820414 39747,$$

the torsion points

$$O, \ (9109543899208458360 20349, -45547719496042291801 0175),$$
$$(-544872729119082402823 0629/4, \ 544872729119082402823 0625/8),$$
$$(45122743287686017103 7309, -225613716438430085518655),$$

and twelve independent points of infinite order

$$P_1 = (15885093250064960913 4809, 5783347758167145246162762217 04042845),$$
$$P_2 = (3511040172007843863 92209, 3098979669449451161946241983325 93845),$$
$$P_3 = (-4277226602909288139 83135, -10485766455261115281091856299487 86727),$$
$$P_4 = (9545007819393757627 42909, 22532600886334522054307161878 3370945),$$
$$P_5 = (4236795982596765919 90909, 15482981095954785259333298763596 6145),$$
$$P_6 = (15358084490958180942 07905, 14014214440804983803697855336 16999513),$$
$$P_7 = (4448018874220560215 35383, 73569216148613399817347986859 758945),$$
$$P_8 = (-12060060158710442786 78751, -740210245609217615143269452335454375),$$
$$P_9 = (-1925622924386935236 17091, -911556889640548767064630159456313855),$$
$$P_{10} = (10508879668527356682 921249, 33851800053181168926568362825476385625),$$
$$P_{11} = (9515144107333695556 70349, 21667652092127680529970331 1439049825),$$
$$P_{12} = (-7355680099955426717481581/81,$$
$$- 6057056719332256026906514463906338491 25/729).$$

Let us also mention that for $t_1 = \frac{44}{29}$, $t_2 = \frac{17}{42}$, $t_3 = \frac{3}{44}$, i.e. $a = \frac{815848}{164547}$, $b = \frac{1512524}{1810017}$, $c = \frac{32060}{201113}$, we get the curve

$$y^2 = x^3 + x^2 - 19393636089646994677 2176x$$
$$+ 294536412537181305061362295224 16740$$

of rank 10, the curve with the smallest conductor among all known curves of rank 10 and torsion group $\mathbb{Z}/2\mathbb{Z} \times \mathbb{Z}/2\mathbb{Z}$. This curve was found by a "brute force" search over all t_1, t_2, t_3 with small numerators and denominators.

3.6.3   Torsion Group $\mathbb{Z}/2\mathbb{Z} \times \mathbb{Z}/4\mathbb{Z}$

In this subsection, we will describe the construction of elliptic curves induced by Diophantine triples with torsion group $\mathbb{Z}/2\mathbb{Z} \times \mathbb{Z}/4\mathbb{Z}$ and rank 4 over $\mathbb{Q}(t)$, and rank 9 over $\mathbb{Q}$. These results were obtained in 2014 by Dujella and Peral [147] and later supplemented with an additional example of rank 9 by the same authors [151]. They still represent record ranks for curves with that torsion group (whether induced by a rational Diophantine triple or not). Another rank 9 curve was found, by other methods, in 2020 by Klagsbrun [163].

In Sect. 2.4, we saw that every elliptic curve with torsion group $\mathbb{Z}/2\mathbb{Z} \times \mathbb{Z}/4\mathbb{Z}$ is isomorphic to a curve of the form

$$y^2 = x(x + x_1^2)(x + x_2^2), \quad x_1, x_2 \in \mathbb{Q} \setminus \{0\}, \ x_1 \neq \pm x_2. \tag{3.31}$$

The point $(x_1 x_2, x_1 x_2 (x_1 + x_2))$ is a rational point on the curve of order 4.

The elliptic curve $y^2 = (ax + 1)(bx + 1)(cx + 1)$, induced by the rational Diophantine triple $\{a, b, c\}$, can be transformed in the standard way into $y^2 = (x + ab)(x + ac)(x + bc)$, and then, by translation, into the form

$$y^2 = x(x + ac - ab)(x + bc - ab). \tag{3.32}$$

By comparing Eqs. (3.31) and (3.32), we see that if we manage to find a rational Diophantine triple $\{a, b, c\}$ such that $ac - ab$ and $bc - ab$ are squares of rational numbers, then the elliptic curve induced by $\{a, b, c\}$ will have a torsion subgroup isomorphic to $\mathbb{Z}/2\mathbb{Z} \times \mathbb{Z}/4\mathbb{Z}$. We can expect that curve to have positive rank since (3.32) contains the point (ab, abc). A simple way to satisfy these two conditions is to choose a and b such that $ab = -1$. Then $ac - ab = ac + 1$ and $bc - ab = bc + 1$ will be squares according to the definition of a Diophantine triple. It remains to find a rational number c such that $\{a, -1/a, c\}$ is a rational Diophantine triple. We have the following system of equations

$$ac + 1 = p^2, \quad -\frac{c}{a} + 1 = q^2. \tag{3.33}$$

If we insert $ac + 1 = p^2$ in $-\frac{c}{a} + 1 = q^2$, we get

$$1 - p^2 + a^2 = z^2, \tag{3.34}$$

where $z = aq$. We can parametrize the rational solutions of equation (3.34) by putting

$$1 - p^2 + a^2 = (-a + (p - 1)\tau)^2.$$

We get a more "symmetric" parametrization if we put

$$p = \frac{\tau + \alpha}{\tau - \alpha},$$

and we obtain that

$$a = \frac{\alpha\tau + 1}{\tau - \alpha}.$$

If we insert this in (3.32), after a bit of rearranging, we get

$$y^2 = x^3 + 2(\alpha^2 + \tau^2 + 4\alpha^2\tau^2 + \alpha^4\tau^2 + \alpha^2\tau^4)x^2$$
$$+ (\tau + \alpha)^2(\alpha\tau - 1)^2(\tau - \alpha)^2(\alpha\tau + 1)^2 x. \tag{3.35}$$

As before, we try to get additional points on the curve of the form $y^2 = x^3 + Ax^2 + Bx$ by observing the factors of B. In this case, we want $x = (\tau + \alpha)^2(\alpha\tau - 1)(\alpha\tau + 1)$ to satisfy Eq. (3.35), which gives us the condition

$$\tau^2 + \alpha^2 + 2 = \square. \tag{3.36}$$

On the other hand, if we want that $x = (\tau + \alpha)(\alpha\tau - 1)^2(\tau - \alpha)$ satisfies Eq. (3.35), we get the condition

$$\alpha^2\tau^2 + 2\alpha^2 + 1 = \square. \tag{3.37}$$

We are looking for a parametric solution to the system of equations (3.36) and (3.37). According to the construction which led to this system, these parametric solutions should give a family of elliptic curves of rank at least 3. However, we will see that, with some luck, we get a family of rank 4.

Parametric solutions of equation (3.36) are known. We will take the solution given in [54, §10]:

$$\tau = \frac{r^2 - s^2 - 2t^2 + 2v^2}{2(rt + sv)}, \qquad \alpha = \frac{rs - 2tv}{rt + sv}. \tag{3.38}$$

By the homogeneity, it is clear that if we use rational parameters, we can eliminate one parameter. When we search for individual curves with a large rank, it will be convenient to work with four integer parameters. However, searching for a parametric family of high rank will be easier if (motivated by some experimental data) we assume that $v = 0$, $r = s + t + 1$. If we now insert (3.38) in (3.37), we get the quartic in the variable s:

$$(12t^2 + 8t + 4)s^4 + (12t^3 + 20t^2 + 12t + 4)s^3$$
$$+ (13t^4 + 12t^3 + 10t^2 + 4t + 1)s^2 + (8t^5 + 8t^4)s + 4t^6 + 8t^5 + 4t^4 = g^2. \tag{3.39}$$

Since it contains a rational point $(0, 2t^3 + 2t^2)$, it can be transformed in a standard way into an elliptic curve over $\mathbb{Q}(t)$:

$$
\begin{aligned}
h^2 = w^3 &+ (13t^4 + 12t^3 + 10t^2 + 4t + 1)w^2 \\
&+ (-96t^8 - 256t^6 - 256t^7 - 128t^5 - 32t^4)w \\
&- 1152t^{12} - 3840t^{11} - 5504t^{10} - 4608t^9 - 2432t^8 - 768t^7 - 128t^6.
\end{aligned}
\tag{3.40}
$$

By checking the factors of

$$
\begin{aligned}
&1152t^{12} + 3840t^{11} + 5504t^{10} + 4608t^9 + 2432t^8 + 768t^7 + 128t^6 \\
&= 128t^6(t + 1)^2(3t^2 + 2t + 1)^2
\end{aligned}
$$

as possible w-coordinates of points on (3.40), we find that the point $(4t^2(3t^2 + 2t + 1), 4t^2(t - 1)(3t + 1)(3t^2 + 2t + 1))$ lies on the elliptic curve given by (3.40). Transforming this point back to quartic (3.39), we find that

$$
s = -\frac{7t^3 + 9t^2 + 3t + 1}{t^2 + 6t + 3}.
$$

Then we easily compute:

$$
\tau = \frac{(3t^2 + 6t + 1)(5t^2 + 2t - 1)}{4t(t - 1)(3t + 1)(t + 1)},
$$

$$
\alpha = -\frac{(t + 1)(7t^2 + 2t + 1)}{t(t^2 + 6t + 3)},
$$

$$
a = -\frac{(t + 1)(31t^4 + 52t^3 + 22t^2 - 4t - 1)(3t^2 + 2t + 1)}{t(11t^4 + 12t^3 + 2t^2 - 4t - 1)(9t^2 + 14t + 7)},
$$

$$
b = \frac{t(11t^4 + 12t^3 + 2t^2 - 4t - 1)(9t^2 + 14t + 7)}{(t + 1)(31t^4 + 52t^3 + 22t^2 - 4t - 1)(3t^2 + 2t + 1)},
$$

$$
c = \big(16(t - 1)(3t + 1)(t + 1)t(t^2 + 6t + 3)(3t^2 + 6t + 1)(5t^2 + 2t - 1)(7t^2 + 2t + 1)\big)/
$$

$$
\big((11t^4 + 12t^3 + 2t^2 - 4t - 1)(9t^2 + 14t + 7)(31t^4 + 52t^3 + 22t^2 - 4t - 1)(3t^2 + 2t + 1)\big).
$$

Now we claim that the induced elliptic curve

$$
E : \quad y^2 = x^3 + A(t)x^2 + B(t)x,
$$

where

$$
\begin{aligned}
A(t) = 2(&87671889t^{24} + 854321688t^{23} + 3766024692t^{22} + 9923033928t^{21} \\
&+ 17428851514t^{20} + 21621621928t^{19} + 19950275060t^{18}
\end{aligned}
$$

$$+ 15200715960t^{17} + 11789354375t^{16} + 10470452464t^{15} + 8925222696t^{14}$$

$$+ 5984900048t^{13} + 2829340620t^{12} + 820299856t^{11} + 59930952t^{10}$$

$$- 66320528t^{9} - 35768977t^{8} - 9381000t^{7} - 1017244t^{6} + 262760t^{5}$$

$$+ 159130t^{4} + 41096t^{3} + 6468t^{2} + 600t + 25),$$

$$B(t) = (t^2 - 2t - 1)^2(69t^4 + 148t^3 + 78t^2 + 4t + 1)^2(13t^2 - 2t - 1)^2$$

$$\times (9t^4 + 28t^3 + 18t^2 + 4t + 1)^2(11t^4 + 12t^3 + 2t^2 - 4t - 1)^2$$

$$\times (9t^2 + 14t + 7)^2(31t^4 + 52t^3 + 22t^2 - 4t - 1)^2(3t^2 + 2t + 1)^2,$$

has rank ≥ 4 over $\mathbb{Q}(t)$. Indeed, it contains four points with x-coordinates

$$X_1 = (9t^4 + 28t^3 + 18t^2 + 4t + 1)^2(11t^4 + 12t^3 + 2t^2 - 4t - 1)^2$$

$$\times (69t^4 + 148t^3 + 78t^2 + 4t + 1)^2,$$

$$X_2 = (3t^2 + 2t + 1)(9t^2 + 14t + 7)^2(13t^2 - 2t - 1)$$

$$\times (9t^4 + 28t^3 + 18t^2 + 4t + 1)(11t^4 + 12t^3 + 2t^2 - 4t - 1)^2$$

$$\times (31t^4 + 52t^3 + 22t^2 - 4t - 1),$$

$$X_3 = (3t^2 + 2t + 1)(9t^2 + 14t + 7)^2(13t^2 - 2t - 1)$$

$$\times (9t^4 + 28t^3 + 18t^2 + 4t + 1)^2(11t^4 + 12t^3 + 2t^2 - 4t - 1)$$

$$\times (69t^4 + 148t^3 + 78t^2 + 4t + 1),$$

$$X_4 = -(3t^2 + 2t + 1)^2(9t^2 + 14t + 7)^2(11t^4 + 12t^3 + 2t^2 - 4t - 1)^2$$

$$\times (31t^4 + 52t^3 + 22t^2 - 4t - 1)^2.$$

Let us observe that the point X_4 corresponds to point $(-1, -c)$ on curve (3.32). Other points are found by looking for points on E with x-coordinates that are divisors of polynomial $B(t)$. An appropriate specialization, e.g. $t = 2$, proves that the points P_1, P_2, P_3, P_4, with x-coordinates X_1, X_2, X_3, X_4, are independent points of infinite order. Therefore, we obtained an elliptic curve over $\mathbb{Q}(t)$ with the torsion group $\mathbb{Z}/2\mathbb{Z} \times \mathbb{Z}/4\mathbb{Z}$ and rank ≥ 4. Let us mention that the previous records with curves of rank ≥ 3 were found by Lecacheux [250], Elkies [162], and Eroshkin [165].

Using the Gusić-Tadić algorithm [202], it can be shown that $\mathrm{rank}(E(\mathbb{Q}(t))) = 4$ and moreover that the points P_1, P_2, P_3, P_4 are free generators of $E(\mathbb{Q}(t))$. For this purpose, we write the equation of the curve E in the form

$$y^2 = (x - e_1)(x - e_2)(x - e_3), \quad e_1, e_2, e_3 \in \mathbb{Z}[t],$$

we observe the square-free factors of $(e_1 - e_2)(e_1 - e_3)$, $(e_2 - e_1)(e_2 - e_3)$ and $(e_3 - e_1)(e_3 - e_2)$, and look for $t_0 \in \mathbb{Q}$ which satisfies the conditions of Theorem 3.6.1. Then the specialization homomorphism $E(\mathbb{Q}(t)) \to E(t_0)(\mathbb{Q})$ will be injective. Furthermore, if $|E(t_0)(\mathbb{Q})_{\text{tors}}| = 8$ and there are points $Q_1, \ldots, Q_r \in E(\mathbb{Q}(t))$ such that $Q_1(t_0), \ldots, Q_r(t_0)$ are free generators of $E(t_0)(\mathbb{Q})$, then the specialization homomorphism $E(\mathbb{Q}(t)) \to E(t_0)(\mathbb{Q})$ is an isomorphism. Therefore, $E(\mathbb{Q}(t))$ and $E(t_0)(\mathbb{Q})$ have the same rank r and $Q_1, \ldots, Q_r$ are free generators of $E(\mathbb{Q}(t))$. It is not difficult to verify that $t_0 = 15$ satisfies the conditions of Theorem 3.6.1. Using `mwrank`, we get that $\text{rank}(E(15)(\mathbb{Q})) = 4$, and thus we proved that

$$\text{rank}(E(\mathbb{Q}(t))) = 4.$$

Furthermore, `mwrank` also finds free generators R_1, R_2, R_3, R_4 of $E(15)(\mathbb{Q})$. If we express $P_1(15), P_2(15), P_3(15), P_4(15)$ in the basis R_1, R_2, R_3, R_4 (modulo torsion), we get that the determinant of the transformation matrix is equal to -1. Therefore, $P_1(15), P_2(15), P_3(15), P_4(15)$ also form a Mordell-Weil basis for $E(15)(\mathbb{Q})$, and we conclude that indeed P_1, P_2, P_3, P_4 are free generators of $E(\mathbb{Q}(t))$.

Let us now show how to obtain a curve induced by a rational Diophantine triple with the torsion group $\mathbb{Z}/2\mathbb{Z} \times \mathbb{Z}/4\mathbb{Z}$ and rank 9 over $\mathbb{Q}$. Previously, such curves of rank 7 were known [110], and were obtained, with the above notation, for $\alpha = 2$. To obtain curves of rank 9, we will search for curves corresponding to τ and α of the form (3.38). As we have already mentioned, those curves of rank 9 do not only represent the record among curves induced by rational Diophantine triples, but also among all curves with the torsion group $\mathbb{Z}/2\mathbb{Z} \times \mathbb{Z}/4\mathbb{Z}$. Previous records of rank 8 were found, by other methods, by Elkies, Eroshkin, and Eroshkin and Dujella (see [118]).

In [147], Dujella and Peral searched over all integer parameters r, s, t, v in the range $|r| + |s| + |t| + |v| \leq 420$. Previously described methods for "sieving" and finding suitable candidates for a high rank were used, which include computing the Mestre-Nagao sums, the 2-Selmer rank and Mestre's conditional upper bound for the rank. Then the rank for these candidates was computed using `mwrank`. In this way, five curves of rank 8 were found, corresponding to the parameters $(r, s, t, v) = (20, -11, 25, 68)$, $(82, 9, 73, 30)$, $(55, 31, 142, 15)$, $(91, 55, 33, 104)$, $(157, 127, 43, 12)$, and one curve of rank 9, which corresponds to the parameters $(r, s, t, v) = (155, 54, 96, 106)$. That curve is induced by the Diophantine triple

$$\left\{ \frac{301273}{556614}, \; -\frac{556614}{301273}, \; -\frac{535707232}{290125899} \right\} .$$

Its minimal Weierstrass equation is

$$y^2 = x^3 + x^2 - 6141005737705911671519806644217969840x$$

$$+ 5857433177348803158586285785929631477808095171159063188,$$

the torsion points are

$$O,\ (-2861469472720778854,\ 0),$$

$$(1431017969855150171,\ 0),\ (1430451502865628682,\ 0),$$

$$(1381707195787460036,\ -100990010591667129753450630),$$

$$(1381707195787460036,\ 100990010591667129753450630),$$

$$(1480328743922840306,\ -103337259355706972940063720),$$

$$(1480328743922840306,\ 103337259355706972940063720),$$

while nine independent points of infinite order are

$$(-612695149795875652,\ 3064309824349077381027308358),$$

$$(-431590874944672564,\ 2903005768083873104158859430),$$

$$(187501554154394546,\ 2170847073897415394832351000),$$

$$(-1383500708967173302,\ 3421314943163837774567917408),$$

$$(1428519047239049546,\ 4551549120021779137548000),$$

$$(1430248713837731282,\ 8182260000869154831593640),$$

$$(1429305792931194266,\ 2901212522992755483557760),$$

$$(103900694057898826,\ 2284841365124562079087206240),$$

$$(1429854291102331316,\ 1726936504767203175719910).$$

The same curve is obtained for the parameters $(r, s, t, v) = (82, -19, 87, 14)$, i.e.
for the rational Diophantine triple

$$\left\{ -\frac{126555}{2686},\ \frac{2686}{126555},\ -\frac{9107022944}{249946125} \right\}.$$

In [149], the search was extended for various special values of parameters
r, s, t, v outside the previously mentioned range. Thus, five more curves of rank
8 were found, corresponding to the parameters $(r, s, t, v) = (131, -29, 49, 96)$,
$(186, -57, 62, 199)$, $(107, 107, 149, 430)$, $(103, 103, 168, 725)$, and $(749, 749,
138, 245)$.

A slightly different construction was used in [151] to obtain several new curves
of rank 8 and one new curve of rank 9. To satisfy condition (3.36), we may take

$$\tau^2 + \alpha^2 + 2 = \left(\tau + \frac{1}{\tau} + \left(\alpha - \frac{1}{\tau} \right) w \right)^2,$$

so we get

$$\alpha = \frac{-2\tau^2 w + 1 - 2w + w^2}{\tau(w^2 - 1)}.$$

Motivated by some experiments, $\tau = 1/7$ is taken as a possible good choice for obtaining curves with higher rank. By searching for $w = w_1/w_2$ in the range $\max(|w_1|, |w_2|) < 3000$, seven examples of rank equal to 8 (for $w = 201/170$, $245/138$, $-800/459$, $1610/1417$, $1955/1754$, $2254/2215$, $2301/1670$), and one example of rank equal to 9 (for $w = 900/781$) were found. The last curve is induced by the Diophantine triple

$$\left\{ \frac{181800}{127673}, \; -\frac{127673}{181800}, \; -\frac{996869751703}{2072406375000} \right\}.$$

Its minimal Weierstrass equation is

$$y^2 + xy = x^3 - 144370584236849299130167544556987839128626 0x$$

$$+ 5185245341269541163221532255116623986091375865258779782416196 00.$$

3.6.4   *Torsion Group* $\mathbb{Z}/2\mathbb{Z} \times \mathbb{Z}/6\mathbb{Z}$

Elliptic curves induced by rational Diophantine triples with torsion group $\mathbb{Z}/2\mathbb{Z} \times \mathbb{Z}/6\mathbb{Z}$ were observed for the first time in [110]. A curve over $\mathbb{Q}(t)$ of rank ≥ 1 and a curve over $\mathbb{Q}$ of rank 4 were constructed. Current records for the rank of elliptic curves with torsion group $\mathbb{Z}/2\mathbb{Z} \times \mathbb{Z}/6\mathbb{Z}$ (regardless of whether the curves are induced by rational Diophantine triples) are rank 2 over $\mathbb{Q}(t)$ and rank 6 over $\mathbb{Q}$. We will show that it is possible to obtain them using elliptic curves induced by rational Diophantine triples (although in the case of record curves over $\mathbb{Q}$, they were not initially discovered in that way).

We will use the idea from Sect. 3.2, where we saw that the condition that the point S is of order 3 leads to Eq. (3.2), which can be written as

$$- a^2 b^2 c^4 + (2a^3 b^2 + 2a^2 b^3)c^3 + (6ab + 12a^2 b^2 - a^4 b^2 - a^2 b^4 + 2a^3 b^3)c^2$$

$$+ (4a + 4b + 6a^2 b + 6ab^2)c + 3 + 4ab = 0. \tag{3.41}$$

Let a and b be such that $c = 0$ is one of the solutions of equation (3.41). We get $4ab + 3 = 0$, i.e. $b = -\frac{3}{4a}$. If we insert this in the condition that $bc + 1$ is a square, we get that $a(4a - 3c)$ is a square, say $a(4a - 3c) = (2a + t)^2$, i.e.

$$c = -\frac{t(4a + t)}{3a}.$$

By inserting the obtained expression for b and c, we see that Eq. (3.41) will be satisfied if

$$36a^2t + (-24 + 24t^2)a + 4t^3 - 9t = 0.$$

From the condition that the discriminant of this equation (in a) is a square, we get that $t^2 + 4$ is a square, i.e. $t = \frac{4u}{u^2-1}$. We obtain

$$a = \frac{u^3 - 9u}{6(u^2 - 1)},$$

$$b = -\frac{18(u^2 - 1)}{4(u^3 - 9u)},$$

and three non-zero solutions of the equation for c:

$$c_1 = \frac{9 - 12u - 18u^2 + 4u^3 + u^4}{6u(u^2 + 2u - 3)},$$

$$c_2 = \frac{9 + 12u - 18u^2 - 4u^3 + u^4}{6u(u^2 - 2u - 3)},$$

$$c_3 = -\frac{16u(u^2 - 3)}{3(u^4 - 10u^2 + 9)}.$$

It is easy to check that $\{a, b, c_1, c_2, c_3\}$ is a rational Diophantine quintuple. Since the triples $\{a, b, c_1\}$, $\{a, b, c_2\}$ and $\{a, b, c_3\}$ satisfy Eq. (3.41), the elliptic curves induced by these triples have the torsion group $\mathbb{Z}/2\mathbb{Z} \times \mathbb{Z}/6\mathbb{Z}$ and rank ≥ 1 over $\mathbb{Q}(u)$. These three curves are birationally equivalent, so in what follows, we will only consider the curve induced by the triple $\{a, b, c_1\}$. This curve can be written in the form $y^2 = x^3 + A(u)x^2 + B(u)x$, where

$$A(u) = 81 - 162u + 54u^2 + 162u^3 + 162u^4 - 54u^5 + 6u^6 + 6u^7 + u^8,$$

$$B(u) = -27(-1 + u)^3 u^3 (3 + u)^3 (6 - 3u + u^2)(3 + 3u + 2u^2).$$

A point of infinite order is

$$P(u) = (-27(-1 + u)^2 u^2 (3 + u)^2,$$

$$27(-3 + u)(-1 + u)^2 u^2 (1 + u)(3 + u)^2 (-3 + 6u + u^2)).$$

In order to increase the rank, we set the condition that

$$\frac{4(-1 + u)u^2(3 + u)^2(6 - 3u + u^2)(3 + 3u + 2u^2)}{(-3 + u)^2}$$

is the x-coordinate of a new point on the curve, and we get the condition that $4 - u + u^2$ is a square, which we can parametrized with $u = -\frac{(-2+w)(2+w)}{1+2w}$. Thus, we obtained a curve of rank ≥ 2 over $\mathbb{Q}(w)$. We can write the curve in the form $y^2 = x^3 + A_1(w)x^2 + B_1(w)x$, where

$$A_1(w) = 185257 + 401184w + 530914w^2 + 1218012w^3$$
$$+ 1041238w^4 - 925272w^5$$
$$- 1369658w^6 + 53676w^7 + 519130w^8$$
$$+ 93984w^9 - 59978w^{10} - 12924w^{11}$$
$$+ 3286w^{12} + 792w^{13} - 14w^{14} - 12w^{15} + w^{16},$$

$$B_1(w) = 27(-7+w)^3(-2+w)^3(-1+w)^3(1+w)^3(2+w)^3(3+w)^3(1+2w)^3$$
$$\times (10 + 19w^2 + 6w^3 + w^4)(47 + 36w - 7w^2 - 6w^3 + 2w^4).$$

Two independent points of infinite order have x-coordinates

$$x_1 = -27(-7+w)^2(-2+w)^2(-1+w)^2(1+w)^2(2+w)^2(3+w)^2(1+2w)^2,$$

$$x_2 = -\frac{4}{(-1+6w+w^2)^2}(-7+w)^2(-2+w)^2(-1+w)(1+w)^2(2+w)^2$$
$$\times (3+w)(1+2w)(10+19w^2+6w^3+w^4)$$
$$\times (47 + 36w - 7w^2 - 6w^3 + 2w^4).$$

The specialization $w_0 = 57$ satisfies the conditions of Theorem 3.6.1, which shows that the rank over $\mathbb{Q}(w)$ is equal to 2.

In this family of curves, we can find examples of curves of rank 5 over $\mathbb{Q}$ for the following values of the parameter: $w = \frac{1}{7}, \frac{12}{11}, \frac{33}{14}$.

Another way to obtain a curve of rank 2 from the initial curve of rank 1 is to set the condition that

$$\frac{4(-1+u)u^3(3+u)^3(3+3u+2u^2)}{(1+u)^2}$$

be the x-coordinate of a new point on the curve. This gives the condition that $2 + u + u^2$ is a square, i.e. $u = \frac{2-w^2}{-1+2w}$. Thus we get a new curve with the torsion group $\mathbb{Z}/2\mathbb{Z} \times \mathbb{Z}/6\mathbb{Z}$ and rank ≥ 2 over $\mathbb{Q}(w)$. The coefficients of that curve are

$$A_2(w) = 3517 - 26568w + 74462w^2 - 102612w^3 + 138610w^4 - 283680w^5$$
$$+ 346730w^6 - 107316w^7 - 122138w^8 + 84648w^9$$
$$+ 7418w^{10} - 15084w^{11}$$
$$+ 1690w^{12} + 576w^{13} + 14w^{14} - 12w^{15} + w^{16},$$

$$B_2(w) = 27(-1+w)^3(3+w)^3(-1+2w)^3(-2+w^2)^3(1-6w+w^2)^3$$

$$\times\,(16-36w+17w^2+6w^3+w^4)(5+7w^2-6w^3+2w^4),$$

while two independent points of infinite order have the following x-coordinates:

$$x_1 = -27(-1+w)^2(3+w)^2(-1+2w)^2(-2+w^2)^2(1-6w+w^2)^2,$$

$$x_2 = -\frac{4}{(-1-2w+w^2)^2}(-1+w)(3+w)(-1+2w)(-2+w^2)^3$$

$$\times\,(1-6w+w^2)^3(5+7w^2-6w^3+2w^4).$$

The specialization $w_0 = 18$ satisfies the conditions of Theorem 3.6.1, which shows that the rank over $\mathbb{Q}(w)$ is equal to 2.

For $w = \frac{7}{19}$, the corresponding curve has rank 6 over $\mathbb{Q}$. This curve is isomorphic to the curve previously found by Elkies in 2006 by searching within the general family of curves with torsion group $\mathbb{Z}/2\mathbb{Z} \times \mathbb{Z}/6\mathbb{Z}$, and represents the record rank for this torsion group. We see that this record curve can also be obtained by rational Diophantine triples. More precisely, it is induced by the rational Diophantine triple

$$\left\{\frac{31269599}{31628160},\ -\frac{23721120}{31269599},\ \frac{1461969791}{7144352640}\right\}.$$

Additional examples of curves with torsion group $\mathbb{Z}/2\mathbb{Z} \times \mathbb{Z}/6\mathbb{Z}$ and rank 6 over $\mathbb{Q}$ were found by Dujella, Peral, and Tadić in 2015 and Dujella and Peral in 2020, so four such curves are known today (see [118] for details on these curves).

We will obtain the third family of curves with torsion group $\mathbb{Z}/2\mathbb{Z} \times \mathbb{Z}/6\mathbb{Z}$ and rank 2 by starting with the rational Diophantine triple $\{a, b, c\}$ from the parametric family of rational Diophantine sextuples from Sect. 3.2. That is the triple $\{a, b, c\}$, where

$$a = \frac{18t(t-1)(t+1)}{(t^2-6t+1)(t^2+6t+1)},$$

$$b = \frac{(t-1)(t^2+6t+1)^2}{6t(t+1)(t^2-6t+1)},$$

$$c = \frac{(t+1)(t^2-6t+1)^2}{6t(t-1)(t^2+6t+1)}.$$

We know that it satisfies the condition that the point S is of order 3, and therefore the elliptic curve induced by the triple $\{a, b, c\}$ has torsion group $\mathbb{Z}/2\mathbb{Z} \times \mathbb{Z}/6\mathbb{Z}$ and rank ≥ 1 over $\mathbb{Q}(t)$. We can write that curve in the form $y^2 = x^3 + A_3(t)x^2 + B_3(t)x$, where

$$A_3(t) = t^8 - 212t^6 - 5184t^5 - 4314t^4 - 5184t^3 - 212t^2 + 1,$$

$$B_3(t) = -3456t^3(t^2 - 18t + 1)(t^2+1)(t^2+6t+1)^3,$$

To increase the rank, we look for the x-coordinate of an additional point in the form $kt(t+1)^2(t^2+6t+1)^2$, where k is a constant. We find candidates for k as roots of the discriminant of the expression obtained by inserting x into the cubic polynomial from the equation of the elliptic curve (after eliminating the quadratic factors). So we find that a suitable candidate is $k=27$ (because the factor $(k-27)^7$ appears in the factorization of the discriminant), and we get the condition that t^2-3t+1 is a square, i.e. $t=-\frac{(-1+w)(1+w)}{3+2w}$. After rearranging, the equation of the curve becomes $y^2=x^3+A_4(w)x^2+B_4(w)x$, where

$$A_4(w) = -1899008 - 5996544w - 1469440w^2 + 13980672w^3 + 14383360w^4$$

$$- 6990336w^5 - 16274176w^6 - 3491328w^7 + 5554240w^8 + 2814336w^9$$

$$- 219616w^{10} - 160416w^{11} + 120808w^{12} + 38928w^{13} - 856w^{14} + w^{16},$$

$$B_4(w) = 3456(-1+w)^3(1+w)^3(3+2w)^3(-2+w^2)^3(-14-12w+w^2)^3$$

$$\times (10+12w+2w^2+w^4)(-44-24w+56w^2+36w^3+w^4),$$

while the x-coordinates of two independent points of infinite order are

$$x_1 = \frac{108(-1+w)^2(1+w)^2(3+2w)^2(-2+w^2)^3(-14-12w+w^2)^3(2+2w+w^2)^2}{(10+12w+2w^2+w^4)^2},$$

$$x_2 = -27(-1+w)(1+w)(3+2w)(-2+w^2)^2$$

$$\times (-14-12w+w^2)^2(-4-2w+w^2)^2.$$

The specialization $w_0=14$ satisfies the conditions of Theorem 3.6.1, which shows that the rank over $\mathbb{Q}(w)$ is equal to 2.

For $w=-39$, we obtain again the previously mentioned Elkies' curve of rank 6 with the equation

$$Y^2 + XY = X^3 - 3768095670099922608026398200571309 0640\,X$$

$$- 36992898397926078743894505902555362159162611772488902400,$$

this time induced by the rational Diophantine triple

$$\left\{\frac{7567037280}{7833785281},\ \frac{4161669360289}{569762123040},\ \frac{1359453258559}{948852707040}\right\}.$$

Similarly as in Sect. 3.6.1, we can obtain an infinite family of elliptic curves with the torsion group $\mathbb{Z}/2\mathbb{Z}\times\mathbb{Z}/6\mathbb{Z}$ and rank ≥ 3 as the intersection of two curves over $\mathbb{Q}(t)$ of rank ≥ 2. To find the intersection of the curves $y^2=x^3+A_1(w)x^2+B_1(w)x$

and $y^2 = x^3 + A_2(w)x^2 + B_2(w)x$, we have to solve the equation $-\frac{(-2+w_1)(2+w_1)}{1+2w_1} = \frac{2-w_2^2}{-1+2w_2}$, i.e.

$$(2w_2 - 1)w_1^2 + (-2w_2^2 + 4)w_1 - w_2^2 - 8w_2 + 6 = 0. \tag{3.42}$$

The discriminant of this equation should be a perfect square, i.e.

$$4(w_2^4 + 2w_2^3 + 11w_2^2 - 20w_2 + 10) = \square.$$

Since this quartic has a rational point (e.g. the point at infinity), it can be transformed into the elliptic curve $y^2 = x^3 - x^2 - 120x + 432$ of rank equal to 1 (a generator is $(2, 14)$, which corresponds to $w_1 = -\frac{3}{2}$, $w_2 = \frac{9}{4}$, $u = -\frac{7}{8}$). This implies the existence of infinitely many rational solutions of (3.42), and the independence of the corresponding points shows that there exist infinitely many elliptic curves with the torsion group $\mathbb{Z}/2\mathbb{Z} \times \mathbb{Z}/6\mathbb{Z}$ and rank ≥ 3. Note that the heuristic from [302] predicts that this is the best possible result for this torsion group (see Table 2.4).

3.6.5   Torsion Group $\mathbb{Z}/2\mathbb{Z} \times \mathbb{Z}/8\mathbb{Z}$

In Theorem 2.4.11, we proved that every elliptic curve over $\mathbb{Q}$ with the torsion group $\mathbb{Z}/2\mathbb{Z} \times \mathbb{Z}/8\mathbb{Z}$ is induced by some rational Diophantine triple. More precisely, it is induced by a triple of the form

$$\left\{ \frac{2t}{t^2 - 1}, \frac{1 - t^2}{2t}, \frac{6t^2 - t^4 - 1}{2t(t^2 - 1)} \right\}. \tag{3.43}$$

The elements of this triple obviously have mixed signs since the product of the first two elements is equal to -1. However, the same curve can be induced by several different triples (explicit examples are given in Sect. 3.6.4). Thus, we may ask if there is a curve with torsion group $\mathbb{Z}/2\mathbb{Z} \times \mathbb{Z}/8\mathbb{Z}$, induced by a rational Diophantine triple that has all elements positive. We will show that the answer is affirmative.

Let us consider the family of rational Diophantine triples

$$\left\{ \frac{2(2w^4 + 4w^3 + 2w^2 + 1)(2w + 1)}{w(2w^2 + 2w - 1)(w + 2)(w - 1)(w + 1)}, -\frac{2(w + 1)^2(w - 1)^2}{w(w + 2)(2w^2 + 2w - 1)}, \right.$$
$$\left. \frac{2w^2(w + 2)^2}{(2w^2 + 2w - 1)(w - 1)(w + 1)} \right\}. \tag{3.44}$$

The family is obtained by factorization of the difference of the j-invariant of the curve induced by the general three-parameter rational Diophantine triple and j-invariant of the curve induced by $\{T, -1/T, T - 1/T\}$ which lacks only the condition that $T^2 + 1$ is a square in order to have the torsion group $\mathbb{Z}/2\mathbb{Z} \times \mathbb{Z}/8\mathbb{Z}$. By comparing the j-invariant of the elliptic curve induced by (3.44) with the j-invariant of the curve induced by (3.43), we obtain the condition $2w^4 + 4w^3 + 2w^2 + 1 = \square$. The curve defined by this equation can be transformed into the elliptic curve $Y^2 = X^3 - X^2 - 9X + 9$, which has rank 1 and whose generator is $P = (0, 3)$. By taking appropriate multiples of the point P, for example, $3P, 5P, 8P$, we get rational Diophantine triples with positive elements. For example, let us take the point $3P = (-360/169, -8211/2197)$. If we transform it back to the quartic, we get $w = 26/89$, and by inserting this value for w in the above family of Diophantine triples, we get the following rational Diophantine triple with positive elements

$$\{a, b, c\} = \left\{ \frac{37471518967}{1381254420}, \frac{5832225}{571948}, \frac{6251648}{1562505} \right\}$$

which induces an elliptic curve with torsion group $\mathbb{Z}/2\mathbb{Z} \times \mathbb{Z}/8\mathbb{Z}$ and rank 1. In this way, we get an infinite family of elliptic curves with torsion group $\mathbb{Z}/2\mathbb{Z} \times \mathbb{Z}/8\mathbb{Z}$ and positive rank (since the point $(0, abc)$ is of infinite order). It is the best-known result of this type (it is not known if there is an infinite family with this torsion group and rank ≥ 2), and this family is equivalent to the family found by Lecacheux in 2004 [251]. We will consider whether there is a rational Diophantine triple with positive elements that induces an elliptic curve with torsion group $\mathbb{Z}/2\mathbb{Z} \times \mathbb{Z}/8\mathbb{Z}$ and rank 0 in the next section.

As for elliptic curves over $\mathbb{Q}$ with torsion group $\mathbb{Z}/2\mathbb{Z} \times \mathbb{Z}/8\mathbb{Z}$, it is known that there exist such elliptic curves of rank 3, and it is not known whether such a curve of rank ≥ 4 exists. In total, there are 34 known curves of rank 3. The first was found by Connell and Dujella in 2000, and the last three curves by AttarBashi, Fisher, Rathbun, and Voznyy in 2022 (see [118] for details). According to Theorem 2.4.11, all such curves are induced by some rational Diophantine triple. For example, the first mentioned curve, whose equation is

$$y^2 + xy = x^3 - 15745932530829089880x$$

$$+ 24028219957095969426339278400,$$

and independent points of infinite order are

$$(2188064030, -7124272297330),$$

$$(-2815745040, -214568724545880),$$

$$(3643261410, -122557804465830),$$

is induced by the rational Diophantine triple

$$\left\{ \frac{408}{145}, -\frac{145}{408}, -\frac{145439}{59160} \right\},$$

while one of the last found curves, whose equation is

$$y^2 + xy = x^3 - 125773582955215019659843433869443803612902341280x$$
$$+ 1654578042723445369075561816231366519388305306314247517581394 2418662400,$$

and independent points of infinite order are

$$(17717758251775551568432211692130/11826721,$$
$$726472136272254107745712870940674199960845481630/40672093519),$$

$$(2526555996702832707895123084069857683030588014705814971872515 4949$$
$$874724706640/79079121352075128626675914808286270045501129127855369,$$
$$2106765868893950319477043345162195244251673901514805587788796 83355$$
$$13789472048810172905201438061214086802058655 82560/$$
$$222378477653494414029323540501238325182919858451 9671241$$
$$44731936278076277714476 53),$$

$$(7044063247449150935860150533337469755962712091387727069976796 07700$$
$$57102728566271816968/$$
$$4587645686689656711878602031324194555260270870245087225111060 89,$$
$$2871377212968706356475083631022485212953145012423036589862235 99143$$
$$593344666205275813631008027047008967522227301908578787419498 104/$$
$$9826181994367085850590085023465427811589979116750497045847031 84396$$
$$189343294840644633470910 8437),$$

(to find such huge generators, the 4-descent and 8-descent methods, implemented in
Magma, were used) is induced by the rational Diophantine triple

$$\left\{ \frac{1266720}{710719},\ -\frac{710719}{1266720},\ \frac{1099458061439}{900281971680} \right\}.$$

3.7 Rank Zero Elliptic Curves Induced by Rational Diophantine Triples

In the previous section, we saw how rational Diophantine triples can be used to find
elliptic curves of high rank. We will now consider the opposite problem and ask how
small the rank of an elliptic curve induced by a rational Diophantine triple can be. In

particular, can the rank of such a curve be 0? We will see that it is pretty easy to find such curves if we allow the elements of the triples to have mixed signs. However, for triples with positive elements, the problem becomes much more difficult, and currently, only one such curve is known (found in 2020 by Dujella and Mikić [145]), whose construction will be described in this section.

Let $\{a, b, c\}$ be a rational Diophantine triple and $ab + 1 = r^2$, $ac + 1 = s^2$, $bc + 1 = t^2$. We know that the curve E given by the equation $y^2 = (ax + 1)(bx + 1)(cx + 1)$ has three points of order 2, $A = (-1/a, 0)$, $B = (-1/b, 0)$, $C = (-1/c, 0)$, and the points $P = (0, 1)$ and $S = (1/abc, 1/rst)$ which are generally independent points of infinite order. So, if we want to find a curve of rank 0, the points P and S have to be of finite order. Recall that $S = 2R$, where $R = ((rs + rt + st + 1)/(abc), ((r + s)(r + t)(s + t))(abc))$, and that the triple $\{a, b, c\}$ is regular, i.e. $c = a + b \pm 2r$, if and only if $S = \mp 2P$ (see Sect. 3.1).

From Mazur's theorem and the fact that $S \in 2E(\mathbb{Q})$, it follows that we have the following possibilities for the points P and S to be of finite order:

- $mP = O$, $m = 3, 4, 6, 8$;
- $mS = O$, $m = 2, 3, 4$.

In particular, since the point P cannot be of order 2, we conclude that it is not possible to have simultaneously rank 0 and the torsion group $\mathbb{Z}/2\mathbb{Z} \times \mathbb{Z}/2\mathbb{Z}$.

By the standard transformations $x \mapsto \frac{x}{abc}$, $y \mapsto \frac{y}{abc}$, from the curve E we get the isomorphic curve

$$E' : \quad y^2 = (x + ab)(x + ac)(x + bc), \tag{3.45}$$

and the points A, B, C, P and S on E correspond to the points $A' = (-bc, 0)$, $B' = (-ac, 0)$, $C' = (-ab, 0)$, $P' = (0, abc)$ and $S' = (1, rst)$ on E'. In the next lemma, we examine all the possibilities for the point S to be of finite order. Note that we have already used part (ii) (without proof) in the construction of rational Diophantine sextuples in Sect. 3.2 and the construction of curves with the torsion group $\mathbb{Z}/2\mathbb{Z} \times \mathbb{Z}/6\mathbb{Z}$ in Sect. 3.6.4.

Lemma 3.7.1

(i) *The condition* $2S = O$ *is equivalent to*

$$(ab + 1)(ac + 1)(bc + 1) = 0.$$

(ii) *The condition* $3S = O$ *is equivalent to*

$$3 + 4(ab + ac + bc) + 6abc(a + b + c) + 12(abc)^2$$
$$- (abc)^2(a^2 + b^2 + c^2 - 2ab - 2ac - 2bc) = 0.$$

(iii) *The point S is of order* 4 *if and only if*

$$((ab + 1)^2 - ab(c - a)(c - b))((ac + 1)^2$$
$$- ac(b - a)(b - c))((bc + 1)^2 - bc(a - b)(a - c)) = 0.$$

Proof

(i) We will work with the isomorphic curve E'. The condition $2S' = O$ implies $rst = -rst$, i.e. $rst = 0$ and

$$(ab + 1)(ac + 1)(bc + 1) = 0.$$

(ii) From $3S' = O$, i.e. $x(2S') = x(-S') = x(S')$, and the formula for the doubling of points on an elliptic curve, we get

$$3 + (ab + ac + bc)$$
$$= \frac{9 + 4(ab + ac + bc)^2 + (abc(a + b + c))^2 + 12(ab + ac + bc)}{4r^2s^2t^2}$$
$$+ \frac{6abc(a + b + c) + 4abc(ab + ac + bc)(a + b + c)}{4r^2s^2t^2}.$$

Hence, we obtain

$$4((abc)^2 + abc(a + b + c) + (ab + ac + bc) + 1)(3 + ab + ac + bc)$$
$$= 9 + 12(ab + ac + bc) + (6abc(a + b + c) + 4(ab + ac + bc)^2)$$
$$+ 4abc(ab + ac + bc)(a + b + c) + (abc(a + b + c))^2,$$

which is equivalent to

$$3 + 4(ab + ac + bc) + 6abc(a + b + c) + 12(abc)^2$$
$$- (abc)^2(a^2 + b^2 + c^2 - 2ab - 2ac - 2bc) = 0.$$

(iii) The condition that the point S' is of order 4 is equivalent to $2S' \in \{A', B', C'\}$. Assume that $2S' = C'$ (the remaining two cases are completely analogous). From the duplication formula, we get

$$2 + (bc + ac)$$
$$= \frac{9 + 4(ab + ac + bc)^2 + (abc(a + b + c))^2 + 12(ab + ac + bc)}{4r^2s^2t^2}$$
$$+ \frac{6abc(a + b + c) + 4abc(ab + ac + bc)(a + b + c)}{4r^2s^2t^2},$$

which is equivalent to

$$(1 + 2ab - abc(c - a - b))^2 = 0,$$

i.e.

$$(ab + 1)^2 = ab(c - a)(c - b).$$

$\square$

Let us now consider the three possibilities for $mS = O$. First assume that $2S = O$. According to Lemma 3.7.1(i), we have $(ab + 1)(ac + 1)(bc + 1) = 0$, and we conclude that a, b, c cannot have the same sign. If we allow mixed signs, we can take $b = -1/a$. In Sect. 3.6.3, we saw that all rational Diophantine triples of the form $\{a, -1/a, c\}$ can be parameterized in the following way:

$$a = \frac{ut + 1}{t - u}, \quad b = \frac{u - t}{ut + 1}, \quad c = \frac{4ut}{(ut + 1)(t - u)}.$$

To find examples of curves of rank 0, assume that the triple $\{a, -1/a, c\}$ is regular. This assumption gives us the condition $(u^2 - 1)(t^2 - 1) = 0$, so we can take $u = 1$. If we now take, for example, $t = 2$, we get a curve with the torsion group $\mathbb{Z}/2\mathbb{Z} \times \mathbb{Z}/4\mathbb{Z}$ and rank 0 which is induced by the triple

$$\left\{ 3, -\frac{1}{3}, \frac{8}{3} \right\}.$$

Let us now assume that $3S = O$. If it were also $3P = O$, then we would have $P = \pm S$, which is a contradiction. Therefore, if the point P has finite order, the only possibility is that P is a point of order 6. This implies that $2P = \pm S$ and $c = a + b \mp 2r$. Inserting $b = (r^2 - 1)/a$ and $c = a + b + 2r$ in the condition from Lemma 3.7.1(ii), we get

$$(2ar - 1 + 2r^2)(-a + 2ar^2 - 2r + 2r^3)(2a^2r - a - 2r + 4ar^2 + 2r^3) = 0.$$

From here, we have

$$a = \frac{-2r(r^2 - 1)}{-1 + 2r^2}, \quad \text{or } a = \frac{-(-1 + 2r^2)}{2r}, \quad \text{or } a = \frac{1 - 4r^2 \pm \sqrt{1 + 8r^2}}{4r}.$$

Let us take

$$(a, b, c) = \left(\frac{-2r(r - 1)(r + 1)}{-1 + 2r^2}, \frac{-(-1 + 2r^2)}{2r}, \frac{(-1 + 2r)(2r + 1)}{2(-1 + 2r^2)r} \right). \tag{3.46}$$

Then the condition $ab > 0$ is equivalent to $r > 1$ or $r < -1$, while the condition $bc > 0$ is equivalent to $-1/2 < r < 1/2$. Therefore, a, b, c cannot have the same sign.

The case

$$(a, b, c) = \left(\frac{-(-1+2r^2)}{2r}, \frac{-2r(r-1)(r+1)}{-1+2r^2}, \frac{(-1+2r)(2r+1)}{2(-1+2r^2)r} \right)$$

is the same as the previous one, just by interchanging the roles of a and b.

Finally, let $8r^2 + 1 = (2rt + 1)^2$ (so that we get rid of the square root in the third case). We get that $r = \frac{-t}{-2+t^2}$, so

$$(a, b, c) = \left(\frac{-t(t-2)(t+2)}{2(-2+t^2)}, \frac{2(t-1)(t+1)}{(-2+t^2)t}, \frac{-(-2+t^2)}{2t} \right)$$

(or by interchanging a and b). The condition $ac > 0$ is equivalent to $t > 2$ or $t < -2$, while the condition $bc > 0$ is equivalent to $-1 < t < 1$. So, in this case, a, b, c cannot have the same sign.

If we allow mixed signs, we can get examples of curves of rank 0. For example, from triples of the form (3.46), for $t = 4$, we get a curve with the torsion group $\mathbb{Z}/2\mathbb{Z} \times \mathbb{Z}/6\mathbb{Z}$ and rank 0, which is induced by the triple

$$\left\{ -\frac{12}{7}, \frac{15}{28}, -\frac{7}{4} \right\}.$$

The last case to be considered is when the point S is of order 4. Then the point R, such that $2R = S$, is of order 8 and therefore, the curve E has the torsion group $\mathbb{Z}/2\mathbb{Z} \times \mathbb{Z}/8\mathbb{Z}$. We saw in Theorem 2.4.11 that every such elliptic curve over $\mathbb{Q}$ is induced by a rational Diophantine triple of the form

$$\left\{ \frac{2t}{t^2-1}, \frac{1-t^2}{2t}, \frac{6t^2 - t^4 - 1}{2t(t^2-1)} \right\}. \tag{3.47}$$

It is clear that the elements of triple (3.47) have mixed signs. If we take, for example $t = 2$, we get a curve with the torsion group $\mathbb{Z}/2\mathbb{Z} \times \mathbb{Z}/8\mathbb{Z}$ and rank 0 which is induced by the triple

$$\left\{ \frac{4}{3}, -\frac{3}{4}, \frac{7}{12} \right\}.$$

It remains to consider whether it is possible to obtain a curve with the torsion group $\mathbb{Z}/2\mathbb{Z} \times \mathbb{Z}/8\mathbb{Z}$ and rank 0, which is induced by a triple with positive elements.

We assume again that the point S is of order 4, and we take $b = (r^2 - 1)/a$, $c = a + b + 2r$. If we insert this in the first factor

$$(ab + 1)^2 - ab(c - a)(c - b)$$

from Lemma 3.7.1(iii), we get the quadratic equation in a:

$$(2r^3 - 2r)a^2 + (4r^4 - 6r^2 + 1)a + 2r^5 - 4r^3 + 2r = 0.$$

Its discriminant,

$$- 4r^4 + 4r^2 + 1,$$

should be a perfect square. This condition defines a quartic that is birationally equivalent to the elliptic curve

$$E_1 : \quad Y^2 = X^3 + X^2 + X + 1,$$

which has rank 1 and a generator $P_1 = (0, 1)$. Therefore, calculating the multiples of the point P_1 on the curve E_1 (adding the torsion point $T_1 = (-1, 0)$ of order 2 only has the effect of replacing r with $-r$), and transferring them back to the quartic, we get candidates for the solution of our problem. However, the condition that all elements of the corresponding triple are positive must still be satisfied (it is sufficient that all the elements of the triple have the same sign because by multiplying all the elements of a rational Diophantine triple by -1 we get a rational Diophantine triple). The first two multiples of P_1 that give triples with positive elements are $6P_1$ and $11P_1$.

The point $6P_1$ gives $r = -\frac{3855558}{3603685}$ and the triple

$$(a, b, c) = \left(\frac{1884586446094351}{25415891646864180}, \; \frac{14442883687791636}{7402559392524605}, \; \frac{60340495895762708555}{14487505263205637124} \right).$$

With the available programs, we could not determine the rank of the induced elliptic curve. Namely, `magma`, `mwrank` and `ellrank` all give that $0 \le \text{rank} \le 2$. Assuming that the parity conjecture holds, the rank should be 0 or 2.

The point $11P_1$ gives $r = \frac{3556951688276668510 6979}{3238381938724095267 2281}$ and the triple (a, b, c), where

$$a = \frac{69705492951192675600645567228019184577147632882703132983}{13201484334991246769290130383656126692130218445953 6763120},$$

$$b = \frac{4782682988007982907580118956394262073206270109554 8790400}{12233666942070950930363744264796639133659 66949698 35459327},$$

$$c = \frac{4798211114664940442174933170939350177779177455854621 79875502 57759801}{154000907539182573640934849105806523907860840550436770208040 56653840}.$$

(Comparing the j-invariants, we get that this curve is induced by the triple (3.47) for $t = \frac{18451786408106133183649}{41916048174422594852689}$.)

For the induced curve, both `mwrank` and the standard `magma` function for computing the rank `MordellWeilShaInformation` give that $0 \le \text{rank} \le 4$. However, `magma` (version `V2.24-7`) function `TwoPowerIsogenyDescentRank Bound`, in which Tom Fisher's algorithm from [170] is implemented, gives that the rank is equal to 0 (this result is given by the fifth of the six steps of the algorithm; it is the step after the 4-descent and before the 8-descent). Thus we found the desired example of a rational Diophantine triple with positive elements which induces an elliptic curve of rank 0.

Let us note that the `magma` function `TwoPowerIsogenyDescentRank Bound` applied to the above curve which corresponds to the point $6P_1$ gives only rank ≤ 2 and does not give a final answer to the question of whether the rank of that curve is 0 or 2. The described construction certainly gives an infinite number of multiples of P_1 which correspond to triples with positive terms (here we use the fact that the set $E_1(\mathbb{Q})$ is dense in $E_1(\mathbb{R})$; by a theorem due to Poincaré and Hurwitz, if the rank of an elliptic curve over $\mathbb{Q}$ is positive, i.e. if the curve has infinitely many rational points, then the set of rational points is dense on the even component of the curve and also on an odd one if it contains at least one rational point, see e.g. [331, p. 78]). However, it is difficult to say something reliable about the distribution of ranks in this family of curves, so we can only speculate that there could be infinitely many curves in this family of rank 0.

3.8   Torsion Groups of Elliptic Curves Induced by Integer Diophantine Triples

So far, we have dealt with elliptic curves induced by rational Diophantine triples. In particular, we have shown that for such curves, each of the four torsion groups allowed by Mazur's theorem for curves with three points of order 2, i.e. $\mathbb{Z}/2\mathbb{Z} \times \mathbb{Z}/2\mathbb{Z}$, $\mathbb{Z}/2\mathbb{Z} \times \mathbb{Z}/4\mathbb{Z}$ $\mathbb{Z}/2\mathbb{Z} \times \mathbb{Z}/6\mathbb{Z}$, and $\mathbb{Z}/2\mathbb{Z} \times \mathbb{Z}/8\mathbb{Z}$, is possible, and moreover appears for infinitely many rational Diophantine triples.

Now we will consider the same question, but for curves induced by (integer) Diophantine triples. We will see that the answer here is different. Namely, the torsion groups $\mathbb{Z}/2\mathbb{Z} \times \mathbb{Z}/4\mathbb{Z}$ and $\mathbb{Z}/2\mathbb{Z} \times \mathbb{Z}/8\mathbb{Z}$ are not possible, while the problem remains open whether the torsion group $\mathbb{Z}/2\mathbb{Z} \times \mathbb{Z}/6\mathbb{Z}$ can appear. We will consider a slightly wider problem; namely, we will consider elliptic curves induced by integer $D(4)$-triples. It is clear that by multiplying the elements of a $D(1)$-triples by 2, we get a $D(4)$-triple, so that we will obtain the desired result for $D(1)$-triples as an immediate corollary of results for $D(4)$-triples. In doing so, we will follow the paper [144] by Dujella and Mikić, in which an error from the proof in [103] was corrected.

Let $\{a, b, c\}$ be a $D(4)$-triple, and let r, s, t be nonnegative integers such that

$$ab + 4 = r^2, \quad ac + 4 = s^2, \quad bc + 4 = t^2.$$

We consider the elliptic curve

$$E : \quad y^2 = (ax + 4)(bx + 4)(cx + 4),$$

which we say that it is induced by the $D(4)$-triple $\{a, b, c\}$.

By transformations $x \mapsto \frac{x}{abc}$, $y \mapsto \frac{y}{abc}$, we bring the curve into the form

$$E' : \quad y^2 = (x + 4bc)(x + 4ac)(x + 4ab).$$

On the curve E', we have three points of order 2:

$$A' = (-4bc, 0), \quad B' = (-4ac, 0), \quad C' = (-4ab, 0),$$

and two additional obvious rational points

$$P' = (0, 8abc), \quad S' = (16, 8rst).$$

It holds that $S' = 2R'$, where

$$R' = (4rs + 4rt + 4st + 16, 8(r + s)(r + t)(s + t)).$$

We will first consider a particular case, which is the only $D(4)$-triple with mixed signs: $\{-1, 3, 4\}$ (and its equivalent $\{-4, -3, 1\}$). The elliptic curve induced by this triple has rank 0 and torsion group $\mathbb{Z}/2\mathbb{Z} \times \mathbb{Z}/4\mathbb{Z}$ (see Example 2.4.5). In this case $B' \in 2E'(\mathbb{Q})$, more precisely $B' = 2P'$, so the point P' is of order 4. Note that in this case, the point R' is also of order 4 since $R' = P' + A'$ and $2R' = 2P'$.

Therefore, we can further assume that a, b, c are positive integers and $a < b < c$.

We will need a precise estimate for the size of the third element c in $D(4)$-triples. Of course, one possibility is that $\{a, b, c\}$ is regular, i.e. $c = a + b + 2r$. We are interested in what can be said about c if the triple is not regular. We will use a construction from [106] that generalizes the construction of the number d_- from the case of $D(1)$-triples. That number extends a $D(1)$-triple to a $D(1)$-quadruple. In the case of $D(n)$-triples, that fourth element will not have the property that the products with it increased by n are squares, but the products will have to be increased by n^2 to get squares. However, in the case of $n = 4$, it will be possible, by dividing by 4, to go from 16 to 4, and get a $D(4)$-quadruple.

Lemma 3.8.1 *If $\{a, b, c\}$ is a $D(n)$-triple and $ab+n = r^2$, $ac+n = s^2$, $bc+n = t^2$, then there are integers e, x, y, z so that*

$$ae + n^2 = x^2, \quad be + n^2 = y^2, \quad ce + n^2 = z^2$$

and

$$c = a + b + \frac{e}{n} + \frac{2}{n^2}(abe + rxy).$$

Proof Let us define

$$e = n(a + b + c) + 2abc - 2rst.$$

Then

$$(ae + n^2) - (at - rs)^2 = an(a + b + c) + 2a^2bc - 2arst + n^2$$
$$- a^2(bc + n) + 2arst - (ab + n)(ac + n) = 0.$$

Therefore, we can take $x = rs - at$, and analogously $y = rt - bs$, $z = cr - st$. We have

$$abe + rxy = abn(a + b + c) + 2a^2b^2c - 2abrst$$
$$+ rst(ab + n) - a(ab + n)(bc + n) - b(ab + n)(ac + n) + abrst$$
$$= -abcn - n^2(a + b) + rstn,$$

and finally

$$a+b+\frac{e}{n}+\frac{2}{n^2}(abe+rxy) = 2a+2b+c+\frac{2abc}{n}-\frac{2rst}{n}-\frac{2abc}{n}-2a-2b+\frac{2rst}{n} = c.$$

$\square$

Lemma 3.8.2 *If $\{a, b, c\}$ is a $D(4)$-triple, then $c = a + b + 2r$ or $c > ab + a + b + 1 > ab$.*

Proof By Lemma 3.8.1, there exists an integer

$$e = 4(a + b + c) + 2(abc - rst) \tag{3.48}$$

and integers x, y, z such that

$$ae + 16 = x^2, \tag{3.49}$$

$$be + 16 = y^2, \tag{3.50}$$

$$ce + 16 = z^2, \tag{3.51}$$

and $c = a + b + \frac{e}{4} + \frac{1}{8}(abe + rxy)$. It is easy to see that x and y from Lemma 3.8.1 for $n = 4$ (and more generally for $n > 0$) are positive. From (3.51), it follows that $e \geq 0$ (the case $e = -1$ implies $c \leq 16$, but the only such $D(4)$-triple is $\{1, 5, 12\}$, which does not satisfy (3.49) and (3.50)). For $e = 0$ we get $c = a + b + 2r$, while for $e \geq 1$ we have that $c > \frac{1}{4}abe + a + b + \frac{e}{4}$.

As we showed in Sect. 1.4, among the numbers a, b, c, there are at most two odd ones. We conclude that $abc - rst$ is even. Therefore, from (3.48), it follows that $e \equiv 0 \pmod 4$, which implies that $e \geq 4$ and $c > ab + a + b + 1$. $\square$

Remark 3.8.3 Lemma 3.8.2 implies that in any case, we have $c \geq a + b + 2r$. Indeed, the inequality $ab+a+b+1 \geq a+b+2r$ is equivalent to $(r-3)(r+1) \geq 0$, which is satisfied for all $D(4)$-triples with positive elements.

Remark 3.8.4 The statement of Lemma 3.8.2 is the best possible, in the sense that the inequality $c > ab$ cannot be replaced by $c > (1 + \varepsilon)ab$ for some fixed $\varepsilon > 0$. Indeed, if for $k \geq 3$ we put $a = k^2 - 4$, $b = k^2 + 2k - 3$, $c = k^4 + 2k^3 - 3k^2 - 4k$, then $\{a, b, c\}$ is a $D(4)$-triple and $\lim_{k \to \infty} \frac{c}{ab} = 1$.

Remark 3.8.5 We can improve the inequality $c > ab+a+b+1$ from Lemma 3.8.2 by a more precise analysis of the term rxy. Indeed, since $e \geq 4$, we have

$$c \geq a + b + 1 + \frac{1}{2}(ab + \sqrt{(ab + 4)(a + 4)(b + 4)}).$$

Furthermore,

$$(ab + 4)(a + 4)(b + 4) - (ab + u(a + b))^2 =$$

$$(4a - 2ua - u^2)b^2 + a((4b - 2ab - u^2)a + 2b(10 - u^2)) + 16a + 16b + 64 > 0$$

for $u = \sqrt{5} - 1$ (since then $4a - 2ua - u^2 = (a - 1)(6 - 2\sqrt{5}) \geq 0$). Therefore, we obtain

$$c \geq ab + \frac{1}{2}(\sqrt{5} + 1)(a + b) + 1.$$

Remark 3.8.6 Let $\{a, b, c\}$ be a $D(1)$-triple. Since $\{2a, 2b, 2c\}$ is a $D(4)$-triple, Lemma 3.8.2 implies that $c = a + b + 2r$ or $c > 2ab + a + b$. However, we can get a stronger result by imitating the proof of Lemma 3.8.2 and Remark 3.8.5. Indeed, we have $c = a + b + d_- + 2abd_- + 2\sqrt{(ab + 1)(ad_- + 1)(bd_- + 1)}$. It is clear that $d_- \geq 0$. If $d_- = 0$, we have $c = a + b + 2r$, while if $d_- \geq 1$, we have $c > a + b + 1 + 2ab + 2(ab + (\sqrt{2} - 1)(a + b)) = 4ab + (2\sqrt{2} - 1)(a + b) + 1$. The family of triples $\{k^2 - 1, k^2 + 2k, 4k^4 + 8k^3 - 4k\}$ shows that the coefficient 4 in $4ab$ is the best possible, while the coefficient $(2\sqrt{2} - 1)$ is not far from the best possible.

Remark 3.8.7 From the proof of Lemmas 3.8.1 and 3.8.2, it follows that any $D(4)$-triple $\{a, b, c\}$ can be extended to a $D(4)$-quadruple with the fourth element $d = a + b + c + \frac{1}{2}(abc \pm rst)$ because the number $abc \pm rst$ is even and $ad + 4 = (\frac{1}{2}(rs \pm at))^2$, $bd + 4 = (\frac{1}{2}(rt \pm bs))^2$, $cd + 4 = (\frac{1}{2}(cr \pm st))^2$.

Now we want to prove that E' has no points of order 4.

Lemma 3.8.8 $A', B', C' \notin 2E'(\mathbb{Q})$

Proof If $A' \in 2E'(\mathbb{Q})$, then Theorem 2.4.9 implies that $4c(a - b)$ is a square. But $4c(a - b) < 0$, so we got a contradiction. Similarly, $B' \notin 2E'(\mathbb{Q})$ because $4a(b - c) < 0$ is not a square.

It remains to consider the point C', and that case is significantly more demanding. If $C' \in 2E'(\mathbb{Q})$, then, by Theorem 2.4.9, we have

$$a(c - b) = X^2, \tag{3.52}$$

$$b(c - a) = Y^2, \tag{3.53}$$

for integers X and Y. We do not have a direct contradiction as in the previous two cases since the numbers $a(c - b)$ and $b(c - a)$ are positive.

By Lemma 3.8.2, we have two possibilities: $c = a + b + 2r$ or $c > ab + a + b + 1$.

Let us first assume that $c = a + b + 2r$. From (3.52) and (3.53), we have that $a = kx^2$, $c - b = ky^2$, $b = lz^2$, $c - a = lu^2$, where k, l, x, y, z, u are positive integers. From here $c = kx^2 + lu^2 = ky^2 + lz^2$, so from $c = a + b + 2r$, we get

$$2r = k(y^2 - x^2) = l(u^2 - z^2). \tag{3.54}$$

By squaring (3.54), we get

$$4r^2 = 16 + 4ab = 16 + 4klx^2z^2 = k^2(y^2 - x^2)^2 = l^2(u^2 - z^2)^2,$$

which implies that $k \in \{1, 2, 4\}$ and $l \in \{1, 2, 4\}$. Since kl is not a perfect square (otherwise, we would have $(2r)^2 = 16 + (2xz\sqrt{kl})^2$, which implies $2r = 5$), without loss of generality, we can assume that $k = 1, l = 2$ or $k = 2, l = 4$. For $k = 1$, $l = 2$, we have $4r^2 = 16 + 8x^2z^2$, which implies $r^2 = 4 + 2x^2z^2$, from where we conclude that r and xz are even. Thus, $r^2 \equiv 4 \pmod{8}$ and $r \equiv 2 \pmod{4}$, but from $2r = 2(u^2 - z^2)$, we get that $u^2 - z^2 \equiv 2 \pmod{4}$, which is a contradiction. If $k = 2, l = 4$, then $4r^2 = 16 + 32x^2z^2$, which implies $r^2 = 4 + 8x^2z^2$. From here $r^2 \equiv 4 \pmod{8}$ and $r \equiv 2 \pmod{4}$, but from $2r = 2(y^2 - x^2)$, we get $y^2 - x^2 \equiv 2 \pmod{4}$, which is a contradiction.

It remains to consider the case when $c > ab + a + b + 1$. Let us write the conditions (3.52) and (3.53) in the form

$$ac - ab = s^2 - r^2 = (s - \alpha)^2, \tag{3.55}$$

$$bc - ab = t^2 - r^2 = (t - \beta)^2, \tag{3.56}$$

where $0 < \alpha < s, 0 < \beta < t$. Then we have

$$r^2 = 2s\alpha - \alpha^2 = 2t\beta - \beta^2. \tag{3.57}$$

From (3.57), we get

$$4(bc + 4)\beta^2 = (ab + 4 + \beta^2)^2$$

and

$$(\beta^2 - 4)^2 = b(4c\beta^2 - a^2b - 2a(4 + \beta^2)). \tag{3.58}$$

From (3.58), we conclude that either $\beta = 1$ or $\beta = 2$, or $\beta^2 \geq \sqrt{b} + 4$.

If $\beta = 1$, then $b(4c - a^2b - 10a) = 9$, which easily leads to a contradiction.

If $\beta = 2$, then from (3.58), we get

$$c = \frac{a^2b + 16a}{16}. \tag{3.59}$$

Now we have

$$s^2 = ac + 4 = \frac{1}{16}(a^3b + 16a^2 + 64) = \frac{1}{16}(a^2r^2 + 12a^2 + 64),$$

so

$$a^2r^2 + 12a^2 + 64 = (ar + n)^2,$$

where n is even and $0 < n < 8$. From $a^2r^2 + 12a^2 + 64 = (ar + 2)^2$, we get $a(r - 3a) = 15$. If $a \leq 3$, then (3.59) implies that $c < b$, a contradiction. The case $a = 5$ gives the triple $\{5, 64, 105\}$ which does not satisfy $c > ab$ (c is equal to $a + b + 2r$), while the case $a = 15$ leads to $15b + 4 = 46^2$, which has no integer solutions. The cases $a^2r^2 + 12a^2 + 64 = (ar + 4)^2$ and $a^2r^2 + 12a^2 + 64 = (ar + 6)^2$ can be treated similarly.

Thus, we are left with the case $\beta^2 \geq \sqrt{b} + 4$, which implies

$$\beta > \max(\sqrt[4]{b}, 2). \tag{3.60}$$

The function $f(\beta) = t^2 - (t - \beta)^2$ is increasing for $0 < \beta < t$. Therefore,

$$ab = t^2 - (t - \beta)^2 - 4 > 2t\sqrt[4]{b} - \sqrt{b} - 4 > 2\sqrt{bc}\sqrt[4]{b} - \sqrt{b} - 4,$$

which implies $ab > \sqrt{bc}\sqrt[4]{b}$, because $\sqrt{b}(\sqrt{c}\sqrt[4]{b} - 1) > 4$ (due to $b \geq 4$ and $c \geq 12$, which follows from the fact that $\{3, 4, 15\}$ and $\{1, 5, 12\}$ are $D(4)$-triples with the smallest values of b and c, respectively). This further gives

$$c < a^2\sqrt{b}. \tag{3.61}$$

Using (3.48), we define the integer d_- as

$$d_- = \frac{e}{4} = a + b + c + \frac{abc - rst}{2}.$$

Then $d_- \neq 0$ (since $c \neq a + b + 2r$) and $\{a, b, c, d_-\}$ is a $D(4)$-quadruple. In particular,

$$ad_- + 4 = \left(\frac{rs - at}{2}\right)^2. \tag{3.62}$$

Furthermore,

$$c = a + b + d_- + \frac{1}{2}(abd_- + \sqrt{(ab + 4)(ad_- + 4)(bd_- + 4)}) > abd_- \tag{3.63}$$

(according to the proof of Lemma 3.8.2). Comparing this with (3.61), we get

$$d_- < \frac{a}{\sqrt{b}}. \tag{3.64}$$

Thus, we have $d_- < a < b$, which implies that b is the largest element in the $D(4)$-triple $\{a, b, d_-\}$. This means that, according to Remark 3.8.3, $b \geq a + d_- + 2\sqrt{ad_- + 4}$ or equivalently $d_- \leq a + b - 2r$. Let us further define

$$c' = a + b + d_- + \frac{1}{2}(abd_- - \sqrt{(ab + 4)(ad_- + 4)(bd_- + 4)}).$$

We have

$$cc' = (a + b + 2r - d_-)(a + b - 2r - d_-) \geq 0.$$

This implies that

$$c < 2(a + b + d_- + \frac{1}{2}abd_-) < 4b + abd_- < 2abd_-. \tag{3.65}$$

Let us denote $p = \frac{rs - at}{2}$. Then $p > 0$ and, by (3.62), we have $ad_- + 4 = p^2$. To estimate p, we define $p' = \frac{rs + at}{2}$. Then we have

$$pp' = \frac{1}{4}(a^2bc + 4ac + 4ab + 16 - a^2bc - 4a^2) = a(b + c - a) + 4,$$

and

$$p < \frac{a(c + b)}{at} < \frac{c + b}{\sqrt{bc}} = \frac{\sqrt{c}}{\sqrt{b}} + \frac{\sqrt{b}}{\sqrt{c}},$$

$$p > \frac{(ac + 4)}{rs} = \frac{s}{r}.$$

Furthermore, we have

$$\frac{\sqrt{c}}{\sqrt{b}} - \frac{s}{r} = \frac{r\sqrt{c} - s\sqrt{b}}{r\sqrt{b}} = \frac{4c - 4b}{r\sqrt{b}(r\sqrt{c} + s\sqrt{b})} < \frac{4c}{2rsb} < \frac{2\sqrt{c}}{ab\sqrt{b}} < \frac{\sqrt{b}}{\sqrt{c}},$$

where we used $c < \frac{ab^2}{2}$, which follows from (3.61). By putting the two obtained estimates for p together, we get

$$\left| p - \frac{\sqrt{c}}{\sqrt{b}} \right| < \frac{\sqrt{b}}{\sqrt{c}}. \tag{3.66}$$

Let us define the integer α by

$$2d_-\beta = p + \alpha.$$

Assume that $\alpha = 0$. Then (3.62) implies that $d_-(4\beta^2 d_- - a) = 4$, so $d_- \in \{1, 2, 4\}$. If $d_- = 1$, then $2\beta = p$, and (3.56) gives

$$r^2 + \frac{p^2}{4} = tp, \tag{3.67}$$

while c satisfies the inequalities

$$ab < ab + a + b + 1 < c < ab + 2a + 2b + 2 < ab + 4b < 2ab$$

(according to Lemma 3.8.2 and (3.63) for $d_- = 1$). The left-hand side of (3.67) is

$$< ab + 4 + \frac{c^2 + 2bc + b^2}{4bc} < ab + 4 + \frac{a}{4} + 1 + \frac{1}{2} + \frac{1}{4a} < ab + \frac{a}{4} + 6.$$

On the other hand, by (3.66) and Remark 3.8.5, the right-hand side of (3.67) is

$$> \sqrt{bc}\left(\frac{\sqrt{c}}{\sqrt{b}} - \frac{\sqrt{b}}{\sqrt{c}}\right) = c - b > ab + \frac{3}{2}a + \frac{1}{2}b + 1.$$

Comparing the two obtained inequalities for (3.67), we get

$$b + \frac{5}{2}a < 10,$$

which is in contradiction with $b \geq 12$ (b is the largest element in the $D(4)$-triple $\{d_-, a, b\}$). The cases $d_- = 2$ and $d_- = 4$ can be solved similarly.

Therefore, we can now assume that $\alpha \neq 0$. We will estimate $2d_-t\beta$ and compare it with c. First, we prove that

$$\beta^2 < \frac{a^2 b}{c}. \tag{3.68}$$

Since $\beta < t$, and the case $\beta = t - 1$ gives $b(c - a) = 1$, which is impossible, we conclude that $t \geq \beta + 2$. This implies $t\beta \geq \beta^2 + 2\beta$ and $ab - t\beta \geq 2\beta - 4 > 0$ due to (3.60). Therefore, we get that $t\beta < ab$, which obviously implies (3.68).

Hence,

$$0 < d_-\beta^2 < \frac{d_- a^2 b}{c} < a.$$

From $2t\beta = r^2 + \beta^2 > ab + 4$, it follows that $2d_-t\beta > abd_- + 4d_-$. On the other hand,

$$2d_-t\beta = (r^2 + \beta^2)d_- < abd_- + 4d_- + a.$$

By combining these two estimates, we obtain

$$abd_- + 4d_- < 2d_-t\beta < abd_- + 4d_- + a. \tag{3.69}$$

Comparing (3.69) with (3.63) and (3.65), we conclude that

$$|2d_-t\beta - c| < 4b. \tag{3.70}$$

Combining estimate (3.66) for p with the trivial estimate $|\alpha| \geq 1$, we get

$$\left| 2d_-\beta - \frac{\sqrt{c}}{\sqrt{b}} \right| = \left| p + \alpha - \frac{\sqrt{c}}{\sqrt{b}} \right| \geq 1 - \frac{\sqrt{b}}{\sqrt{c}}.$$

Let us notice that $ad_- > 26$. Indeed, the $D(4)$-pairs such that $ad_- \leq 26$ are $\{1, 5\}$, $\{1, 12\}$, $\{1, 21\}$, $\{2, 6\}$, $\{3, 4\}$ and $\{3, 7\}$. From the first three pairs, taking into account (3.63) and (3.64), we find triples

$$\{5, 12, 96\}, \{12, 21, 320\}, \{12, 96, 1365\}, \{21, 32, 780\}, \{21, 320, 7392\},$$

which satisfy neither (3.52) nor (3.53). From the last three pairs, we cannot get a $D(4)$-triple because of (3.64).

Finally, we get

$$
|2d_- t\beta - c| = \left| 2d_- t\beta - t\frac{\sqrt{c}}{\sqrt{b}} + t\frac{\sqrt{c}}{\sqrt{b}} - c \right| \geq t\left| 2d_-\beta - \frac{\sqrt{c}}{\sqrt{b}} \right| - \left| t\frac{\sqrt{c}}{\sqrt{b}} - c \right|
$$

$$
= t\left| 2d_-\beta - \frac{\sqrt{c}}{\sqrt{b}} \right| - \left(t\frac{\sqrt{c}}{\sqrt{b}} - c \right) \geq t\left(1 - \frac{\sqrt{b}}{\sqrt{c}} \right) - \left(t\frac{\sqrt{c}}{\sqrt{b}} - c \right)
$$

$$
= t\left(1 - \frac{\sqrt{b}}{\sqrt{c}} \right) - c\left(\sqrt{1 + \frac{4}{bc}} - 1 \right) > \sqrt{bc} - b - c\left(\sqrt{1 + \frac{4}{bc}} - 1 \right)
$$

$$
> \sqrt{ab^2 d_-} - b - \frac{2}{b} \geq b\left(\sqrt{ad_-} - 1 - \frac{1}{72} \right) > 4b,
$$

which contradicts (3.70). $\square$

Theorem 3.8.9 $E'(\mathbb{Q})_{\mathrm{tors}} \cong \mathbb{Z}/2\mathbb{Z} \times \mathbb{Z}/2\mathbb{Z}$ *or* $\mathbb{Z}/2\mathbb{Z} \times \mathbb{Z}/6\mathbb{Z}$.

Proof According to Mazur's theorem, since E' has three points of order 2, the only possibilities for $E'(\mathbb{Q})_{\mathrm{tors}}$ are $\mathbb{Z}/2\mathbb{Z} \times \mathbb{Z}/2k\mathbb{Z}$ for $k = 1, 2, 3, 4$. However, Lemma 3.8.8 shows that cases $k = 2$ and 4 are not possible for elliptic curves induced by $D(4)$-triples with positive elements. $\square$

Corollary 3.8.10 *Let* $\{a, b, c\}$ *be a* $D(1)$-*triple. Then the torsion group of the elliptic curve* $y^2 = (ax+1)(bx+1)(cx+1)$ *is either* $\mathbb{Z}/2\mathbb{Z} \times \mathbb{Z}/2\mathbb{Z}$ *or* $\mathbb{Z}/2\mathbb{Z} \times \mathbb{Z}/6\mathbb{Z}$.

Remark 3.8.11 Note that the analogue of Theorem 3.8.9 and Corollary 3.8.10 does not hold for general $D(n^2)$-triples and their induced elliptic curves

$$
y^2 = (ax + n^2)(bx + n^2)(cx + n^2).
$$

If in the examples from Sect. 3.6 of elliptic curves with various torsion groups induced by rational Diophantine $D(1)$-triples we multiply all elements by the least common multiple of their denominators, let us call it n, we will get a $D(n^2)$-triple with the same torsion group. As a simple concrete example, let us present the $D(9)$-triple $\{8, 54, 104\}$, which induces an elliptic curve with the torsion group $\mathbb{Z}/2\mathbb{Z} \times \mathbb{Z}/4\mathbb{Z}$.

Let us mention that no example of a $D(1)$ or $D(4)$-triple is known that induces an elliptic curve with torsion group $\mathbb{Z}/2\mathbb{Z} \times \mathbb{Z}/6\mathbb{Z}$. It is known that this torsion group cannot appear for some concrete families of $D(1)$-triples. As we have already argued, such examples exist for general $D(n^2)$-triples. For example, the $D(294^2)$-triple $\{32, 539, 1215\}$ induces an elliptic curve with the torsion group $\mathbb{Z}/2\mathbb{Z} \times \mathbb{Z}/6\mathbb{Z}$.

As we mentioned, there are several results showing that the torsion group $\mathbb{Z}/2\mathbb{Z} \times \mathbb{Z}/6\mathbb{Z}$ cannot appear for elliptic curves induced by certain families of Diophantine triples. To exclude this torsion group, the following result from [300] (see also [244]) is often very useful.

Lemma 3.8.12 *Let M, N be integers such that $MN(M-N) \neq 0$. An elliptic curve E given by the equation $y^2 = x(x+M)(x+N)$ has the torsion group $\mathbb{Z}/2\mathbb{Z} \times \mathbb{Z}/6\mathbb{Z}$ if and only if there exist integers α, β, δ such that $\frac{\alpha}{\beta} \notin \{-2, -1, -\frac{1}{2}, 0, 1\}$ and*

$$M = \delta^2(\alpha^4 + 2\alpha^3\beta), \quad N = \delta^2(2\alpha\beta^3 + \beta^4).$$

Proof Since the curve E has three points of order 2, it will have the torsion group $\mathbb{Z}/2\mathbb{Z} \times \mathbb{Z}/6\mathbb{Z}$ if and only if it has some rational point of order 3. The condition for the point $Q = (x, y)$ to be of order 3 is

$$3x^4 + (4M + 4N)x^3 + 6x^2MN - M^2N^2 = 0. \tag{3.71}$$

Solving Eq. (3.71) for N, we get that $x(x + M)$, and thus also $(M + x)/x$, must be a square. So let us put $\frac{M}{x} + 1 = (1 + t)^2$. Inserting it into (3.71), we get that $\frac{N}{x} = \frac{2t+1}{t^2}$. Let $t = \frac{\alpha}{\beta}$, where α, β are relatively prime. Then

$$\frac{M}{x} = \frac{\alpha(2\beta + \alpha)}{\beta^2}, \quad \frac{N}{x} = \frac{\beta(2\alpha + \beta)}{\alpha^2}. \tag{3.72}$$

Since, according to the Lutz-Nagell theorem, x is an integer, we conclude that x is of the form $x = d\alpha^2\beta^2$. Plugging this into (3.72), we get that $M = d\alpha^3(\alpha + 2\beta)$, $N = d\beta^3(2\alpha + \beta)$. Finally, by inserting the obtained expressions for M, N, x into the equation of the curve, we get that d is a perfect square, say $d = \delta^2$, which gives us the formulas for M and N from the statement of the theorem. The required point of order 3 has coordinates $Q = (\delta^2\alpha^2\beta^2, \pm\delta^3\alpha^2\beta^2(\alpha + \beta)^2)$. $\qquad \square$

We now illustrate an application of Lemma 3.8.12 on the simple parametric family of triples $\{k - 1, k + 1, 4k\}$ studied in [93]. Later, we will apply it to the family of triples containing the pair $\{1, 3\}$ in Sect. 4.8.2. Other similar results can be found in [98, 278, 293].

Proposition 3.8.13 *Let $k \geq 2$ be an integer and E_k the elliptic curve induced by the Diophantine triple $\{k - 1, k + 1, 4k\}$. Then*

$$E_k(\mathbb{Q})_{\text{tors}} \cong \mathbb{Z}/2\mathbb{Z} \times \mathbb{Z}/2\mathbb{Z}.$$

Proof By Theorem 3.8.9, it remains to show that $E_k(\mathbb{Q})_{\text{tors}}$ cannot be $\mathbb{Z}/2\mathbb{Z} \times \mathbb{Z}/6\mathbb{Z}$. We will apply Lemma 3.8.12 on the equivalent curve

$$E_k' : \quad y^2 = x(x + 3k^2 - 4k + 1)(x + 3k^2 + 4k + 1).$$

By Lemma 3.8.12, E_k' has torsion $\mathbb{Z}/2\mathbb{Z} \times \mathbb{Z}/6\mathbb{Z}$ if and only if there are integers α, β, δ such that $\frac{\alpha}{\beta} \notin \{-2, -1, -\frac{1}{2}, 0, 1\}$ and

$$3k^2 - 4k + 1 = \delta^2(\alpha^4 + 2\alpha^3\beta), \quad 3k^2 + 4k + 1 = \delta^2(2\alpha\beta^3 + \beta^4).$$

By summing these two equations, we get

$$6k^2 + 2 = \delta^2((\alpha^2 + \alpha\beta + \beta^2)^2 - 3\alpha^2\beta^2), \tag{3.73}$$

which is impossible since the left-hand side of (3.73) is $\equiv 2 \pmod 3$, while the right-hand side is $\equiv 0$ or $1 \pmod 3$. $\square$

3.9   Elliptic Curves Induced by Diophantine Triples over Quadratic Fields

In this section, we study torsion groups of elliptic curves induced by Diophantine triples over quadratic fields, i.e. we will suppose that a, b, c are elements of some quadratic field $\mathbb{K} = \mathbb{Q}(\sqrt{d}) = \{\alpha + \beta\sqrt{d} : \alpha, \beta \in \mathbb{Q}\}$, where $d \neq 1$ is a square-free integer, and $ab + 1$, $ac + 1$, and $bc + 1$ are squares in $\mathbb{K}$. Since we again have three 2-torsion points, according to [226, 232] (see (2.41)), possible torsion groups for such curves which do not appear over $\mathbb{Q}$ are $\mathbb{Z}/2\mathbb{Z} \times \mathbb{Z}/10\mathbb{Z}$, $\mathbb{Z}/2\mathbb{Z} \times \mathbb{Z}/12\mathbb{Z}$, and $\mathbb{Z}/4\mathbb{Z} \times \mathbb{Z}/4\mathbb{Z}$. The last one can only appear over $\mathbb{Q}(i)$. In 2017, Dujella, Jukić Bokun, and Soldo [132] showed that all these torsion groups indeed appear over some quadratic fields and that there are infinitely many Diophantine triples over quadratic fields which induce elliptic curves with these torsion groups. They also presented examples of elliptic curves over quadratic fields with reasonably large ranks. Here we will sketch the main results of that paper.

As before, we consider the curve

$$E : \quad y^2 = (x + ab)(x + bc)(x + ac) \tag{3.74}$$

with three rational points of order 2: $A = (-bc, 0)$, $B = (-ac, 0)$, $C = (-ab, 0)$, and other two obvious rational points $P = (0, abc)$ and $S = (1, rst)$, where $ab + 1 = r^2$, $ac + 1 = s^2$, $bc + 1 = t^2$.

To obtain the torsion group $\mathbb{Z}/2\mathbb{Z} \times \mathbb{Z}/10\mathbb{Z}$, we will construct a triple such that the point $P = (0, abc)$ is of order 5. Triples are sought among regular triples, e.g. triples of the form $\{a, b, c\}$, where

$$b = \frac{r^2 - 1}{a}, \quad c = a + b + 2r.$$

We have $x(2P) = 1$ (since $2P = -S$ for regular triples) and

$$x(3P) = \frac{1}{a^2}4r(r - 1)(1 + r)(ra + r^2 - 1)(a + r + 1)(a + r)(a + r - 1).$$

The condition $5P = O$ can be written in the form $x(3P) - x(2P) = 0$, i.e.

$$(-4r^2 + 4r^4)a^4 + (4r - 20r^3 + 16r^5)a^3 + (-1 + 16r^2 - 40r^4 + 24r^6)a^2$$
$$+ (-4r + 24r^3 - 36r^5 + 16r^7)a - 4r^2 + 12r^4 - 12r^6 + 4r^8 = 0.$$
$$(3.75)$$

If $r^2 - 1$ is a perfect square, i.e. $r = \frac{w^2+1}{2w}$, then the polynomial on the left-hand side of (3.75), after multiplying by $64t^8$, factorizes as the product of two quadratic factors in a:

$$((4w^6 - 4w^2)a^2 + (4w^7 + 4w^5 + 4w^3 - 4t)a + w^8 - 2w^6 + 2w^2 - 1)$$
$$\times ((4w^6 - 4w^2)a^2 + (4w^7 - 4w^5 - 4w^3 - 4w)a + w^8 - 2w^6 + 2w^2 - 1).$$

The roots of the factors are elements of the quadratic fields $\mathbb{Q}(\sqrt{w^4 + w^2 - 1})$, respectively $\mathbb{Q}(\sqrt{-w^4 + w^2 + 1})$. Note that $w^4 + w^2 - 1$ and $-w^4 + w^2 + 1$ are not rational squares, since both conditions $w^4 + w^2 - 1 = z^2$ and $-w^4 + w^2 + 1 = z^2$ are birationally equivalent to the elliptic curve

$$E_1 : \quad y^2 = x^3 + x^2 + 4x + 4,$$

with the torsion group $\mathbb{Z}/6\mathbb{Z}$ and the rank equal to 0 over $\mathbb{Q}$ (and the torsion points correspond to $w = \pm 1$ for the first curve and $w = 0, \pm 1$ for the second curve, which are excluded values since $w = \pm 1$ gives $r = \pm 1$).

Let $u \in \mathbb{Q} \setminus \{-1, -\frac{2}{3}, 0, 1\}$. It is easy to check that the elliptic curve E_1 over the quadratic field $\mathbb{Q}(v)$, $v = \sqrt{(2u^2 + 2u + 1)/(1 - u^2)}$, contains the point

$$P_1 = \left(\frac{-6u - 4}{u - 1}, \frac{10v(u + 1)}{u - 1} \right).$$

It can be shown (see [132] for details) that for all but finitely many rationals u, the point P_1 is a point of infinite order on the curve E_1 over $\mathbb{Q}(v)$. Every multiple mP_1 generates a Diophantine triple such that the induced elliptic curve E over $\mathbb{Q}(v)$ has torsion group $\mathbb{Z}/2\mathbb{Z} \times \mathbb{Z}/10\mathbb{Z}$.

We conclude that there exist infinitely many quadratic fields $\mathbb{K}$ such that for each of them, there exist infinitely many Diophantine triples which induce elliptic curves with torsion group $\mathbb{Z}/2\mathbb{Z} \times \mathbb{Z}/10\mathbb{Z}$ over $\mathbb{K}$.

We may try to find a curve within the considered family with a reasonably large rank. Since our curves have rational coefficients, to determine their rank over a quadratic field $\mathbb{Q}(\sqrt{d})$ we can use the formula

$$\operatorname{rank}(E(\mathbb{Q}(\sqrt{d}))) = \operatorname{rank}(E(\mathbb{Q})) + \operatorname{rank}(E^{(d)}(\mathbb{Q})),$$

where $E^{(d)}$ denotes the d-quadratic twist of E (see [38]). If E is given by the equation $y^2 = f(x)$, then $E^{(d)}$ has the equation $dy^2 = f(x)$.

By taking $u = 3$, i.e. $v^2 = -\frac{25}{8}$, in the above construction, we get a curve of rank 3 and torsion group $\mathbb{Z}/2\mathbb{Z} \times \mathbb{Z}/10\mathbb{Z}$ over $\mathbb{Q}(\sqrt{-2})$. In this case, $P_1 = (-11, 25\sqrt{-2})$, $w = -\frac{5}{4}\sqrt{-2}$ and $a = \frac{475}{561} + \frac{12737}{22440}\sqrt{-2}$. Hence, the curve is induced by the Diophantine triple

$$\left\{ \frac{475}{561} + \frac{12737}{22440}\sqrt{-2}, \; -\frac{475}{561} + \frac{12737}{22440}\sqrt{-2}, \; \frac{160}{561}\sqrt{-2} \right\}$$

and the equation is

$$y^2 = x^3 - \frac{505764769}{251776800}x^2 + \frac{15937}{14450}x - \frac{2178}{7225}.$$

Let us mention that the current record for the rank for arbitrary elliptic curves over quadratic fields with torsion group $\mathbb{Z}/2\mathbb{Z} \times \mathbb{Z}/10\mathbb{Z}$ is 4 (5 under the parity conjecture, see [5, 46, 354]).

The starting point in constructing elliptic curves with torsion group $\mathbb{Z}/2\mathbb{Z} \times \mathbb{Z}/12\mathbb{Z}$ is the family of Diophantine triples $\{a, b, c\}$ from Sect. 3.2:

$$
\begin{aligned}
a &= \frac{18t(t^2 - 1)}{(t^2 - 6t + 1)(t^2 + 6t + 1)}, \\
b &= \frac{(t - 1)(t^2 + 6t + 1)^2}{6t(t + 1)(t^2 - 6t + 1)}, \\
c &= \frac{(t + 1)(t^2 - 6t + 1)^2}{6t(t - 1)(t^2 + 6t + 1)}.
\end{aligned}
\tag{3.76}
$$

We know that the triples $\{a, b, c\}$, for rationals $t \neq -1, 0, 1$, induce elliptic curves with torsion $\mathbb{Z}/2\mathbb{Z} \times \mathbb{Z}/6\mathbb{Z}$ and positive rank over $\mathbb{Q}$ ($P = (0, abc)$ is a point of infinite order). In order to obtain a point of order 12 on that curve, we consider the point P_6 of order 6 with x-coordinate

$$x(P_6) = \frac{2t^4 + 3t^3 - 14t^2 + 3t + 2}{3t(t^2 - 6t + 1)} \tag{3.77}$$

and search for a point P_{12} such that $2P_{12} = P_6$. According to Theorem 2.4.9, we obtain the conditions

$$(t^2 - 6t + 1)(t^2 + 18t + 1) = u^2, \tag{3.78}$$

$$6t(t^2 + 1) = v^2. \tag{3.79}$$

The condition (3.78) leads to the elliptic curve

$$y^2 = x^3 - x^2 - 225x - 1215, \tag{3.80}$$

which has torsion group $\mathbb{Z}/2\mathbb{Z}$ and rank equal to 1 over $\mathbb{Q}$. A point of infinite order is $P_\infty = (27, -108)$.

For every parameter t which is generated by multiples mP_∞ ($m \geq 2$), from (3.79), we get an elliptic curve over the quadratic field $\mathbb{Q}(\sqrt{6t(1+t^2)})$. Therefore, over this quadratic field the Diophantine triple (3.76) induces an elliptic curve with torsion group $\mathbb{Z}/2\mathbb{Z} \times \mathbb{Z}/12\mathbb{Z}$.

We conclude that there exist infinitely many quadratic fields $\mathbb{K}$ such that for each of them, there exist infinitely many Diophantine triples which induce elliptic curves with torsion $\mathbb{Z}/2\mathbb{Z} \times \mathbb{Z}/12\mathbb{Z}$ over $\mathbb{K}$.

By taking multiples $2P_\infty$ and $3P_\infty$, we get $t = 6/35$ and $t = 41615/426$, and in both cases, we obtain curves of rank between 1 and 3. For larger multiples, the coefficients of the curves are too large, and we are not able to compute reasonable bounds for the rank. For $t = 41615/426$, we can conclude that the rank of the corresponding curve over $\mathbb{Q}(\sqrt{5117449349905165})$ is equal to 3, assuming the parity conjecture. For constructing a curve with (unconditional) rank 2, using different parametrization of triples, see [132]. The current record for the rank for arbitrary elliptic curves over quadratic fields with torsion group $\mathbb{Z}/2\mathbb{Z} \times \mathbb{Z}/12\mathbb{Z}$ is 5 (see [354]).

Finally, we consider the torsion group $\mathbb{Z}/4\mathbb{Z} \times \mathbb{Z}/4\mathbb{Z}$ for which we know that it can appear only over $\mathbb{Q}(i)$, and we show that there exist infinitely many Diophantine triples which induce elliptic curves with torsion $\mathbb{Z}/4\mathbb{Z} \times \mathbb{Z}/4\mathbb{Z}$ over $\mathbb{Q}(i)$. Moreover, there exists a Diophantine triple which induces an elliptic curve with torsion $\mathbb{Z}/4\mathbb{Z} \times \mathbb{Z}/4\mathbb{Z}$ and positive rank over $\mathbb{Q}(i)(t)$.

Our starting point is the two-parameter family of Diophantine triples $\{a, b, c\}$, where

$$a = \frac{tu + 1}{t - u},$$

$$b = -\frac{1}{a}, \tag{3.81}$$

$$c = \frac{4tu}{(tu + 1)(t - u)},$$

for which we saw in Sect. 3.6.3 that they induce elliptic curves with torsion group $\mathbb{Z}/2\mathbb{Z} \times \mathbb{Z}/4\mathbb{Z}$ over $\mathbb{Q}$, for admissible rationals t, u.

In order to get the torsion group $\mathbb{Z}/4\mathbb{Z} \times \mathbb{Z}/4\mathbb{Z}$, we apply Theorem 2.4.9 to the point $B = (-ac, 0)$ of order 2 to get a point B_4 of order 4 such that $2B_4 = B$. By noticing that -1 is a square in $\mathbb{Q}(i)$, we obtain the condition

$$tu(u^2 + 1)(t^2 + 1) = n^2. \tag{3.82}$$

From (3.82), by substitutions $U = (t^3 + t)u$, $N = (t^3 + t)n$, we get the elliptic curve

$$N^2 = U^3 + (t^3 + t)^2 U,$$

with the point $P_t = (t^2+1, (t^2+1)^2)$. The point P_t does not generate an appropriate Diophantine triple, but using its multiple $2P_t$, we obtain

$$u = \frac{(t^2 - 1)^2}{4t(t^2 + 1)}. \tag{3.83}$$

By inserting (3.83) into (3.81), we obtain the parametric family of Diophantine triples $\{a, b, c\}$:

$$a = \frac{t(t^4 + 2t^2 + 5)}{3t^4 + 6t^2 - 1},$$

$$b = \frac{-3t^4 - 6t^2 + 1}{t(t^4 + 2t^2 + 5)},$$

$$c = \frac{16t(t - 1)^2(t + 1)^2(t^2 + 1)}{(t^4 + 2t^2 + 5)(3t^4 + 6t^2 - 1)},$$

which induces an elliptic curve with torsion $\mathbb{Z}/4\mathbb{Z} \times \mathbb{Z}/4\mathbb{Z}$ over $\mathbb{Q}(i)(t)$. Moreover, by taking an appropriate specialization (e.g. $t = 2$), we can check that the point $P = (0, abc)$ is a point of infinite order.

By inserting $t = 4/3$, we obtain a curve of rank 6 and torsion group $\mathbb{Z}/4\mathbb{Z} \times \mathbb{Z}/4\mathbb{Z}$ over $\mathbb{Q}(i)$. The curve is induced by the Diophantine triple

$$\left\{ \frac{3796}{4653}, \; -\frac{4653}{3796}, \; \frac{78400}{490633} \right\}$$

and has the equation

$$y^2 = x^3 - \frac{2308406209001}{2166486666201} x^2 + \frac{86600502800}{2166486666201} + \frac{6146560000}{240720740689}.$$

The current record for the rank over $\mathbb{Q}(i)$ for arbitrary elliptic curves with torsion group $\mathbb{Z}/4\mathbb{Z} \times \mathbb{Z}/4\mathbb{Z}$ is 8 [304]. The current record ranks for all 26 possible torsion groups over quadratic fields can be found on the web page [119].

3.10 Elliptic Curves Induced by Rational Diophantine Quadruples

In previous sections, we considered elliptic curves induced by rational Diophantine triples. In particular, in Sect. 3.6, we showed which torsion groups over $\mathbb{Q}$ can appear for such elliptic curves. In this section, we will consider similar questions but for elliptic curves induced by rational Diophantine quadruples. Such curves were considered in 2000 by Dujella [102] and in 2022 by Dujella and Soydan [159].

Let $\{a, b, c, d\}$ be a rational Diophantine quadruple. To extend it to a rational Diophantine quintuple, we need to find a rational number X such that $aX + 1$, $bX + 1$, $cX + 1$ and $dX + 1$ are rational squares. As in the case of the extension of a triple to quadruples, we multiply these four conditions. We obtain the equation

$$Y^2 = (aX + 1)(bX + 1)(cX + 1)(dX + 1),$$

which defines a genus 1 curve. By the substitution

$$y = \frac{Y(d - a)(d - b)(d - c)}{(dX + 1)^2}, \qquad x = \frac{(aX + 1)(d - b)(d - c)}{dX + 1},$$

we obtain the following elliptic curve

$$E : \quad y^2 = x(x + (b - a)(d - c))(x + (c - a)(d - b)). \tag{3.84}$$

We say that this elliptic curve is *induced by the rational Diophantine quadruple* $\{a, b, c, d\}$.

We consider the question of which torsion groups are possible for elliptic curves induced by rational Diophantine quadruples. We will show that, as in the case of elliptic curves induced by rational Diophantine triples, all four torsion groups that are allowed by Mazur's theorem, i.e. $\mathbb{Z}/2\mathbb{Z} \times \mathbb{Z}/k\mathbb{Z}$ for $k = 2, 4, 6, 8$, are possible, and in fact, they can be achieved for infinitely many rational Diophantine quadruples. We will present curves with moderately large rank in each of the four cases.

There are three non-trivial rational 2-torsion points on E:

$$A = (0, 0), \quad B = (-(b - a)(d - c), 0), \quad C = (-(c - a)(d - b), 0),$$

and another two obvious rational points:

$$P = ((b - a)(c - a), \; (b - a)(c - a)(d - a)),$$

$$Q = ((ad + 1)(bc + 1), \; \sqrt{(ab + 1)(ac + 1)(ad + 1)(bc + 1)(bd + 1)(cd + 1)}).$$

The points P and Q will play an important role in our constructions. In particular, we will be interested in the question under which assumptions these points may have finite order.

Proposition 3.10.1 *A rational Diophantine quadruple* $\{a, b, c, d\}$ *is regular if and only if* $2P = \pm Q$.

Proof The first coordinate of $2P$ is $\frac{1}{4}(a - b - c + d)^2$. We have seen in Sect. 1.3 that the regularity condition can be written in the form

$$(a - b - c + d)^2 = 4(ad + 1)(bc + 1), \tag{3.85}$$

so it is equivalent to $x(2P) = x(Q)$. $\qquad\qquad\square$

Assume that the rational Diophantine quadruple $\{a, b, c, d\}$ can be extended to a rational Diophantine quintuple $\{a, b, c, d, e\}$. Then there is another rational point on E:

$$Z = \left(\frac{(de + 1)(b - a)(c - a)}{ae + 1}, \ \frac{(b - a)(c - a)(d - a)}{ae + 1} \sqrt{\frac{(be + 1)(ce + 1)(de + 1)}{ae + 1}} \right).$$

Proposition 3.10.2 *The Diophantine quintuple* $\{a, b, c, d, e\}$ *is regular if and only if* $Z \pm P = \pm Q$.

Proof A straightforward computation shows that the condition that the first coordinate of the point Q is equal to the first coordinate of $Z + P$ or $Z - P$ is equivalent to the regularity condition (3.9) written in terms of a, b, c, d, e. $\qquad\square$

Propositions 3.10.1 and 3.10.2 suggest that if an irregular Diophantine quadruple $\{a, b, c, d\}$ is contained in an irregular quintuple $\{a, b, c, d, e\}$, then we may expect that the rank of the induced curve is ≥ 3. This idea was applied in [102], were subquadruples of rational Diophantine sextuples found by Gibbs [195] were analyzed, and several elliptic curves of rank 8 were found. In [159], the same method applied to new examples of sextuples given in [196] produced several curves of rank 9. One of these curves corresponds to the following quadruple:

$$\{a, b, c, d\} = \left\{ \frac{1218560}{31752611}, \ \frac{111}{77}, \ \frac{34191}{19712}, \ \frac{1155}{16} \right\}.$$

To find curves of rank 10, in [159], the authors considered quadruples which contain two regular triples. We already considered such quadruples in Sect. 3.6.1. If we put $n = t$, $a = at$ in (3.19), we obtain a two-parameter family of rational Diophantine quadruples $\{a, b, c, d\}$, where

$$b = \frac{(t^2 + 2at + 4t + 3)(t^2 + 2at - 4t + 3)}{16t^2 a},$$

$$c = \frac{(t - 1)(t + 3)(t - 3)(t + 1)}{4t^2 a},$$

$$d = \frac{(t^2 - 2at - 4t + 3)(t^2 - 2at + 4t + 3)}{16t^2 a},$$

and the corresponding family of elliptic curves, which is equivalent to

$$y^2 = x^3 + A_1 x^2 + B_1 x,$$

where

$$A_1 = 6t^8 - 48t^6 a^2 + 96t^4 a^4 - 120t^6 + 992t^4 a^2$$

$$+ 708t^4 - 432t^2 a^2 - 1080t^2 + 486,$$

$$B_1 = (t^2 + 2at - 1)(t^2 - 6at - 9)(3t^2 + 2at - 3)(t^2 - 2at - 9)$$

$$\times (t^2 + 6at - 9)(t^2 - 2at - 1)(t^2 + 2at - 9)(3t^2 - 2at - 3).$$

To further increase the likelihood of finding curves with large rank, we can force $a^2 + 1$ to be a perfect square, since then four quadratic factors appearing in the factorization of B_1 have square discriminant with respect to t, and thus factorize into two linear factors. This increases the upper bound for the rank (2.33) and indicates that the rank indeed could be larger. Thus we take $a = \frac{v^2-1}{2v}$, and search for high-rank curves with reasonably small numerators and denominators of the parameters t and v. For $t = \frac{142}{53}$, $v = \frac{142}{23}$, we obtain a curve of rank 10, corresponding to the quadruple

$$\{a, b, c, d\} = \left\{ \frac{19635}{6532}, \frac{84196064}{50458067}, -\frac{1144273}{8775316}, -\frac{46592463}{201832268} \right\}.$$

The above examples all have torsion group $\mathbb{Z}/2\mathbb{Z} \times \mathbb{Z}/2\mathbb{Z}$. Now we will consider how the torsion group can be enlarged and all possible torsion groups allowed by Mazur's theorem constructed. We start with the torsion group $\mathbb{Z}/2\mathbb{Z} \times \mathbb{Z}/4\mathbb{Z}$.

Consider the curve

$$E : \quad y^2 = x(x + p_1)(x + p_2),$$

where $p_1 = (b - a)(d - c)$, $p_2 = (c - a)(d - b)$. By Theorem 2.4.9, the point Q is in $2E(\mathbb{Q})$, because $x_1 = (ad + 1)(bc + 1)$, $x_1 + p_1 = (ac + 1)(bd + 1)$ and $x_1 + p_2 = (ab + 1)(cd + 1)$ are perfect squares. The point Q will be of order 2 if $ad + 1 = 0$, i.e. if $d = -1/a$. In that case, the point R such that $2R = Q$ will be of order 4, and $E(\mathbb{Q})$ will have a subgroup $\mathbb{Z}/2\mathbb{Z} \times \mathbb{Z}/4\mathbb{Z}$.

Thus we need to find rational Diophantine quadruples which contain a subtriple of the form $\{a, -1/a, b\}$. However, this was already done in Sect. 3.6.3. We have

$$a = \frac{ut + 1}{t - u},$$

$$b = \frac{4tu}{(tu + 1)(t - u)}.$$

It remains to find the fourth element c of the quadruple. We may take c such that $\{a, b, c, d\}$ is a regular quadruple,

$$(a + b - c - d)^2 = 4(ab + 1)(cd + 1).$$

In that way, we get

$$c = \frac{(u - 1)(u + 1)(t - 1)(t + 1)}{(ut + 1)(t - u)}.$$

Hence, we showed that there are infinitely many rational Diophantine quadruples which induce curves with torsion group $\mathbb{Z}/2\mathbb{Z} \times \mathbb{Z}/4\mathbb{Z}$.

Other possibility is to take c to be $\frac{8(d-a-b)(a+d-b)(b+d-a)}{(a^2+b^2+d^2-2ab-2ad-2bd)^2}$ (see Exercise 9 in Sect. 1.6), so we get

$$c = \frac{8(t-u)(ut+1)(-4ut+t^2+u^2+u^2t^2+1)(u-1)(u+1)(t-1)(t+1)(t^2+4ut+u^2+u^2t^2+1)}{(1 - 8ut - 12u^2t^2 + 2u^2 + 2t^2 - 8u^3t^3 + u^4t^4 + 8ut^3 + t^4 + u^4 + 8tu^3 + 2u^2t^4 + 2u^4t^2)^2}.$$

Within this two-parameter family of quadruples, we can find examples of curves of rank equal to 6, e.g. for $(t, u) = (3, 1/12)$, corresponding to the quadruple

$$\left\{ \frac{3}{7}, \frac{48}{175}, -\frac{625153729200}{363378690481}, -\frac{7}{3} \right\}.$$

To achieve the torsion group $\mathbb{Z}/2\mathbb{Z} \times \mathbb{Z}/6\mathbb{Z}$, we will find quadruples which satisfy the condition $3Q = O$, since then the point R such that $2R = Q$ will be a point of order 6. To simplify the condition $3Q = O$, we assume that $\{a, b, c, d\}$ is a regular quadruple. Then we may take $R = \pm P$ by Proposition 3.10.1. There are several known parametrizations of Diophantine triples $\{a, b, c\}$. We will use parametrization (3.7), followed by the substitutions (which were for the first time used in [157]):

$$t_1 = \frac{k}{t_2 t_3}, \quad t_2 = m - \frac{1}{t_3}.$$

From the regularity equation (3.85), we get

$$d = \frac{-2(1 - t_1 + t_3 t_1)(-t_3 + t_2 t_3 + 1)(-t_2 + 1 + t_1 t_2)(-1 + t_1 t_2 t_3)}{(1 + t_1 t_2 t_3)^3}.$$

Now the condition $3Q = O$ becomes a complicated algebraic equation in terms of t_3, k and m. Motivated by experimentally found numerical solutions to this equation, we take

$$t_3 = -\frac{2k^2 + 1}{k(k^2 + 2)}.$$

By inserting this in $3Q = O$, and factorizing the obtained expression, we get a factor

$$4k^4 m - 4k^3 m + 6k^2 m - 2km + 2m - 3k^4 + 6k^3 - 3k^2 + 6k,$$

and thus we can express m in terms of k as

$$m = \frac{3k(k^3 - 2k^2 + k - 2)}{2(2k^2 + 1)(k^2 - k + 1)}.$$

Finally, we can express a, b, c, d in terms of k:

$$a = \frac{-2k(k^2 + 2)(3k^3 - 2k^2 + 2k - 2)}{(k + 1)(k - 1)(2k^2 + 1)(2k^2 + k + 2)},$$

$$b = \frac{-k(k + 1)(k - 1)(2k^2 + k + 2)(4k^2 - k + 4)}{2(k^2 + 2)(k^2 - k + 1)^2(2k^2 + 1)},$$

$$c = \frac{2(2k^2 + 1)(2k^3 - 2k^2 + 2k - 3)}{(k + 1)(k - 1)(2k^2 + k + 2)(k^2 + 2)},$$

$$d = \frac{(2k^2 + 1)(k + 1)(k^2 + 2)(k - 1)}{2(k^2 - k + 1)^2(2k^2 + k + 2)}.$$

Hence, we obtained an infinite family of quadruples, which induces curves with torsion group $\mathbb{Z}/2\mathbb{Z} \times \mathbb{Z}/6\mathbb{Z}$.

Within this family, we can find a curve of rank equal to 3 for $k = 23$, corresponding to the quadruple

$$\left\{ -\frac{16051953}{11214104}, \; -\frac{170244712}{1784519841}, \; \frac{914623}{5622936}, \; \frac{5498328}{10310521} \right\}.$$

Finally, we would like to obtain the torsion group $\mathbb{Z}/2\mathbb{Z} \times \mathbb{Z}/8\mathbb{Z}$. Let us start with the quadruple $\{a, b, c, d\}$, where

$$a = \frac{ut + 1}{t - u},$$

$$b = \frac{4ut}{(ut + 1)(t - u)},$$

$$c = \frac{(u - 1)(u + 1)(t - 1)(t + 1)}{(ut + 1)(t - u)},$$

$$d = -\frac{t - u}{ut + 1},$$

for which we know that the corresponding curve has the torsion group which contains $\mathbb{Z}/2\mathbb{Z} \times \mathbb{Z}/4\mathbb{Z}$. The point $Q = (0, 0)$ of order 2 is in $2E(\mathbb{Q})$, so there is a point R such that $2R = Q$. To get the torsion group $\mathbb{Z}/2\mathbb{Z} \times \mathbb{Z}/8\mathbb{Z}$, we have to force the point R to be in $2E(\mathbb{Q})$, i.e. $R = 2S$, for a point S in $E(\mathbb{Q})$. Then the point S will be of order 8. The point R has the following coordinates:

$$\left(\frac{(u+t)^2(ut-1)^2}{(ut+1)^2(t-u)^2}, \ \frac{(u^2+1)(t^2+1)(u+t)^2(ut-1)^2}{(ut+1)^3(t-u)^3} \right).$$

We want that for $x_1 = \frac{(u+t)^2(ut-1)^2}{(ut+1)^2(t-u)^2}$, the numbers x_1, $x_1 + p_1$ and $x_1 + p_2$ are squares. However, x_1 is already a square, while $x_1 + p_1 = \square$ and $x_1 + p_2 = \square$ both lead to the same condition: $(u^2+1)(t^2+1)$ is a square. By putting

$$(u^2+1)(t^2+1) = (u^2+1+(t-u)v)^2,$$

we get

$$t = -\frac{-2u^2v - 2v + v^2u + u^3 + u}{u^2 + 1 - v^2},$$

and finally,

$$a = \frac{(u-v+1)(u-v-1)}{2(u-v)},$$

$$b = -\frac{2(u^2+1-v^2)u(-2u^2v - 2v + v^2u + u^3 + u)}{(u^2+1)^2(u-v)(u-v+1)(u-v-1)},$$

$$c = \frac{(-2u^2v-2v+v^2u+u^3+u-u^2-1+v^2)(-2u^2v-2v+v^2u+u^3+u+u^2+1-v^2)(u+1)(u-1)}{2(u^2+1)^2(u-v)(u-v+1)(u-v-1)},$$

$$d = -\frac{2(u-v)}{(u-v+1)(u-v-1)}.$$

The general form of an elliptic curve with torsion group $\mathbb{Z}/2\mathbb{Z} \times \mathbb{Z}/8\mathbb{Z}$ is

$$y^2 = x\left(x + \left(\frac{2T}{T^2-1}\right)^2\right)\left(x + \left(\frac{T^2-1}{2T}\right)^2\right).$$

By comparing the j-invariant of this curve with that of our curve E, we find that the j-invariants coincide for $T = \frac{v}{vu-u^2-1}$. This indicates that every curve with torsion group $\mathbb{Z}/2\mathbb{Z} \times \mathbb{Z}/8\mathbb{Z}$ can be obtained from a rational Diophantine quadruple.

For example, the quadruple

$$\left\{ \frac{1804}{1197}, \ -\frac{226796}{539847}, \ \frac{303199}{239932}, \ -\frac{1197}{1804} \right\},$$

obtained for $(u, v) = (2, -25/19)$, induces a curve with torsion $\mathbb{Z}/2\mathbb{Z} \times \mathbb{Z}/8\mathbb{Z}$ and rank 3 (equivalent to the curve found by Connell and Dujella in 2000, see Sect. 3.6.5), which is the largest known rank for curves with this torsion group.

3.11 Exercises

1. Find a rational number $e \neq 0$ with the property that $\left\{\frac{1}{16}, \frac{33}{16}, \frac{17}{4}, \frac{105}{16}, e\right\}$ is a rational Diophantine quintuple.
2. Let $\{x_1, x_2, x_3, x_4\}$ be a regular rational Diophantine quadruple. Show that one of the values of x_5 in Theorem 3.1.3 is equal to 0, while the other is equal to

$$\frac{4 y_{12} y_{13} y_{14} y_{23} y_{24} y_{34}}{(x_1 x_2 x_3 x_4 - 1)^2}.$$

3. Let $\{x_1, x_2, x_3, x_4\}$ be an (integer) Diophantine quadruple. Show that x_5 in Theorem 3.1.3 is less than 1 (and therefore it is not a positive integer).
4. Find a rational Diophantine quadruple $\{x_1, x_2, x_3, x_4\}$ such that both values of x_5 in Theorem 3.1.3 are equal to 0.
5. Let $\{a, b, c, d\}$ be a rational Diophantine quadruple such that $abcd = 1$ and let

$$e = \frac{(a + b - c - d)^2 - 4(ab + 1)(cd + 1)}{4d(ab + 1)(ac + 1)(bc + 1)}.$$

   Prove that $\{a, b, c, d, e\}$ is a rational Diophantine quintuple.
6. Let $a = \frac{33}{280}, b = \frac{77}{270}, c = \frac{384}{385}, d = \frac{2625}{88}$. Check that $\{a, b, c, d\}$ is a rational Diophantine quadruple satisfying $abcd = 1$. Find positive rationals e and f such that $\{a, b, c, d, e, f\}$ is a rational Diophantine sextuple (see [196]).
7. Let $\{a, b, c\}$ be a rational Diophantine triple and

$$d = \frac{8(c - a - b)(a + c - b)(b + c - a)}{(a^2 + b^2 + c^2 - 2ab - 2ac - 2bc)^2}.$$

   Write the point with the x-coordinate d in terms of points $P = (0, 1)$ and $S = (\frac{1}{abc}, \frac{rst}{abc})$ on the elliptic curve

$$y^2 = (ax + 1)(bx + 1)(cx + 1).$$

8. Find positive rational numbers t_i, $i = 1, 2, 3, 4, 5, 6$, such that the rational Diophantine sextuple from Sect. 3.2, for $t = t_i$, contains exactly i negative numbers.
9. Find a positive integer $t \neq 31$ with the property that there exists a rational point (x, y) on (3.5) such that the curve E'' defined by (3.6) has no rational points of order 2. Determine the torsion group and rank of $E''(\mathbb{Q})$. Find a prime p such that E'' has additive reduction at p.
10. Show that by taking $v = \frac{P_k - P_{k-1}}{P_{k+1} - P_k}$ in (3.16), we get sextuples with one integer element.
11. Let $T = (4, 10)$ be a free generator and $Z = (0, 0)$ the 2-torsion point on the curve $E : y^2 = x^3 + 9x$. Compute the sextuples that correspond to the points $2T, 2T + Z, 3T$ and $3T + Z$ under the construction from Sect. 3.5.

12. Show that there are infinitely many rational Diophantine sextuples $\{a, b, c, d, e, f\}$ with the property that $a^2 + 1$ is a perfect square.

13. Find a rational number k with the property that the elliptic curve

$$y^2 = ((k+1)x + 1)(4kx + 1)((16k^3 - 4k)x + 1)$$

has rank equal to 9 [354].

14. Let $\{a, b, c\}$ be an (integer) Diophantine triple. Show that the elliptic curve $y^2 = (ax + 1)(bx + 1)(cx + 1)$ has positive rank.

15. Find all $D(16)$-triples $\{a, b, c\}$, such that $0 < a < b$ and $c < 0$. For each such triple, determine the torsion group and rank of the elliptic curve given by the equation $y^2 = (ax + 16)(bx + 16)(cx + 16)$.

16. Prove that the elliptic curve induced by the rational Diophantine triple $\{3/4, -4/3, -7/12\}$ has the torsion group $\mathbb{Z}/2\mathbb{Z} \times \mathbb{Z}/8\mathbb{Z}$.

17. Prove that there are infinitely many rational numbers w with the property that $2w^4 + 4w^3 + 2w^2 + 1$ is a perfect square and all elements of triple (3.44) have the same sign.

18. Let

$$c_k = \frac{1}{6}\left((2 + \sqrt{3})(7 + 4\sqrt{3})^k + (2 - \sqrt{3})(7 - 4\sqrt{3})^k - 4\right)$$

for a positive integer k. Show that $\{3, c_k, c_{k+1}\}$ is a $D(1)$-triple and compute the limit $\dfrac{c_{k+1} - 12c_k}{3 + c_k}$ as k tends to ∞ (compare the result with Remark 3.8.6).

19. Let $\{a, b, a + b + 2r\}$ be a regular Diophantine triple. Assume that at least one of the integers a, b is odd. Prove that the torsion group of the elliptic curve induced by this triple is $\mathbb{Z}/2\mathbb{Z} \times \mathbb{Z}/2\mathbb{Z}$.

20. Find a $D(48)$-triple $\{a, b, c\}$ such that the elliptic curve $y^2 = (ax + 48)(bx + 48)(cx + 48)$ has the torsion group $\mathbb{Z}/2\mathbb{Z} \times \mathbb{Z}/4\mathbb{Z}$.

21. Find a $D(144)$-triple $\{a, b, c\}$ (mixed signs allowed) such that the elliptic curve $y^2 = (ax + 144)(bx + 144)(cx + 144)$ has the torsion group $\mathbb{Z}/2\mathbb{Z} \times \mathbb{Z}/6\mathbb{Z}$.

22. Find a rational number k such that the elliptic curve induced by the rational Diophantine triple $\{k - 1, k + 1, 4k\}$ has a rational point of order 4. Show that there are infinitely many rationals with this property.

23. Show that the elliptic curve induced by the triple $\left\{-\frac{35}{36}, \frac{27}{35}, \frac{161}{180}\right\}$ has the torsion group $\mathbb{Z}/2\mathbb{Z} \times \mathbb{Z}/12\mathbb{Z}$ and rank equal to 2 over $\mathbb{Q}(\sqrt{-155})$.

24. Let

$$E : \quad y^2 + xy = x^3 - 1070x + 7812.$$

(a) Show that E has torsion group $\mathbb{Z}/2\mathbb{Z} \times \mathbb{Z}/8\mathbb{Z}$ and rank 0 over $\mathbb{Q}$.

(b) Find a rational Diophantine triple which induces an elliptic curve isomorphic to E.

(c) Find a rational Diophantine quadruple which induces an elliptic curve isomorphic to E.

Chapter 4
Integer Points on Elliptic Curves

4.1 Preliminaries on Diophantine Equations

The problem of extending a Diophantine pair $\{a, b\}$ to a Diophantine triple $\{a, b, c\}$, by eliminating c from the system of equations $ac + 1 = x^2$ and $bc + 1 = y^2$, leads to the quadratic Diophantine equation $bx^2 - ay^2 = b - a$, a so-called generalized Pellian equation. Thus, in this section, we collect needed preliminaries concerning Pell's and Pellian equations, as well as the continued fractions, which are used in the standard methods for solving such equations and their generalizations. We also give initial information on Baker's theory of linear forms in logarithms of algebraic numbers, which will be an important tool in solving Thue equations and systems of generalized Pellian equations in later sections. These problems are closely related to the problem of extending a Diophantine triple to a quadruple and the problem of finding integer points on the corresponding elliptic curves. We will end this section with some results on the simultaneous Diophantine approximations of algebraic numbers, as this is another useful tool for solving systems of generalized Pellian equations.

4.1.1 Pell's Equation

A Diophantine equation of the form

$$x^2 - dy^2 = 1, \tag{4.1}$$

where d is a positive integer which is not a perfect square, is called *Pell's equation*. The case when d is a perfect square is excluded because in that case equation (4.1) has only trivial solutions $x = \pm 1$, $y = 0$. Indeed, if $d = \delta^2$, then from $(x - \delta y)(x + \delta y) = 1$, it follows that $x - \delta y = x + \delta y = \pm 1$. The

A. Dujella, *Diophantine m-tuples and Elliptic Curves*,
Developments in Mathematics 79, https://doi.org/10.1007/978-3-031-56724-7_4

equation is named after the English mathematician John Pell, to whom Euler, it seems mistakenly, attributed a solution method. Some particular equations of this type can be found in texts of ancient Greek mathematicians (Archimedes—"cattle problem" and Diophantus—although he was mainly interested in rational solutions). The first systematic treatment was made by Indian mathematicians: Brahmagupta in the seventh century and Bhāskara in the twelfth century ("chakravāla"—"the cyclic process"). Among the European mathematicians, methods for solving Pell's equations were given by Brouncker, Fermat, Euler and Lagrange, who was also the first to provide a rigorous proof of the correctness of the proposed method. Since Pell had very little to do with Pell's equation, some authors call it the Brahmagupta-Pell equation to correct this historical misattribution.

The books [11, 27, 223] are devoted to Pell's equations, and we can also recommend the textbooks [52, 116, 292, 323, 358], which provide detailed approaches to this topic. An overview of the history of this equation can be found in [357, Chapter 1.9].

As the first step in studying Pell's equation, we will prove that it has infinitely many solutions in positive integers. We will use the Dirichlet theorem on Diophantine approximations which says that for any irrational number α, there exist infinitely many rational numbers $\frac{p}{q}$ with the property

$$\left| \alpha - \frac{p}{q} \right| < \frac{1}{q^2}. \tag{4.2}$$

The idea of its proof is, for a given positive integer Q, to consider $Q + 1$ numbers

$$0, \ 1, \ \{\alpha\} = \alpha - \lfloor \alpha \rfloor, \ \{2\alpha\}, \ \ldots, \ \{(Q-1)\alpha\}$$

from the segment $[0, 1]$ and subdivision of this segment into Q disjoint subintervals of width $1/Q$, and by the Dirichlet box principle, conclude that there is a subinterval containing at least two of these $Q + 1$ numbers.

Lemma 4.1.1 *Let d be a positive integer which is not a perfect square. Then there exists an integer k, $|k| < 1 + 2\sqrt{d}$, such that the equation*

$$x^2 - dy^2 = k \tag{4.3}$$

has infinitely many solutions in positive integers.

Proof By the mentioned Dirichlet theorem (or properties of continued fractions, see (4.8) below), there are infinitely many pairs of positive integers (x, y) with the property

$$\left| \sqrt{d} - \frac{x}{y} \right| < \frac{1}{y^2}, \quad \text{i.e. } |x - y\sqrt{d}| < \frac{1}{y}.$$

For each such pair (x, y), we have

$$|x + y\sqrt{d}| = |x - y\sqrt{d} + 2y\sqrt{d}| < \frac{1}{y} + 2y\sqrt{d} \le (1 + 2\sqrt{d})y,$$

so that

$$|x^2 - dy^2| = |x - y\sqrt{d}| \cdot |x + y\sqrt{d}| < 1 + 2\sqrt{d}.$$

Since there are infinitely many pairs (x, y) with this property and there are only finitely many integers with absolute value less than $1 + 2\sqrt{d}$, we conclude that there exists an integer k, such that $|k| < 1 + 2\sqrt{d}$, for which Eq. (4.3) has infinitely many solutions. $\qquad\square$

Theorem 4.1.2 *Pell's equation* $x^2 - dy^2 = 1$ *has at least one solution in positive integers* x *and* y.

Proof Infinitely many solutions of Eq. (4.3) can be divided into k^2 classes by putting solutions (x_1, y_1) and (x_2, y_2) in the same class if and only if $x_1 \equiv x_2$ (mod k) and $y_1 \equiv y_2$ (mod k). One of these classes contains at least two (in fact, infinitely many) distinct solutions (x_1, y_1), (x_2, y_2) (x_1, x_2 are distinct positive integers). Put

$$x = \frac{x_1 x_2 - dy_1 y_2}{k}, \quad y = \frac{x_1 y_2 - x_2 y_1}{k}$$

(we divide the solution $x_2 + y_2\sqrt{d}$ by $x_1 + y_1\sqrt{d}$ and rationalize the denominator). We claim that $x, y \in \mathbb{Z}$, $y \ne 0$ and $x^2 - dy^2 = 1$. We have: $x_1 x_2 - dy_1 y_2 \equiv x_1^2 - dy_1^2 = k \equiv 0$ (mod k), $x_1 y_2 - x_2 y_1 \equiv x_1 y_1 - x_1 y_1 \equiv 0$ (mod k), so that $x, y \in \mathbb{Z}$. Assume that $y = 0$, i.e. $x_1 y_2 = x_2 y_1$. Then

$$k = x_2^2 - dy_2^2 = x_2^2 - d \cdot \frac{x_2^2 y_1^2}{x_1^2} = \frac{x_2^2}{x_1^2}(x_1^2 - dy_1^2) = \frac{x_2^2}{x_1^2} \cdot k,$$

i.e. $x_1^2 = x_2^2$, contradicting the assumption that x_1 and x_2 are distinct positive integers. Finally,

$$x^2 - dy^2 = \frac{1}{k^2}\left((x_1 x_2 - dy_1 y_2)^2 - d(x_1 y_2 - x_2 y_1)^2\right)$$

$$= \frac{1}{k^2}(x_1^2 x_2^2 + d^2 y_1^2 y_2^2 - dx_1^2 y_2^2 - dx_2^2 y_1^2)$$

$$= \frac{1}{k^2}(x_1^2 - dy_1^2)(x_2^2 - dy_2^2) = \frac{1}{k^2} \cdot k \cdot k = 1.$$

$\qquad\square$

The smallest solution (x, y) in positive integers of Pell's equation (4.1) is called its *fundamental solution*. It is denoted by (x_1, y_1), and often also as $x_1 + y_1\sqrt{d}$ (note that due to the irrationality of $\sqrt{d}$, the representation in the form $a + b\sqrt{d}, a, b \in \mathbb{Q}$, is unique).

Theorem 4.1.3 *Pell's equation $x^2 - dy^2 = 1$ has infinitely many solutions. If (x_1, y_1) is the fundamental solution, then all solutions of this equation in positive integers are given by the formula*

$$x_n + y_n\sqrt{d} = (x_1 + y_1\sqrt{d})^n, \quad n \in \mathbb{N}, \tag{4.4}$$

i.e.

$$x_n = x_1^n + \binom{n}{2} dx_1^{n-2}y_1^2 + \binom{n}{4} d^2 x_1^{n-4} y_1^4 + \cdots,$$

$$y_n = nx_1^{n-1}y_1 + \binom{n}{3} dx_1^{n-3}y_1^3 + \binom{n}{3} d^2 x_1^{n-5} y_1^5 + \cdots.$$

Proof From (4.4), it follows $x_n - y_n\sqrt{d} = (x_1 - y_1\sqrt{d})^n$, and by multiplication, we obtain

$$x_n^2 - dy_n^2 = (x_1^2 - dy_1^2)^n = 1,$$

which shows that (x_n, y_n) are indeed solutions (and they give infinitely many solutions).

Assume now that (s, t) is a solution not of the form (x_n, y_n), $n \in \mathbb{N}$. Since $x_1 + y_1\sqrt{d} > 1$ and $s + t\sqrt{d} > 1$, we conclude that there exists $m \in \mathbb{N}$ such that

$$(x_1 + y_1\sqrt{d})^m < s + t\sqrt{d} < (x_1 + y_1\sqrt{d})^{m+1}. \tag{4.5}$$

Multiplying (4.5) by $(x_1 + y_1\sqrt{d})^{-m} = (x_1 - y_1\sqrt{d})^m$, we get

$$1 < (s + t\sqrt{d})(x_1 - y_1\sqrt{d})^m < x_1 + y_1\sqrt{d}.$$

Let us define $a, b \in \mathbb{Z}$ by $a + b\sqrt{d} = (s + t\sqrt{d})(x_1 - y_1\sqrt{d})^m$. We have $a^2 - db^2 = (s^2 - dt^2)(x_1^2 - dy_1^2)^m = 1$. From $a + b\sqrt{d} > 1$, it follows $0 < a - b\sqrt{d} < 1$, which implies $a > 0$ and $b > 0$. Hence, (a, b) is a solution in positive integers of the equation $x^2 - dy^2 = 1$ and $a + b\sqrt{d} < x_1 + y_1\sqrt{d}$, which is a contradiction. $\quad\square$

Remark 4.1.4 From (4.4), it easily follows that the sequences (x_n) and (y_n) satisfy the recurrences

$$x_{n+2} = 2x_1 x_{n+1} - x_n, \quad y_{n+2} = 2x_1 y_{n+1} - y_n, \quad n \geq 0,$$

where (x_1, y_1) is the fundamental solution of (4.1), while $(x_0, y_0) = (1, 0)$ is the "trivial solution".

Apart from the equation $x^2 - dy^2 = 1$, the equations $x^2 - dy^2 = -1, \pm 4$ are also often called Pell's equations. The basic properties of these equations and the connections of all four equations with the problem of finding the units (invertible elements in the corresponding ring of integers) in real quadratic fields $\mathbb{Q}(\sqrt{d})$ can be found in [116, Chapters 10.3 and 12.1].

4.1.2   Continued Fractions

Theorem 4.1.3 and Remark 4.1.4 show us how we can generate all solutions of Pell's equation $x^2 - dy^2 = 1$ if we know its fundamental (smallest) solution. Thus, there remains a problem of how to find the fundamental solution. In some cases, we can find it by simply inserting $y = 1, 2, 3, \ldots$ and checking whether $dy^2 + 1$ is a perfect square. However, even for relatively small values of d, the fundamental solution can be large, e.g. for $d = 94$, the fundamental solution is $2143295 + 221064\sqrt{94}$. Therefore, we need a more efficient algorithm for finding fundamental solutions.

A reasonably efficient algorithm follows from connections between Pell's equation and Diophantine approximation, in particular with continued fractions. Indeed, any non-trivial solution of the equation $x^2 - dy^2 = 1$ induces very good rational approximation of the irrational number $\sqrt{d}$:

$$\left| \sqrt{d} - \frac{x}{y} \right| - \frac{1}{y|x + y\sqrt{d}|} < \frac{1}{2\sqrt{d}y^2}. \tag{4.6}$$

It is known that all very good rational approximations of a real number can be obtained from its continued fraction expansion.

Thus, in this subsection, we will briefly mention basic definitions and results concerning continued fractions and their connections with rational approximations of real numbers, in particular of quadratic irrationals. We will omit most of the proofs. They can be found, e.g. in [233, 303, 309], [317, Chapter 1], and [116, Chapter 8].

Let $\alpha \in \mathbb{R}$. An expression of the form

$$\alpha = a_0 + \cfrac{1}{a_1 + \cfrac{1}{a_2 + \ddots}},$$

where $a_0 \in \mathbb{Z}$, and $a_1, a_2, \ldots \in \mathbb{N}$, is called *a (simple) continued fraction expansion* of α. We will often denote this continued fraction as $[a_0; a_1, a_2, \ldots]$. The numbers

$a_0, a_1, a_2, \ldots$ are called *partial quotients* and can be defined as follows:

$$a_0 = \lfloor \alpha \rfloor, \quad \alpha = a_0 + \frac{1}{\alpha_1}, \quad a_1 = \lfloor \alpha_1 \rfloor, \quad \alpha_1 = a_1 + \frac{1}{\alpha_2}, \quad a_2 = \lfloor \alpha_2 \rfloor, \ldots.$$

We continue the procedure as long as $a_k \neq \alpha_k$. The continued fraction expansion of α is finite if and only if α is a rational number. If $\alpha = \frac{m}{n}$, the numbers $a_0, a_1, a_2, \ldots$ are in fact the quotients from the Euclidean algorithm applied to the integers m and n.

The rational numbers

$$\frac{p_k}{q_k} = a_0 + \cfrac{1}{a_1 + \cfrac{1}{a_2 + \cfrac{\ddots}{+ \cfrac{1}{a_k}}}} = [a_0; a_1, \ldots, a_k]$$

are called *convergents of the continued fraction expansion*. The numerators and denominators of the convergents satisfy the following recurrences:

$$p_{n+2} = a_{n+2}p_{n+1} + p_n, \quad p_0 = a_0, \quad p_1 = a_0 a_1 + 1, \quad (p_{-1} = 1, \; p_{-2} = 0),$$

$$q_{n+2} = a_{n+2}q_{n+1} + q_n, \quad q_0 = 1, \quad q_1 = a_1, \quad (q_{-1} = 0, \; q_{-2} = 1).$$

The following important relation which connects convergents with consecutive indices can be proved easily by induction:

$$q_n p_{n-1} - p_n q_{n-1} = (-1)^n. \tag{4.7}$$

Relation (4.7) implies that $\frac{p_{2k}}{q_{2k}} \leq \alpha$ and $\alpha \leq \frac{p_{2k+1}}{q_{2k+1}}$ for all k. Furthermore, it can be shown that the following inequalities hold:

$$\frac{p_0}{q_0} < \frac{p_2}{q_2} < \frac{p_4}{q_4} < \cdots \leq \alpha \leq \cdots < \frac{p_5}{q_5} < \frac{p_3}{q_3} < \frac{p_1}{q_1}.$$

If α is an irrational number, then $\lim_{n \to \infty} \frac{p_n}{q_n} = \alpha$.

We may ask how well the convergents approximate the number α. The answer is given by the following inequalities:

$$\frac{1}{q_n(q_n + q_{n+1})} < \left| \alpha - \frac{p_n}{q_n} \right| < \frac{1}{q_n^2}. \tag{4.8}$$

Some sort of a converse of this fact also holds (Legendre's theorem): if $\frac{p}{q}$ is a rational number which satisfies the inequality $\left| \alpha - \frac{p}{q} \right| < \frac{1}{2q^2}$, then $\frac{p}{q}$ is a convergent of α. We will prove the following more general statement due to Worley [359] (see also [109]).

Theorem 4.1.5 *Let α be a real number and c a positive real number. If a rational number $\frac{p}{q}$ satisfies the inequality*

$$\left| \alpha - \frac{p}{q} \right| < \frac{c}{q^2}, \tag{4.9}$$

then

$$\frac{p}{q} = \frac{r p_{k+1} \pm s p_k}{r q_{k+1} \pm s q_k},$$

for some $k \geq -1$ and non-negative integers r, s such that $rs < 2c$ (and some choice of the sign).

Proof We will assume that $\alpha < \frac{p}{q}$. In the case $\alpha > \frac{p}{q}$, the proof is analogous. We will also assume that α is irrational (for α rational, a small modification of the proof is needed). Let k be the largest odd integer such that

$$\alpha < \frac{p}{q} \leq \frac{p_k}{q_k}.$$

(If $\frac{p}{q} > \frac{p_1}{q_1}$, then we take $k = -1$.) Let us define numbers r and s by

$$p = r p_{k+1} + s p_k,$$
$$q = r q_{k+1} + s q_k.$$

By formula (4.7), the determinant of this system is ± 1, so r, s are integers, and since $\frac{p_{k+1}}{q_{k+1}} < \frac{p}{q} \leq \frac{p_k}{q_k}$, we have $r \geq 0$ and $s > 0$.

By the maximality of k, we have

$$\left| \frac{p_{k+2}}{q_{k+2}} - \frac{p}{q} \right| < \left| \alpha - \frac{p}{q} \right| < \frac{c}{q^2}.$$

Furthermore,

$$\left| \frac{p_{k+2}}{q_{k+2}} - \frac{p}{q} \right| = \frac{(a_{k+2} q_{k+1} + q_k)(r p_{k+1} + s p_k) - (a_{k+2} p_{k+1} + p_k)(r q_{k+1} + s q_k)}{q q_{k+2}}$$

$$= \frac{s a_{k+2} - r}{q q_{k+2}}.$$

Hence,

$$q(sa_{k+2} - r) < cq_{k+2} = \frac{c}{s}((sa_{k+2} - r)q_{k+1} + q),$$

i.e.

$$(sa_{k+2} - r)(q - \frac{c}{s}q_{k+1}) < \frac{c}{s}q.$$

Thus, we have

$$\frac{1}{sa_{k+2} - r} > \frac{q - \frac{c}{s}q_{k+1}}{\frac{c}{s}q} = \frac{s}{c} - \frac{1}{r + \frac{sq_k}{q_{k+1}}} \geq \frac{s}{c} - \frac{1}{r}.$$

Therefore, we obtained the following quadratic inequality for r:

$$r^2 - sra_{k+2} + ca_{k+2} > 0. \qquad (4.10)$$

We now distinguish two cases:

1) $s^2 a_{k+2} \geq 4c$

   In this case, we have $s^4 a_{k+2}^2 - 4cs^2 a_{k+2} \geq (s^2 a_{k+2} - 4c)^2$, so the solutions of inequality (4.10) satisfy either

   $$r < \frac{1}{2s}\left(s^2 a_{k+2} - \sqrt{s^4 a_{k+2}^2 - 4cs^2 a_{k+2}}\right) \leq \frac{2c}{s}$$

   or

   $$r > \frac{1}{2s}\left(s^2 a_{k+2} + \sqrt{s^4 a_{k+2}^2 - 4cs^2 a_{k+2}}\right) \geq \frac{1}{s}(s^2 a_{k+2} - 2c).$$

   From the first possibility, it follows that $rs < 2c$. If the second possibility occurs, we introduce the substitution $t = sa_{k+2} - r$. The number t is a positive integer. Now we have

   $$p = rp_{k+1} + sp_k = (sa_{k+2} - t)p_{k+1} + sp_k = sp_{k+2} - tp_{k+1},$$

   $$q = sq_{k+2} - tq_{k+1}$$

   and $st = s^2 a_{k+2} - rs < 2c$.

2) $s^2 a_{k+2} < 4c$

   If $r < \frac{1}{2}sa_{k+2}$, then $rs < \frac{1}{2}s^2 a_{k+2} < 2c$. If $\frac{1}{2}sa_{k+2} \leq r < sa_{k+2}$, then we again define $t = sa_{k+2} - r$ and we get $st \leq \frac{1}{2}s^2 a_{k+2} < 2c$.

$$\qquad\qquad\qquad\qquad\qquad\qquad\qquad\qquad\qquad\qquad\qquad\qquad\qquad\qquad \square$$

From inequality (4.6) and Theorem 4.1.5 (in fact, already from Legendre's theorem, which is a special case of Theorem 4.1.5 for $c = 1/2$), we conclude that for any solution of Pell's equation $x^2 - dy^2 = 1$, $\frac{x}{y}$ is a convergent of $\sqrt{d}$. The number $\sqrt{d}$ is a quadratic irrational number, and hence its continued fraction expansion is periodic. Moreover, the number $\sqrt{d} + \lfloor\sqrt{d}\rfloor$ is reduced (it is greater than 1, and its conjugate $-\sqrt{d} + \lfloor\sqrt{d}\rfloor$ is from the interval $(-1, 0)$), and thus its expansion is purely periodic. This implies that $\sqrt{d}$ has continued fraction expansion of the form

$$\sqrt{d} = [a_0; \overline{a_1, a_2, \ldots, a_{\ell-1}, 2a_0}],$$

where $a_0 = \lfloor\sqrt{d}\rfloor$, while the bar over the numbers $a_1, a_2, \ldots, a_{\ell-1}$ denotes that this block of numbers repeats infinitely. Moreover, it can be shown that it satisfies the "palindromic property": $a_1 = a_{\ell-1}, a_2 = a_{\ell-2}, \ldots$.

Here we give an algorithm for finding continued fraction expansion of quadratic irrationals. Let α be a quadratic irrational number. We write it in the form $\alpha = \frac{s_0 + \sqrt{d}}{t_0}$, where $d, s_0, t_0 \in \mathbb{Z}$, $t_0 \neq 0$, $d \neq \square$ and $t_0 | (d - s_0^2)$. If $\alpha = \sqrt{d}$, then we have $s_0 = 0$, $t_0 = 1$. The numbers a_i (partial quotients) are computed recursively by the following formulas:

$$a_i = \left\lfloor \frac{s_i + \sqrt{d}}{t_i} \right\rfloor, \quad s_{i+1} = a_i t_i - s_i, \quad t_{i+1} = \frac{d - s_{i+1}^2}{t_i}. \tag{4.11}$$

It can be shown that the sequences of integers (s_i) and (t_i) are bounded. More precisely, for sufficiently large index i,

$$|s_i| < \sqrt{d}, \quad 0 < t_i < s_i + \sqrt{d} < 2\sqrt{d}.$$

In that way, we see that the expansion has to be periodic (since there exist distinct indices j and k such that $(s_j, t_j) = (s_k, t_k)$).

Note that, under the assumption that $t_i > 0$, we have $a_i = \left\lfloor \frac{s_i + \sqrt{d}}{t_i} \right\rfloor = \left\lfloor \frac{s_i + \lfloor\sqrt{d}\rfloor}{t_i} \right\rfloor$ (see [116, Chapter 8.6]). Thus, algorithm (4.11) works only with integers (and does not require a precise approximation for the irrational number $\sqrt{d}$).

From the above estimates for the size of elements of sequences (s_i) and (t_i) (which in the case $\alpha = \sqrt{d}$ hold for any i), we obtain an estimate for the length of the period of the continued fraction expansion of $\sqrt{d}$, namely $\ell(d) < 2\sqrt{d} \cdot 2\sqrt{d} = 4d$. By more precise analysis of the relations between s_i and t_i (in particular, the congruence $s_i^2 \equiv d \pmod{t_i}$), we can get the bound $\ell(d) = O(\sqrt{d} \log d)$ (see [309, Chapter 3.2]). The best-known explicit result of this type is $\ell(d) < \frac{7}{2\pi^2}\sqrt{d} \log d$ for sufficiently large d (see [72]), while the conjecture (related to the Riemann hypothesis) is that $\ell(d) = O(\sqrt{d} \log \log d)$.

By comparing the rational and irrational parts in the equation

$$\sqrt{d} = \frac{\frac{s_{n+1}+\sqrt{d}}{t_{n+1}} p_n + p_{n-1}}{\frac{s_{n+1}+\sqrt{d}}{t_{n+1}} q_n + q_{n-1}},$$

we get the relation

$$p_n^2 - dq_n^2 = (-1)^{n+1} t_{n+1}, \quad \text{for all } n \geq -1. \tag{4.12}$$

It shows us that the solutions of Pell's equation $x^2 - dy^2 = 1$ correspond to those indices n for which $(-1)^{n+1} t_{n+1} = 1$ holds. It is not hard to see that $t_i = 1$ if and only if $\ell \mid i$ (ℓ is the length of the period) and that $t_i \neq -1$ always holds. Therefore, we have

Theorem 4.1.6 *Let ℓ be the length of the period of the continued fraction expansion of $\sqrt{d}$.*

If ℓ is even, then the equation $x^2 - dy^2 = -1$ has no solution, and all solutions of the equation $x^2 - dy^2 = 1$ are given by $(x, y) = (p_{n\ell-1}, q_{n\ell-1})$, $n \in \mathbb{N}$. In particular, the fundamental solution is $(p_{\ell-1}, q_{\ell-1})$.

If ℓ is odd, then all solutions of the equation $x^2 - dy^2 = -1$ are given by $(x, y) = (p_{(2n-1)\ell-1}, q_{(2n-1)\ell-1})$, and all solutions of the equation $x^2 - dy^2 = 1$ are given by $(x, y) = (p_{2n\ell-1}, q_{2n\ell-1})$, $n \in \mathbb{N}$. In particular, the fundamental solution of $x^2 - dy^2 = 1$ is $(p_{2\ell-1}, q_{2\ell-1})$.

Example 4.1.7 Find the fundamental solution of $x^2 - 113y^2 = 1$.
Let us first find the continued fraction expansion of $\sqrt{113}$ by algorithm (4.11):

$$a_0 = 10, \ \ s_1 = a_0 t_0 - s_0 = 10, \ \ t_1 = \frac{d - s_1^2}{t_0} = 13;$$

$$a_1 = \left\lfloor \frac{s_1 + 10}{t_1} \right\rfloor = 1, \ \ s_2 = 3, \ \ t_2 = 8; \quad a_2 = 1, \ \ s_3 = 5, \ \ t_3 = 11;$$

$$a_3 = 1, \ \ s_4 = 6, \ \ t_4 = 7; \quad a_4 = 2, \ \ s_5 = 8, \ \ t_5 = 7.$$

We can continue with the algorithm until we get $(s_k, t_k) = (s_1, t_1)$. Nevertheless, we can also note that $t_4 = t_5$. This partial repetition indicates that we have reached the middle of the full period. Namely,

- if $s_n = s_{n+1}$, then $\ell = 2n$;
- if $t_n = t_{n+1}$, then $\ell = 2n + 1$

(proof can be found in [303, Part I, Chapter 3.24]). In our case, we have $\ell = 9$, and the palindromic property and the fact that the last element in the period is $2a_0$ allow us to reconstruct the whole expansion:

$$\sqrt{113} = [10; \overline{1, 1, 1, 2, 2, 1, 1, 1, 20}].$$

Table 4.1 Convergents of $\sqrt{113}$

i	-1	0	1	2	3	4	5	6	7	8
a_i		10	1	1	1	2	2	1	1	1
p_i	1	10	11	21	32	85	202	287	489	776
q_i	0	1	1	2	3	8	19	27	46	73

In order to compute the fundamental solution of the equation $x^2 - 113y^2 = 1$, it suffices to find the fundamental solution of the equation $x^2 - 113y^2 = -1$ (which is (p_8, q_8)) and square it (see, e.g. [116, Theorem 10.14]).

From Table 4.1, we find that the fundamental solution of $x^2 - 113y^2 = 1$ is

$$(776 + 73\sqrt{113})^2 = 1204353 + 113296\sqrt{113}.$$

4.1.3   Pellian Equations

An equation of the form

$$x^2 - dy^2 = N, \tag{4.13}$$

where d is a positive integer which is not a perfect square, and N is a non-zero integer, is called a *Pellian equation*. It is clear that such an equation may not have integer solutions, but if it has a solution, then it has infinitely many solutions. Indeed, if $x + y\sqrt{d}$ is a solution of (4.13) and $u + v\sqrt{d}$ is a solution of the corresponding Pell's equation $x^2 - dy^2 = 1$, then

$$(x + y\sqrt{d})(u + v\sqrt{d}) = (ux + dvy) + (uy + vx)\sqrt{d} \tag{4.14}$$

also satisfies Eq. (4.13), because

$$(ux + dvy)^2 - d(uy + vx)^2 = (x^2 - dy^2)(u^2 - dv^2) = N \cdot 1 = N.$$

Since Pell's equation has infinitely many solutions, from (4.14), it follows that Eq. (4.13) also has infinitely many solutions (if it has at least one).

Two solutions, $x + y\sqrt{d}$ and $x' + y'\sqrt{d}$, of Eq. (4.13) are called *associated* if one can be obtained from the other by multiplying by a solution of Pell's equation as in formula (4.14). It is easy to check that in that way, we introduced an equivalence relation on the set of all solutions of Eq. (4.13) (note that $(u + v\sqrt{d})^{-1} = u - v\sqrt{d}$, which implies that the relation is symmetric). We say that mutually associated solutions form one *class of solutions*. It is easy to see that $x + y\sqrt{d}$ and $x' + y'\sqrt{d}$ are associated if and only if

$$xx' \equiv dyy' \,(\mathrm{mod}\ N), \quad xy' \equiv x'y \,(\mathrm{mod}\ N)$$

(see the proof of Theorem 4.1.2).

Let $\mathbf{K}$ be a class of solutions and let $x_i + y_i\sqrt{d}$, $i = 1, 2, 3, \ldots$ be its elements. Then the class consisting of solutions $x_i - y_i\sqrt{d}$ is denoted by $\bar{\mathbf{K}}$ and called the *conjugate class* of the class $\mathbf{K}$. If $\mathbf{K} = \bar{\mathbf{K}}$ holds, we say that the class $\mathbf{K}$ is *ambiguous*.

Among all elements of a class $\mathbf{K}$, we choose one, denoted by $x^* + y^*\sqrt{d}$, which we call the *fundamental solution of the equation $x^2 - dy^2 = N$ in the class $\mathbf{K}$*. We chose it in such a way that y^* takes the smallest possible non-negative value among all elements $x + y\sqrt{d}$ in the class $\mathbf{K}$. With this requirement, x^* is also uniquely determined, except in the case when $\mathbf{K}$ is ambiguous. If $\mathbf{K}$ is ambiguous, then we choose x^* such that it satisfies the additional condition $x^* \geq 0$. Note that $|x^*|$ has the least possible value within the class $\mathbf{K}$.

The following result of Chebyshev [60] (Nagell's book [292, Section 58] is often cited as a more accessible reference) gives explicit and sharp upper bounds for fundamental solutions of a Pellian equation in terms of the fundamental solution of the corresponding Pell's equation.

Theorem 4.1.8 *Let $u + v\sqrt{d}$ be the fundamental solution of Pell's equation $x^2 - dy^2 = 1$. Then all fundamental solutions $x^* + y^*\sqrt{d}$ of the Pellian equation $x^2 - dy^2 = N$ satisfy the inequalities*

$$0 \leq y^* \leq \frac{v}{\sqrt{2(u + \varepsilon)}}\sqrt{|N|},$$

$$|x^*| \leq \sqrt{\frac{1}{2}(u + \varepsilon)|N|},$$

where $\varepsilon = 1$ if $N > 0$, and $\varepsilon = -1$ if $N < 0$. In particular, there are only finitely many fundamental solutions (and thus only finitely many classes of solutions).

Proof We will prove the claim for $N < 0$. The proof for $N > 0$ is quite similar. Let us define the integers x', y' by $x' + y'\sqrt{d} = (x^* + y^*\sqrt{d})(u - \delta v\sqrt{d})$, where $\delta = 1$ if $x^* \geq 0$, and $\delta = -1$ if $x^* < 0$. Then $x' + y'\sqrt{d}$ belongs to the same class as $x^* + y^*\sqrt{d}$, and thus the minimality of y^* implies that

$$y' = uy^* - \delta vx^* \geq y^*,$$

which gives $v|x^*| \leq (u - 1)y^*$. By squaring, we get

$$v^2(dy^{*2} + N) \leq (u^2 - 2u + 1)y^{*2},$$

i.e. $y^{*2}(2u - 2) \leq |N| \cdot v^2$, and we obtain the desired inequality for y^*. Now we have

$$x^{*2} = dy^{*2} + N \leq \frac{-dNv^2}{2u - 2} + N = \frac{-N(u^2 - 2u + 1)}{2u - 2} = \frac{|N| \cdot (u - 1)}{2}.$$

$\square$

Example 4.1.9 Solve the equation

$$x^2 - 120y^2 = -119. \tag{4.15}$$

This equation appears when we want to extend the Diophantine pair $\{1, 120\}$ to a Diophantine triple. Namely, by eliminating c from the conditions $c + 1 = y^2$ and $120c + 1 = x^2$, we obtain exactly Eq. (4.15).

The fundamental solution of the corresponding Pell equation $x^2 - 120y^2 = 1$ is $11 + \sqrt{120}$. Hence, the fundamental solutions of the equation $x^2 - 120y^2 = -119$ satisfy the inequalities $0 \leq y^* \leq 2$, $|x^*| \leq 24$. It is easy to check that the only solutions which satisfy these inequalities are $\pm 1 + \sqrt{120}$ and $\pm 19 + 2\sqrt{120}$ (and they are not associated). Therefore, all solutions are given by

$$x + y\sqrt{120} = (\pm 1 + \sqrt{120})(11 + \sqrt{120})^n \quad \text{or}$$

$$x + y\sqrt{120} = (\pm 19 + 2\sqrt{120})(11 + \sqrt{120})^n,$$

for $n \in \mathbb{Z}$. It means that we have four sequences of solutions (in positive integers), each satisfying the same recurrence relation as the sequence of solutions of the corresponding Pell's equation:

$$x_0 = 1, \ y_0 = 1; \ x_1 = 131, \ y_1 = 12; \ x_{n+2} = 22x_{n+1} - x_n,$$

$$y_{n+2} = 22y_{n+1} - y_n,$$

$$x_0' = -1, \ y_0' = 1; \ x_1' = 109, \ y_1' = 10; \ x_{n+2}' = 22x_{n+1}' - x_n',$$

$$y_{n+2}' = 22y_{n+1}' - y_n',$$

$$x_0'' = 19, \ y_0'' = 2; \ x_1'' = 449, \ y_1'' = 41; \ x_{n+2}'' = 22x_{n+1}'' - x_n'',$$

$$y_{n+2}'' = 22y_{n+1}'' - y_n'',$$

$$x_0''' = -19, \ y_0''' = 2; \ x_1''' = 31, \ y_1''' = 3; \ x_{n+2}''' = 22x_{n+1}''' - x_n''',$$

$$y_{n+2}''' = 22y_{n+1}''' - y_n'''.$$

By inserting the values of y in $c = y^2 - 1$, we obtain extensions of the pair $\{1, 120\}$ to a triple:

$$c = 3, 8, 99, 143, 1680, 4095, 47960, 69168, 809999, 1974024, 23116863, \ldots.$$

Proposition 4.1.10 *Suppose that* $|N| < \sqrt{d}$. *If* $x + y\sqrt{d}$ *is a solution of the equation* $x^2 - dy^2 = N$, *then* $\frac{x}{y}$ *is a convergent of the continued fraction expansion of* $\sqrt{d}$.

Proof Assume first that $N > 0$. Then $x > y\sqrt{d}$, and so we have

$$0 < \frac{x}{y} - \sqrt{d} = \frac{N}{y(x + y\sqrt{d})} < \frac{N}{2\sqrt{d}y^2} < \frac{1}{2y^2}.$$

From Legendre's theorem (Theorem 4.1.5 for $c = 1/2$), it follows that $\frac{x}{y}$ is an (odd) convergent of $\sqrt{d}$.

Let now $N < 0$. Then $x < y\sqrt{d}$ and so we have

$$0 < \frac{y}{x} - \frac{1}{\sqrt{d}} = \frac{|N|}{x\sqrt{d}(x + y\sqrt{d})} < \frac{|N|}{2\sqrt{d}x^2} < \frac{1}{2x^2}.$$

We conclude that $\frac{y}{x}$ is a convergent of $\frac{1}{\sqrt{d}}$. But if $\frac{y}{x}$ is the i-th convergent of $\frac{1}{\sqrt{d}}$, then $\frac{x}{y}$ is the $(i - 1)$-st convergent of $\sqrt{d}$. $\qquad\square$

By Proposition 4.1.10, the solvability of the equation $x^2 - dy^2 = N$ in coprime integers x, y, for $|N| < \sqrt{d}$, can be decided from the continued fraction expansion of $\sqrt{d}$ by checking whether any of the first 2ℓ convergents (where ℓ is the period) satisfy the relation

$$p_i^2 - dq_i^2 = (-1)^{i+1}t_{i+1} = N.$$

If $|N|$ is not significantly larger than $\sqrt{d}$ (say, e.g. $|N| < 4\sqrt{d}$), then we can use Theorem 4.1.5 instead of Legendre's theorem. In that case, we use the relation

$$(rp_i \pm sp_{i-1})^2 - d(rq_i \pm sq_{i-1})^2 = (-1)^{i+1}(r^2t_{i+1} - s^2t_i \mp 2rss_{i+1}). \qquad (4.16)$$

We say that a solution $x_0 + y_0\sqrt{d}$ is *primitive* if x_0 and y_0 are coprime. If $\gcd(x_0, y_0) = g$, then $\frac{x_0}{g} + \frac{y_0}{g}\sqrt{d}$ is a primitive solution of the equation $x^2 - dy^2 = \frac{N}{g^2}$.

Example 4.1.11 Let k be a positive integer. Find all positive integers N such that $N < 4k$, and the equation

$$x^2 - (k^2 + 1)y^2 = N$$

has a primitive solution.

We have

$$0 < \frac{x}{y} - \sqrt{k^2+1} < \frac{N}{2\sqrt{k^2+1}\,y^2} < \frac{2k}{\sqrt{k^2+1}\,y^2} < \frac{2}{y^2}.$$

By Theorem 4.1.5, the numbers x and y have the form $x = rp_i \pm sp_{i-1}$, $y = rq_i \pm sq_{i-1}$, where $rs < 4$. The number $\sqrt{k^2+1}$ has the following continued fraction expansion

$$\sqrt{k^2+1} = [k; \overline{2k}].$$

Furthermore, $s_i = k$ and $t_i = 1$ for all $i \geq 1$. Thus, it suffices to insert $i = 1$ and $(r, s) = (1, 0), (1, 1), (1, 2), (2, 1), (1, 3), (3, 1)$ in formula (4.16). We get the following values of N satisfying the condition $0 < N < 4k$:

$$N = 1, \ 2k, \ 4k - 3 \quad \text{(for all } k \in \mathbb{N}\text{)},$$

and additionally $N = 10$ for $k = 3$ (which is equal to $6k - 8$).

Remark 4.1.12 When we consider extensions of a $D(-1)$-pair of the form $\{1, k^2 + 1\}$ to a $D(-1)$-triple, we obtain the Pellian equation

$$x^2 - (k^2 + 1)y^2 = k^2. \tag{4.17}$$

Since the right-hand side of (4.17) is significantly larger than $\sqrt{k^2+1}$, we cannot apply the method from the previous example. It is easy to see that for $k \geq 2$, (4.17) has at least three classes of solutions in integers, corresponding to the fundamental solutions $(x_0, y_0) = (k, 0)$, $(k^2 - k + 1, k - 1)$, and $(-(k^2 - k + 1), k - 1)$. For certain values of k, there are two additional classes of solutions. E.g. for $k = 2t^2$, we also have fundamental solutions $(x_0, y_0) = (\pm(2t^3 + t), t)$, and additional solutions occur also for $k = 4s^3 - 4s^2 + 3s - 1$, for $k = F_{2n}$, and some other polynomial or exponential subfamilies. However, there is no known example in which (4.17) has more than five classes of solutions. That motivated the conjecture proposed by Dujella in 2012 that for $k \geq 2$, Eq. (4.17) always has three or five classes of solutions. It is easy to see that each fundamental solution (x_0, y_0) has to satisfy $y_0 < k$, so the conjecture actually says that there is at most one solution of Eq. (4.17) with $0 < y < k - 1$ and $x > 0$. For some partial results concerning this conjecture, see [248, 267, 334].

4.1.4 Linear Forms in Logarithms

The problem of extending Diophantine triples to quadruples can be transformed into solving a system of Pell's or (generalized) Pellian equations. By solving each of

those equations individually and equating the expressions obtained for the common
unknown in the system, we arrive at an equation that can be written in an approximate form (ignoring terms that rapidly tend to zero for large indices/exponents m
and n)

$$\gamma \cdot \alpha^m \approx \delta \cdot \beta^n,$$

where $\alpha, \beta, \gamma, \delta$ are quadratic irrationals. By taking logarithms, we get

$$m \log \alpha - n \log \beta + \log \frac{\gamma}{\delta} \approx 0.$$

From the end of the 1960s and the pioneer work of the English mathematician
Alan Baker (1939–2018) until today, several results are known which indicate that
a linear form in logarithms of algebraic numbers cannot be very close to zero. Such
results give an (explicit) upper bound for m and n in our problem of finding solutions
to the system of generalized Pellian equations, and they can be applied to many other
Diophantine problems.

The development of the theory of linear forms in logarithms was motivated by
Hilbert's seventh problem, which asked to prove that if α is an algebraic number
$\neq 0, 1$ and β is algebraic and irrational, then α^β is a transcendental number. This
claim was proved independently by Gelfond and Schneider in 1934. In 1967, Baker
generalized that result and proved that if $\alpha_1, \ldots \alpha_n$ are algebraic numbers $\neq 0, 1$ and
$\beta_1, \ldots, \beta_n$ are algebraic numbers such that $1, \beta_1, \ldots, \beta_n$ are linearly independent
over $\mathbb{Q}$, then the number $\alpha_1^{\beta_1} \cdot \ldots \cdot \alpha_n^{\beta_n}$ is transcendental.

Thus, if $\alpha_1, \ldots, \alpha_n, \beta_1, \ldots, \beta_n$ are as above, and α_{n+1} is an arbitrary algebraic
number, then $\beta_1 \log \alpha_1 + \cdots + \beta_n \log \alpha_n \neq \log \alpha_{n+1}$. Moreover, Baker showed that
the number

$$|\beta_1 \log \alpha_1 + \cdots + \beta_n \log \alpha_n - \log \alpha_{n+1}|$$

cannot be very small [21, Chapter 2].

For applications to Diophantine equations, a slightly different situation is of
interest, in which β_i are integers. So we look at expressions of the form

$$\Lambda = b_1 \log \alpha_1 + \cdots + b_n \log \alpha_n,$$

where $b_i \in \mathbb{Z}, i = 1, \ldots, n$. In our applications, α_i will be positive real numbers,
so that log will be an ordinary real natural logarithm function. In the general case,
log is the principal branch of the complex logarithm function.

It can be shown that if $\Lambda \neq 0$, then $|\Lambda|$ can be estimated from below by a bound
that depends on absolute values of b_i and the degrees and heights of α_i.

One of the strongest (and easiest to use) known general results of this type is the
following theorem of Baker and Wüstholz from 1993 [25] (see also [26, Chapter
7.2])

Theorem 4.1.13 (Baker-Wüstholz) *Let* $\Lambda = b_1 \log \alpha_1 + \cdots + b_n \log \alpha_n$ *be a linear form in logarithms of algebraic numbers* $\alpha_1, \ldots, \alpha_n$ *with integer coefficients* $b_1, \ldots, b_n$. *If* $\Lambda \neq 0$, *then*

$$\log |\Lambda| \geq -18(n+1)! \, n^{n+1} (32d)^{n+2} \log(2nd) h'(\alpha_1) \cdot \ldots \cdot h'(\alpha_n) \log B,$$

where $d = [\mathbb{Q}(\alpha_1, \ldots, \alpha_n) : \mathbb{Q}]$, $B = \max\{|b_1|, \ldots, |b_n|\}$, $h'(\alpha) = \max\{h(\alpha), \frac{1}{d}|\log \alpha|, \frac{1}{d}\}$, *and* $h(\alpha)$ *is the logarithmic Weil height.*

Let us recall here some definitions of heights and measures of algebraic numbers. Let α be an algebraic number. Then there exists a unique irreducible monic polynomial $g(x)$ with rational coefficients such that $g(\alpha) = 0$. Furthermore, any polynomial $f(x)$ over $\mathbb{Q}$ which satisfies $f(\alpha) = 0$ is divisible by $g(x)$ (see, e.g. [298, Chapter 9.2]). The polynomial $g(x)$ is called the *minimal polynomial* of α. The constant multiple of $g(x)$ with coprime integer coefficients and positive leading coefficient is called the *minimal polynomial over* $\mathbb{Z}$ of α. The *degree* of an algebraic number is the degree of its minimal polynomial.

Let α be an algebraic number, and let $P(x)$ be its minimal polynomial over $\mathbb{Z}$,

$$P(x) = a_d x^d + \cdots + a_1 x + a_0 = a_d \prod_{i=1}^{d} (x - \alpha^{(i)}), \quad \alpha^{(i)} \in \mathbb{C}.$$

The (naïve) *height* of α is $H(\alpha) = \max\{|a_0|, |a_1|, \ldots, |a_d|\}$, while

$$M(\alpha) = |a_d| \prod_{i=1}^{d} \max\{|\alpha^{(i)}|, 1\},$$

is the *Mahler measure*, and $h(\alpha) = \frac{1}{d} \log M(\alpha)$ is the *logarithmic Weil height*.

The result of Baker and Wüstholz was improved in 2000 by Matveev [268]. There are also improvements for the cases (which are most important for applications) $n = 2$ [246, 247] and $n = 3$ [9, 276, 277]. These improvements are very relevant when solving parametric families of Diophantine equations because they usually give the conclusion that for large values of parameters, we only have "trivial" solutions, and they leave a relatively small number of individual equations that need to be examined separately. When it comes to just one concrete equation, then the Baker-Wüstholz theorem (and even the initial Baker's result [19] from 1968), in combination with the reduction methods that we will deal with in the following sections, is usually sufficient.

4.1.5   Simultaneous Diophantine Approximation

Many important problems on Diophantine equations can be reformulated in terms of corresponding problems on Diophantine approximations. In these reformulations, apart from "ordinary" approximations, *simultaneous* Diophantine approximations, i.e. problems of simultaneous approximations of several real numbers, can also appear. A typical example is a system of Pell (or Pellian equations). E.g. consider the system

$$x^2 - 2y^2 = 1, \quad z^2 - 3y^2 = 1,$$

which implies that $\sqrt{2} \approx \frac{x}{y}$, $\sqrt{3} \approx \frac{z}{y}$ (more precisely $\left|\sqrt{2} - \frac{x}{y}\right| < \frac{1}{2y^2}$, $\left|\sqrt{3} - \frac{z}{y}\right| < \frac{1}{2y^2}$). Hence, we see that the irrational numbers $\sqrt{2}$ and $\sqrt{3}$ should have very good rational approximations with a common denominator. Thus, it is natural to ask whether such approximations can exist and, in particular, what is the critical exponent which separates the quality of approximations which have infinitely many solutions from those that have only finitely many solutions.

We will give analogues of the main result from "ordinary" approximations, i.e. the Dirichlet theorem.

Theorem 4.1.14 (Dirichlet) *Let α_{ij} ($i = 1, \ldots, n$; $j = 1, \ldots, m$) be real numbers and $Q > 1$ a positive integer. Then there exist integers $q_1, \ldots, q_m, p_1, \ldots, p_n$ such that*

$$1 \leq \max\{|q_1|, \ldots, |q_m|\} < Q^{n/m},$$

$$|\alpha_{i1}q_1 + \cdots + \alpha_{im}q_m - p_i| \leq \frac{1}{Q}, \quad i = 1, \ldots, n. \tag{4.18}$$

Proof Consider the points

$$(\{\alpha_{11}x_1 + \cdots + \alpha_{1m}x_m\}, \ldots, \{\alpha_{n1}x_1 + \cdots + \alpha_{nm}x_m\}),$$

where x_j, $j = 1, \ldots, m$, are integers satisfying the condition $0 \leq x_j < Q^{n/m}$. There are at least Q^n such points, and each of them lies in the closed unit cube $I^n = \{(t_1, \ldots, t_n) : 0 \leq t_k \leq 1, k = 1, \ldots, n\}$. Furthermore, $(1, 1, \ldots, 1) \in I^n$, so all together there are at least $Q^n + 1$ points in I^n.

We divide I^n into Q^n pairwise disjoint subcubes with sides of length $\frac{1}{Q}$. (The cubes contain some of their faces or edges and not others.) According to Dirichlet's box (pigeonhole) principle, two of the considered points are in the same subcube, say the points

$$(\alpha_{11}x_1 + \cdots + \alpha_{1m}x_m - y_1, \ldots, \alpha_{n1}x_1 + \cdots + \alpha_{nm}x_m - y_n),$$

$$(\alpha_{11}x_1' + \cdots + \alpha_{1m}x_m' - y_1', \ldots, \alpha_{n1}x_1' + \cdots + \alpha_{nm}x_m' - y_n'),$$

with $(x_1, \ldots, x_m) \neq (x'_1, \ldots, x'_m)$. Put $q_j = x_j - x'_j$ for $j = 1, \ldots, m$, and $p_i = y_i - y'_i$ for $i = 1, \ldots, n$. Then (4.18) is obviously satisfied. $\qquad\square$

Corollary 4.1.15 *Assume that at least one of the numbers $\alpha_1, \alpha_2, \ldots, \alpha_n$ is irrational. Then there exist infinitely many n-tuples of rational numbers $\frac{p_1}{q}, \ldots, \frac{p_n}{q}$ satisfying*

$$\left| \alpha_i - \frac{p_i}{q} \right| < \frac{1}{q^{1+1/n}}, \quad i = 1, \ldots, n. \tag{4.19}$$

Proof We apply Theorem 4.1.14 with $m = 1$. We conclude that there exist (coprime) integers $q, p_1, \ldots, p_n$ such that

$$1 \leq q < Q^n \quad \text{and} \quad |\alpha_i q - p_i| \leq \frac{1}{Q}, \ i = 1, \ldots, n. \tag{4.20}$$

It is clear that (4.20) implies (4.19).

Let α_1 be irrational. Then $|\alpha_1 q - p_1| \neq 0$. Thus, for fixed $q, p_1, \ldots, p_n$, inequalities (4.20) can hold only for $Q \leq \frac{1}{|\alpha_1 q - p_1|}$. As $Q \to \infty$, we get infinitely many different solutions. $\qquad\square$

Corollary 4.1.16 *Let $1, \alpha_1, \ldots, \alpha_m$ be real numbers linearly independent over the rationals. Then there exist infinitely many $(m + 1)$-tuples of coprime integers $(q_1, \ldots, q_m, p)$ satisfying*

$$q = \max\{|q_1|, \ldots, |q_m|\} > 0 \quad \text{and} \quad |\alpha_1 q_1 + \cdots + \alpha_m q_m - p| < \frac{1}{q^m}. \tag{4.21}$$

Proof From Theorem 4.1.14 for $n = 1$, it follows that there exist integers $q_1, \ldots, q_m, p$ such that

$$1 \leq \max\{|q_1|, \ldots, |q_m|\} < Q^{1/m} \quad \text{and} \quad |\alpha_1 q_1 + \cdots + \alpha_m q_m - p| \leq \frac{1}{Q}. \tag{4.22}$$

It is clear that (4.22) implies (4.21). By linear independence, we have $|\alpha_1 q_1 + \cdots + \alpha_m q_m - p| \neq 0$. Thus, for fixed $q_1, \ldots, q_m, p$, inequality (4.22) holds only for $Q \leq \frac{1}{|\alpha_1 q_1 + \cdots + \alpha_m q_m - p|}$. Hence, as $Q \to \infty$, we get infinitely many different solutions. $\qquad\square$

Although Dirichlet's theorems on simultaneous approximations are direct generalizations of Dirichlet's theorem for $n = 1$, when we speak about effective algorithms for finding rational numbers whose existence is guaranteed by those theorems, there is a significant difference between cases $n = 1$ and $n \geq 2$. Namely, for $n = 1$, as we have already seen, rational approximations with the required property can be obtained by continued fractions. For $n \geq 2$, the situation is more complex. There are various generalizations of continued fractions and their properties to higher dimensions. One of them is the LLL algorithm, which efficiently (in a polynomial time) provides rational approximations of a similar quality as the ones guaranteed by Dirichlet's theorems (some additional factors will appear compared to the optimal approximations provided by the theory). Additional information on the LLL algorithm and its applications can be found in [116, Section 8.9], [297], [332, Chapters 5 and 6], and [341].

Let $\alpha_1, \ldots, \alpha_n$ be real numbers and $Q > 1$ an integer. By applying the LLL algorithm to the lattice generated by the matrix

$$
B = \begin{pmatrix}
1 & 0 & 0 & \cdots & 0 \\
-\lceil C\alpha_1 \rceil & C & 0 & \cdots & 0 \\
-\lceil C\alpha_2 \rceil & 0 & C & \cdots & 0 \\
\vdots & & 0 & 0 & \ddots & \vdots \\
-\lceil C\alpha_n \rceil & 0 & 0 & \cdots & C
\end{pmatrix},
$$

where $C = Q^{n+1}$ and $\lceil x \rceil$ denotes the nearest integer to a real number x, we find integers $q, p_1, \ldots, p_n$ such that

$$
1 \le q \le 2^{n/4} Q^n \quad \text{and} \quad |\alpha_i q - p_i| \le \frac{\sqrt{5} \cdot 2^{(n-4)/4}}{Q}, \quad i = 1, \ldots, n. \tag{4.23}
$$

For the proof, see [116, Theorem 8.59] and [297, Chapter 6].

We will see in Sect. 4.3, that by Roth's theorem (Theorem 4.3.3), the exponent 2 in Dirichlet's theorem for ordinary Diophantine approximations is the best possible when α is an algebraic number. For simultaneous approximations, i.e. Corollary 4.1.15, an analogous result can be obtained as a consequence of *Schmidt's subspace theorem*. A proof of this theorem and an overview of its various versions, as well as some applications, can be found in [43, Chapter 7], [318, Chapter 5], and [361].

Theorem 4.1.17 (Schmidt) *Let $L_1, \ldots, L_n$ be linearly independent linear forms in n variables with coefficients which are algebraic numbers and let $\delta > 0$. Then all integer points $x = (x_1, \ldots, x_n)$ which satisfy the inequality*

$$
|L_1(x) \cdot \ldots \cdot L_n(x)| < \frac{1}{\|x\|^\delta},
$$

where $\|x\| = \max\{|x_1|, \ldots, |x_n|\}$, lie in finitely many proper subspaces of $\mathbb{Q}^n$.

Corollary 4.1.18 *Let $\alpha_1, \ldots, \alpha_n$ be algebraic numbers such that $1, \alpha_1, \ldots, \alpha_n$ are linearly independent over $\mathbb{Q}$ and let $\delta > 0$. Then there are only finitely many n-tuples $(\frac{p_1}{q}, \ldots, \frac{p_n}{q})$ of rational numbers such that*

$$
\left| \alpha_i - \frac{p_i}{q} \right| < \frac{1}{q^{1+\frac{1}{n}+\delta}}, \quad i = 1, 2, \ldots, n. \tag{4.24}
$$

Proof If we multiply all inequalities in (4.24) and multiply the result by q^{n+1}, we obtain

$$q|\alpha_1 q - p_1| \cdot \ldots \cdot |\alpha_n q - p_n| < \frac{1}{q^\delta}.$$

Let $x = (p_1, \ldots, p_n, q) \in \mathbb{Z}^{n+1}$, and let us consider the linear forms

$$L_i(x_1, \ldots, x_{n+1}) = \alpha_i x_{n+1} - x_i, \quad 1 \le i \le n, \quad L_{n+1}(x_1, \ldots, x_{n+1}) = x_{n+1}.$$

For sufficiently large q, we have

$$|L_1(x) \cdot \ldots \cdot L_{n+1}(x)| < \frac{1}{q^\delta} < \frac{1}{\|x\|^{\delta/2}}.$$

By applying Theorem 4.1.17 for dimension $n + 1$, we conclude that the solutions of this inequality lie in the union of finitely many proper rational subspaces. The elements of such a subspace satisfy an equation of the form

$$c_1 x_1 + \cdots + c_{n+1} x_{n+1} = 0, \quad c_1, \ldots, c_{n+1} \in \mathbb{Q}.$$

For solutions of (4.24), which lie in this subspace, we have

$$(c_1 \alpha_1 + \cdots + c_n \alpha_n + c_{n+1})q = c_1(\alpha_1 q - p_1) + \cdots + c_n(\alpha_n q - p_n).$$

Set $\gamma = |c_1 \alpha_1 + \cdots + c_n \alpha_n + c_{n+1}|$. Then $\gamma > 0$, due to the linear independence of $1, \alpha_1, \ldots, \alpha_n$. For a given subspace, the number γ is fixed. Furthermore,

$$\gamma \cdot q \le |c_1||\alpha_1 q - p_1| + \cdots + |c_n||\alpha_n q - p_n| \le |c_1| + \cdots + |c_n|.$$

Therefore, q is bounded. Hence, in each of the finitely many subspaces, we have only finitely many integers q which satisfy (4.24). $\qquad\square$

Corollary 4.1.18 (as well as Roth's theorem, which can be obtained from Corollary 4.1.18 for $n = 1$) is "ineffective", i.e. it does not provide an explicit upper bound for the size of integers q which satisfy (4.24). There are effective results of this kind for concrete values of α_i's. Let us mention that in 1993, Rickert [307] proved that

$$\max\left(\left|\sqrt{2} - \frac{p_1}{q}\right|, \left|\sqrt{3} - \frac{p_2}{q}\right| \right) > \frac{10^{-7}}{q^{1.913}},$$

for all $p_1, p_2, q \in \mathbb{N}$. As a consequence, it follows that all solutions of the system of Pellian equations

$$x^2 - 2y^2 = u, \quad z^2 - 3y^2 = v$$

satisfy the inequality $\max(|x|, |y|, |z|) \leq (10^7 \cdot \max(|u|, |v|))^{12}$.

Rickert's result was obtained by the hypergeometric method. Baker introduced this method in [17] to obtain effective results for rational approximations to certain algebraic numbers such as $\sqrt[3]{2}$. In [18], the method was adapted to simultaneous rational approximations of certain algebraic numbers. The results and ideas from Rickert's paper have many applications to problems on Diophantine m-tuples. The first such application appeared in [93], where the following theorem from [307] was used in proving that the Diophantine triple $\{k - 1, k + 1, 4k\}$ extends uniquely to a Diophantine quadruple $\{k - 1, k + 1, 4k, d\}$ with $d = 16k^3 - 4k$.

Theorem 4.1.19 *For an integer $k \geq 2$, the numbers*

$$\theta_1 = \sqrt{(k - 1)/k}, \quad \theta_2 = \sqrt{(k + 1)/k}$$

satisfy

$$\max\left(|\theta_1 - p_1/q|, |\theta_2 - p_2/q|\right) > (271k)^{-1} q^{-1-\lambda}$$

for all integers p_1, p_2, q with $q > 0$, where

$$\lambda = \lambda(k) = \frac{\log(12k\sqrt{3} + 24)}{\log(27(k^2 - 1)/32)}.$$

In obtaining an absolute upper bound for the size of Diophantine tuples, the following very useful result of Bennett [30] will play an important role.

Theorem 4.1.20 *If a_i, p_i, q and N are integers for $0 \leq i \leq 2$, with $a_0 < a_1 < a_2$, $a_j = 0$ for some $0 \leq j \leq 2$, $q \neq 0$ and $N > M^9$, where*

$$M = \max_{0 \leq i \leq 2}\{|a_i|\} \geq 3,$$

then we have

$$\max_{0 \leq i \leq 2}\left\{\left|\sqrt{1 + \frac{a_i}{N}} - \frac{p_i}{q}\right|\right\} > (130N\gamma)^{-1} q^{-\lambda},$$

where

$$\lambda = 1 + \frac{\log(32.04\, N\gamma)}{\log\left(1.68\, N^2 \prod_{0 \leq i < j \leq 2}(a_i - a_j)^{-2}\right)}$$

and

$$
\gamma = \begin{cases} \frac{(a_2-a_0)^2(a_2-a_1)^2}{2a_2-a_0-a_1} & \text{if } a_2 - a_1 \geq a_1 - a_0, \\ \frac{(a_2-a_0)^2(a_1-a_0)^2}{a_1+a_2-2a_0} & \text{if } a_2 - a_1 < a_1 - a_0. \end{cases}
$$

The last two theorems are results on simultaneous approximations of square roots of two rationals which are very close to 1. The problem of extending a Diophantine triple $\{a, b, c\}$ to a quadruple leads to the system of Pellian equations $az^2 - cx^2 = a - c$, $bz^2 - cy^2 = b - c$, and its solutions provide good rational approximations with the common denominator z to irrational numbers $\sqrt{a/c}$ and $\sqrt{b/c}$. However, these numbers are generally not close to 1. Nevertheless, by multiplying them by rational numbers $\sqrt{ac+1}/a$ and $\sqrt{bc+1}/b$, respectively, we obtain irrational numbers close to 1.

Therefore, we will apply Theorem 4.1.20 to the numbers

$$
\theta_1 = \frac{s}{a}\sqrt{\frac{a}{c}} \quad \text{and} \quad \theta_2 = \frac{t}{b}\sqrt{\frac{b}{c}},
$$

where $ac + 1 = s^2$, $bc + 1 = t^2$. Namely,

$$
\theta_1 = \sqrt{\frac{(ac+1)a}{a^2c}} = \sqrt{1 + \frac{1}{ac}} = \sqrt{1 + \frac{b}{abc}},
$$

$$
\theta_2 = \sqrt{1 + \frac{1}{bc}} = \sqrt{1 + \frac{a}{abc}},
$$

so that θ_1 and θ_2 are indeed square roots of rationals, which are close to 1. On the other hand, we will see in the next lemma that every solution of our problem induces good simultaneous rational approximations of numbers θ_1 and θ_2.

Lemma 4.1.21 *Let $a < b < c$. All positive integer solutions x, y, z of the simultaneous Pellian equations*

$$
az^2 - cx^2 = a - c,
$$

$$
bz^2 - cy^2 = b - c,
$$

satisfy

$$
\max\left(\left|\theta_1 - \frac{sbx}{abz}\right|, \left|\theta_2 - \frac{tay}{abz}\right|\right) < \frac{c}{2a} \cdot z^{-2}.
$$

Proof We have

$$\left|\frac{s}{a}\sqrt{\frac{a}{c}} - \frac{sbx}{abz}\right| = \frac{s}{az\sqrt{c}}|z\sqrt{a} - x\sqrt{c}| = \frac{s}{az\sqrt{c}} \cdot \frac{c-a}{z\sqrt{a} + x\sqrt{c}}$$

$$< \frac{s(c-a)}{2a\sqrt{ac}z^2} < \frac{c}{2a} \cdot z^{-2},$$

$$\left|\frac{t}{b}\sqrt{\frac{b}{c}} - \frac{tay}{abz}\right| = \frac{t}{bz\sqrt{c}} \cdot \frac{c-b}{z\sqrt{b} + y\sqrt{c}} < \frac{t(c-b)}{2b\sqrt{bc}z^2} < \frac{c}{2b} \cdot z^{-2} < \frac{c}{2a} \cdot z^{-2}.$$

$\square$

The condition $N > M^9$ in Theorem 4.1.20 is restrictive for some applications since it asks for a very big gap between c and b. In this respect, several improvements which slightly relax this condition are known (see [16, 64, 183, 184]).

Let us mention that a similar idea as with the above numbers θ_1 and θ_2 was used by Bugeaud and Dujella in [49] (see also [48, Chapter 5.4]), where they studied k-th power Diophantine m-tuples. Instead of $(\frac{a}{b})^{1/k}$, which is suggested by the equation $ay^k - bz^k = a - b$, they considered the algebraic number $(\frac{a(bc+1)}{b(ac+1)})^{1/k}$, to take advantage of a result of Shorey [324] on linear forms in logarithms of algebraic numbers close to 1.

4.2   Mordell's Equation

In 1923, Mordell proved that a Diophantine equation of the form

$$y^2 = x^3 + ax^2 + bx + c,$$

where the cubic polynomial on the right-hand side of the equation has no multiple roots, has only finitely many integer solutions. In other words, he proved that elliptic curves with integer coefficients have finitely many integer points. Mordell's proof uses the reduction of this equation to finitely many Thue equations, about which more will be said later.

Finding all integer solutions of such equations (i.e. all integer points on elliptic curves) is not a simple task. However, many results are known, especially for *Mordell's equation* of the form

$$y^2 = x^3 + k, \tag{4.25}$$

which were obtained by quite elementary methods—by considering congruences modulo 4 and 8 and properties of quadratic residues. In this section, we will present several such results, and additional examples can be found in [8, Chapter 14.1] and [284, Chapter 26].

Proposition 4.2.1 *Let $k = (4b - 1)^3 - 4a^2$, where a is an integer with no prime factors of the form $4l + 3$. Then the equation $y^2 = x^3 + k$ has no integer solutions.*

Proof We have $k \equiv -1 \pmod 4$, so $y^2 \equiv x^3 - 1 \pmod 4$. Since $y^2 \equiv 0$ or $1 \pmod 4$, x cannot be even or congruent to -1 modulo 4. Therefore, $x \equiv 1 \pmod 4$. Let us write the equation $y^2 = x^3 + (4b - 1)^3 - 4a^2$ in the form

$$y^2 + 4a^2 = x^3 + (4b - 1)^3 = (x + 4b - 1)(x^2 - x(4b - 1) + (4b - 1)^2).$$

The last factor $x^2 - x(4b - 1) + (4b - 1)^2$ is congruent to 3 modulo 4. Therefore, it must have at least one prime factor p, congruent to 3 modulo 4. But that prime factor p can divide the sum of two squares $y^2 + 4a^2$ only if both y and a are divisible by p, and this contradicts the assumption that a has no prime factors of the form $4l + 3$.
$\square$

Some integers k that satisfy the conditions of Proposition 4.2.1 are $k = -73, -65, -57, -37, -17, -5, 11, 23, 87$.

Proposition 4.2.2 *Let $k = 2b^2 - a^3$, where $a \equiv 2$ or $4 \pmod 8$, $b \equiv 1 \pmod 2$, and all prime factors of b are of the form $8l \pm 1$. Then the equation $y^2 = x^3 + k$ has no integer solutions.*

Proof We have $y^2 \equiv x^3 + 2 \pmod 4$, so $x \not\equiv 0 \pmod 2$ and $x \not\equiv 1 \pmod 4$. Therefore, $x \equiv 3 \pmod 4$, i.e. $x \equiv 3$ or $7 \pmod 8$. Furthermore,

$$y^2 - 2b^2 = x^3 - a^3 = (x - a)(x^2 + ax + a^2).$$

If $x \equiv 3 \pmod 8$, then $x^2 + ax + a^2 \equiv 1 + 3a + a^2 \equiv \pm 3 \pmod 8$, so $x^2 + ax + a^2$ has at least one prime factor p of the form $8l \pm 3$. By assumption, p does not divide b, so we get that the Legendre symbol $(\frac{2}{p})$ satisfies

$$\left(\frac{2}{p}\right) = \left(\frac{2b^2}{p}\right) = \left(\frac{y^2}{p}\right) = 1.$$

We got a contradiction because $(\frac{2}{p}) = 1$ if and only if $p \equiv \pm 1 \pmod 8$.

If $x \equiv 7 \pmod 8$, then $x - a \equiv 7 - a \equiv \pm 3 \pmod 8$, so $x - a$ must have at least one prime factor of the form $8l \pm 3$, from which we get a contradiction in exactly the same way as in the previous case.
$\square$

Some values of k that satisfy the conditions of Proposition 4.2.2 are $k = -62, -6, 34, 58, 66, 90$.

Example 4.2.3 Let us show that the equation $y^2 = x^3 + 45$ has no integer solutions.

From $y^2 \equiv x^3 + 5 \pmod 8$, we have that $x \equiv 3$ or $7 \pmod 8$. If it were $x \equiv 0 \pmod 3$, then from $x = 3X$, $y = 3Y$, we would get $Y^2 = 3X^3 + 5 \equiv 2 \pmod 3$, which is a contradiction because 2 is not a quadratic residue modulo 3. Therefore, x is not divisible by 3.

Assume that $x \equiv 3 \pmod 8$. Let us write the equation in the form

$$y^2 - 2 \cdot 6^2 = x^3 - 27 = (x - 3)(x^2 + 3x + 9).$$

We have $x^2 + 3x + 9 \equiv 3 \pmod 8$, so this number must have a prime factor p of the form $8l \pm 3$. Since we checked that $p \neq 3$, p cannot be a factor of $y^2 - 2 \cdot 6^2$ because 2 is not a quadratic residue modulo p.

Assume that $x \equiv 7 \pmod 8$. Now we write the initial equation in the form

$$y^2 - 2 \cdot 3^2 = x^3 + 27 = (x + 3)(x^2 - 3x + 9).$$

We have $x^2 - 3x + 9 \equiv 5 \pmod 8$, so this number must have a prime factor p of the form $8l \pm 3$. However, p cannot be a factor of $y^2 - 2 \cdot 3^2$, so we got a contradiction again.

We can write the equation $y^2 = x^3 + k$ in the form $x^3 = y^2 - k$ and factor it in the field $\mathbb{Q}(\sqrt{k})$:

$$x^3 = (y + \sqrt{k})(y - \sqrt{k}). \tag{4.26}$$

In order to be able to use this factorization to solve the initial equation, we need some information about integers in the quadratic field $\mathbb{K} = \mathbb{Q}(\sqrt{k})$ (see, e.g. [22, Chapter 7] and [298, Chapter 9]). We will assume that k is square-free. First of all, let us recall that the ring of integers $O_{\mathbb{K}}$ in $\mathbb{K}$ is $\{u + v\sqrt{k} : u, v \in \mathbb{Z}\}$ if $k \equiv 2$ or $3 \pmod 4$, and $\{u + v\frac{1+\sqrt{k}}{2} : u, v \in \mathbb{Z}\}$ if $k \equiv 1 \pmod 4$. The next important information is the structure of the set of units (invertible elements) in $O_{\mathbb{K}}$. If $k \geq 2$, then there are infinitely many units in $\mathbb{K}$, and they have the form $\pm\varepsilon_k^n$, $n \in \mathbb{Z}$, where ε_k is the fundamental unit. If k is negative, then $\mathbb{K}$ has units ± 1, and these are the only units, except in the cases $k = -1$ and $k = -3$. The units in $\mathbb{Q}(i)$ are $\pm 1, \pm i$, and in $\mathbb{Q}(\sqrt{-3})$ they are $\pm 1, \frac{\pm 1 \pm \sqrt{-3}}{2}$.

In Eq. (4.26), we have that the product of two numbers is equal to a cube. If it were a product of (ordinary) integers and the factors were relatively prime, then we could conclude that both factors are cubes. In this conclusion, the uniqueness of factorization of integers into prime factors is used. However, that property is not valid for integers in all quadratic fields. More precisely, the only quadratic fields with such property for $k < 0$ are those for $k = -1, -2, -3, -7, -11, -19, -43, -67, -163$, while a conjecture is that there are infinitely many such fields for $k > 0$.

For example, in $\mathbb{Q}(\sqrt{-5})$, we have:

$$21 = 3 \cdot 7 = (1 + 2\sqrt{-5})(1 - 2\sqrt{-5}) = (4 + \sqrt{-5})(4 - \sqrt{-5}) \tag{4.27}$$

and it can be shown that all factors $3, 7, 1 + 2\sqrt{-5}, 1 - 2\sqrt{5}, 4 + \sqrt{-5}, 4 - \sqrt{-5}$ are *irreducible*, i.e. they are divisible only by units and their associated numbers (two numbers are associated if their quotient is a unit).

However, the uniqueness of the factorization can be obtained if we consider ideals instead of numbers because every proper ideal in $O_{\mathbb{K}}$ can be represented in a unique way as a product of prime ideals (see, e.g. [23, Chapter 11.4]).

Example 4.2.4 Let us find all integer solutions to the equation $y^2 = x^3 - 11$.

It is known that $\mathbb{Q}(\sqrt{-11})$ has unique factorization property. Units in $\mathbb{Q}(\sqrt{-11})$ are ± 1. Therefore, from

$$(y + \sqrt{-11})(y - \sqrt{-11}) = x^3$$

and the fact that the factors on the left-hand side of this equation are relatively prime (because their common factor divides $2\sqrt{-11}$, and x cannot be divisible by either 2 or 11), it follows that there exists an algebraic integer $w \in \mathbb{Q}(\sqrt{-11})$ such that

$$y + \sqrt{-11} = \pm w^3.$$

Since $-w^3 = (-w)^3$, we can omit the minus sign. Therefore,

$$y + \sqrt{-11} = w^3, \quad w = a + \frac{b}{2}(1 + \sqrt{-11}) = \left(a + \frac{b}{2}\right) + \frac{b}{2}\sqrt{-11}.$$

Equating the coefficients of $\sqrt{-11}$, we get

$$1 = 3\left(a + \frac{b}{2}\right)^2 \cdot \frac{b}{2} - 11 \cdot \left(\frac{b}{2}\right)^3,$$

so

$$b(3a^2 + 3ab - 2b^2) = 2.$$

From here, we have $b = \pm 1$ or ± 2, so $(a, b) = (0, -1), (1, -1), (1, 2), (-3, 2)$. Now from $y = (a + \frac{b}{2})^3 + 3(a + \frac{b}{2}) \cdot (\frac{b}{2})^2 \cdot (-11)$, it follows $y = \pm 4, \pm 58$, so all solutions of the given equation are $(x, y) = (3, \pm 4), (15, \pm 58)$.

4.3   Thue Equations

In the next section, we will show how the problem of finding integer points on an elliptic curve can be reduced to the problem of solving a finite number of Thue equations. So let us first say something about Thue equations.

Let

$$F(x, y) = a_0 x^n + a_1 x^{n-1} y + \cdots + a_n y^n$$

be a binary form with integer coefficients, irreducible over $\mathbb{Q}$, of degree $n \geq 3$. Note that the form F cannot be irreducible over $\mathbb{C}$. Namely, according to the fundamental theorem of algebra,

$$F(x, 1) = a_0(x - \theta_1) \cdot \ldots \cdot (x - \theta_n),$$

where $\theta_1, \ldots, \theta_n$ are algebraic numbers of degree n, so

$$F(x, y) = y^n F\left(\frac{x}{y}, 1\right) = a_0(x - \theta_1 y) \cdot \ldots \cdot (x - \theta_n y).$$

Irreducibility over $\mathbb{Q}$ implies that $f(x) = F(x, 1)$ does not have multiple roots, i.e. that θ_i are mutually distinct. Indeed, if $f(x)$ had multiple roots, then $\gcd(f, f') \in \mathbb{Q}[x]$ would be a non-trivial factor of f, and then $f(x)$, and also $F(x, y)$, would not be irreducible over $\mathbb{Q}$. Let $m \neq 0$ be an integer. Then the Diophantine equation $F(x, y) = m$ is called a *Thue equation*. In 1909, the Norwegian mathematician Axel Thue (1863–1922), using his famous result from Diophantine approximations, proved that such an equation has only finitely many integer solutions. Let us first prove a simple special case.

Theorem 4.3.1 *If the polynomial $F(x, 1)$ has no real roots, then the equation $F(x, y) = m$ has only finitely many integer solutions. More precisely, all solutions satisfy the inequality*

$$|y| \leq \frac{\sqrt[n]{|m|/|a_0|}}{\min_{1 \leq i \leq n} |\operatorname{Im}(\theta_i)|},$$

where $n \geq 3$ is the degree of the polynomial F, and θ_i are the roots of the polynomial $F(x, 1)$.

Proof Assume that (x, y) is a solution of the equation $F(x, y) = m$ and take θ_k so that $|x - \theta_k y| = \min_{1 \leq i \leq n} |x - \theta_i y|$. Then it is obvious that $|y||\operatorname{Im}(\theta_k)| = |\operatorname{Im}(\theta_k y)| \leq |x - \theta_k y| \leq \sqrt[n]{|m|/|a_0|}$, so we get the statement of the theorem. $\qquad\square$

Let us also present an alternative proof of Theorem 4.3.1, with a slightly different upper bound for $|y|$, according to [284, Chapter 22]. For $y = 0$, we can have at most two solutions, so suppose $y \neq 0$. Since the polynomial $F(x, 1)$ has no real roots, we conclude that there exists a real number $c > 0$ such that $|F(z, 1)| > c$ for every real number z. Now from

$$|m| = |F(x, y)| = |y|^n \left|F\left(\frac{x}{y}, 1\right)\right| > c|y|^n,$$

it follows that $|y| < \sqrt[n]{|m|/c}$.

An example of the equation

$$F(x, y) = x^n + (x - y)^2(2x - y)^2 \cdot \ldots \cdot \left(\frac{n}{2}x - y\right)^2 = 1,$$

for an even number n, where the polynomial $F(x, 1)$ obviously has no real roots, and which has solutions $\pm(1, 1), \pm(1, 2), \ldots, \pm(1, n/2)$, indicates that the bound $|y| < \sqrt[n]{|m|/c}$ cannot be significantly improved.

Thue's result from Diophantine approximations concerns rational approximations of algebraic numbers. The first important result of this type was Liouville's theorem.

Theorem 4.3.2 (Liouville) *Let α be a real algebraic number of degree d. Then there is a constant $c(\alpha) > 0$ such that*

$$\left|\alpha - \frac{p}{q}\right| > \frac{c(\alpha)}{q^d}$$

for all rational numbers $\frac{p}{q}$, where $q > 0$ and $\frac{p}{q} \neq \alpha$.

Proof Let $P(x)$ be the minimal polynomial of α over $\mathbb{Z}$. Without loss of generality, we can assume that $|\alpha - \frac{p}{q}| \leq 1$ (otherwise, we can put $c(\alpha) = 1$). If we expand $P(x)$ into the Taylor series around α, we get

$$\left|P\left(\frac{p}{q}\right)\right| = \left|\sum_{i=1}^{d}\left(\frac{p}{q} - \alpha\right)^i \frac{1}{i!}P^{(i)}(\alpha)\right| < \frac{1}{c(\alpha)} \cdot \left|\alpha - \frac{p}{q}\right|, \tag{4.28}$$

where $c(\alpha) = \left(2\sum_{i=1}^{d}\frac{1}{i!}|P^{(i)}(\alpha)|\right)^{-1}$.

Since the polynomial $P(x)$ is irreducible, we have $P(\frac{p}{q}) \neq 0$. Therefore, the number $q^d|P(\frac{p}{q})|$ is a positive integer, so $|P(\frac{p}{q})| \geq \frac{1}{q^d}$. If we compare this with (4.28), we get the statement of the theorem. $\qquad\square$

Let α be a real algebraic number of degree $d \geq 2$. Liouville's theorem implies that the inequality

$$\left|\alpha - \frac{p}{q}\right| < \frac{1}{q^\mu} \tag{4.29}$$

has only finitely many rational solutions $\frac{p}{q}$ if $\mu > d$.

In 1909, Thue proved that (4.29) has only finitely many solutions if $\mu > \frac{d}{2} + 1$, Siegel proved that the same statement holds if $\mu > 2\sqrt{d}$, while Dyson and Gelfond proved the statement for $\mu > \sqrt{2d}$. Finally, in 1955, Roth proved that inequality (4.29) has only finitely many solutions if $\mu > 2$. Klaus Roth was awarded the Fields medal in 1958 for this result.

Theorem 4.3.3 (Roth) *Let α be a real algebraic number of degree $d \geq 2$. Then for every $\delta > 0$ the inequality*

$$\left| \alpha - \frac{p}{q} \right| < \frac{1}{q^{2+\delta}} \tag{4.30}$$

has only finitely many solutions in the rational numbers $\frac{p}{q}$.

The idea of proving Roth's theorem (and previous improvements of Liouville's theorem) is to try to modify the basic steps in the proof of Liouville's theorem. The proof of Lioville's theorem starts from the minimal polynomial of α. In his improvement of Liouville's theorem, Thue used a polynomial of the form $x_2 Q(x_1) - P(x_1)$, Siegel used a more general polynomial $P(x_1, x_2)$ in two variables, while Roth used a polynomial $P(x_1, \ldots, x_m)$ in several variables. The main difficulty in this approach occurs when checking the last step from the proof of Liouville's theorem. For polynomials in one variable, the conclusion that $P(\frac{p}{q}) \neq 0$ was quite simple. However, with polynomials in several variables, the set of solutions to the equation $P(x_1, \ldots, x_m) = 0$ is an algebraic variety in $\mathbb{R}^m$, and it is very difficult to show that $P(\frac{p}{q}, \ldots, \frac{p}{q}) \neq 0$. An attempt to overcome this obstacle is by using m-tuples $\frac{p_1}{q_1}, \ldots, \frac{p_m}{q_m}$ of different rational approximations and trying to prove that $P(\frac{p_1}{q_1}, \ldots, \frac{p_m}{q_m}) \neq 0$. It turns out that the denominators $q_1 < q_2 < \cdots < q_m$ have to grow fast. For example, in the case of $m = 2$, we need two good approximations $\frac{p_1}{q_1}, \frac{p_2}{q_2}$ of α such that q_2 is much larger than q_1. This is the reason why one good approximation does not give any contradiction, and Roth's theorem, as well as all other improvements to Liouville's theorem obtained by this method, is "ineffective" in the sense that it does not provide any bound for the size of the denominator q solutions of (4.30).

However, for $d \geq 3$, from the proof of Roth's theorem, Davenport and Roth [76] deduced an explicit upper bound for the number of solutions of (4.30), in terms of δ, d and $H(\alpha)$. The proof of Roth's theorem can be found in [254, Part II, Chapter 4] and [317, Chapter 5].

Theorem 4.3.4 (Thue)   *A Thue equation has only finitely many integer solutions.*

Proof Let $F(x, y) = m$ be the given Thue equation. With the above-introduced notation, we can write

$$a_0(x - \theta_1 y) \cdot \ldots \cdot (x - \theta_n y) = m. \tag{4.31}$$

We may assume that $y \neq 0$ because for $y = 0$, we have at most two solutions. Let us divide both sides of (4.31) by y^n and take the absolute values. Thus, we get

$$|a_0| \cdot \left| \theta_1 - \frac{x}{y} \right| \cdot \ldots \cdot \left| \theta_n - \frac{x}{y} \right| = \left| \frac{m}{y^n} \right|. \tag{4.32}$$

As in the proof of Theorem 4.3.1, let us take θ_k such that

$$|x - \theta_k y| = \min_{1 \le i \le n} |x - \theta_i y|,$$

i.e.

$$\left| \theta_k - \frac{x}{y} \right| = \min_{1 \le i \le n} \left| \theta_i - \frac{x}{y} \right|.$$

Let $\gamma = \frac{1}{2} \min_{i \ne j} |\theta_i - \theta_j| > 0$. For y large enough, both sides of (4.32) can be made arbitrarily small. This is especially true for the smallest factor on the left-hand side, i.e. $|\theta_k - \frac{x}{y}|$. Thus, there exists $y_0 > 0$ such that for $y \ge y_0$, $|\theta_k - \frac{x}{y}| < \gamma$. For $i \ne k$, we have

$$\left| \theta_i - \frac{x}{y} \right| \ge |\theta_i - \theta_k| - \left| \theta_k - \frac{x}{y} \right| > 2\gamma - \gamma = \gamma.$$

Therefore, it follows from (4.32) that

$$\left| \theta_k - \frac{x}{y} \right| < \left| \frac{m}{a_0 y^n \gamma^{n-1}} \right| = \frac{c}{|y|^n}. \tag{4.33}$$

Since $n \ge 3$, Roth's theorem (actually already Thue's theorem, but not Liouville's theorem) implies that inequality (4.33) has only finitely many solutions, which had to be proved. $\square$

Remark 4.3.5 From the properties of linear Diophantine equations and Pell's equations (Sect. 4.1.1), we know that the statement of Theorem 4.3.4 is not valid if the degree of F is equal to 1 or 2. On the other hand, the statement of Theorem 4.3.4 is valid if the assumption that the polynomial $F(x, y)$ is irreducible over $\mathbb{Q}$ is replaced by the assumption that the polynomial $F(x, 1)$ has at least three different roots.

Indeed, suppose that the polynomial $F(x, y)$ is reducible over $\mathbb{Q}$. If F has at least two different irreducible factors F_1 and F_2, then we get finitely many systems of Diophantine equations $F_1(x, y) = m_1$, $F_2(x, y) = m_2$. Each of these systems has finitely many (complex) solutions (according to Bézout's theorem (see [330, Appendix A.4]), the number of solutions is not larger than the product of the degrees of F_1 and F_2). It remains to consider the case $F(x, y) = aG(x, y)^k$ where the polynomial G is irreducible over $\mathbb{Q}$. If $\deg G \ge 3$, then from Theorem 4.3.4, it follows that the equation $F(x, y) = m$ has finitely many solutions. So, the only cases when the equation $F(x, y) = m$, where F is a binary form, may have infinitely many solutions are equations of the form

$$F(x, y) = a(bx + cy)^n \quad \text{or} \quad F(x, y) = a(bx^2 + cxy + dy^2)^{n/2},$$

and these are precisely the cases when $F(x, 1)$ has less than three different roots.

4.4   Transformation of Elliptic Curves to Thue Equations

Let us consider the general equation of the form

$$y^2 = x^3 + ax^2 + bx + c,$$

where the coefficients a, b, c are integers, and the cubic polynomial on the right-hand side has no multiple roots. We will describe Mordell's argument from 1923 by which he showed that such an equation has only finitely many integer solutions. At the same time, this is an important step in one of the general methods for finding all integer points on an elliptic curve. A more general result, that the number of integer points on any non-singular cubic curve with integer coefficients is finite, was proved in 1929 by Siegel. It is important to note that although birational transformations preserve the structure of the set of rational points on the curve, this is not true for the set of integer points, which essentially depends on the chosen model (equation) of the curve.

The idea is to factorize the polynomial

$$f(x) = x^3 + ax^2 + bx + c = (x - \vartheta_1)(x - \vartheta_2)(x - \vartheta_3). \tag{4.34}$$

Thus, we obtain fields $\mathbb{Q}(\vartheta_i)$ in which we consider Eq. (4.34). Three cases are possible:

(1) all three roots of f are rationals (and then necessarily integers);
(2) one root of f is rational, and the other two are quadratic irrationals;
(3) f is irreducible over $\mathbb{Q}$; its roots are algebraic integers of degree 3.

Recall that in $\mathbb{Z}$ we have: if $XY = Z^\ell$ and $\gcd(X, Y) = 1$, then there exist $U, V \in \mathbb{Z}$ so that $X = \pm U^\ell$, $Y = \pm V^\ell$, $Z = \pm UV$. The generalization of this result to the integers of a number field $\mathbb{K}$ is given by the following lemma (the proof can be found in [284, Chapter 15]).

Lemma 4.4.1 *All solutions* (X, Y, Z) *in* $O_\mathbb{K}$ *of the equation*

$$XY = cZ^\ell,$$

where the ideal $\langle X, Y \rangle$ *divides* δ, *for a given ideal* δ, *have the form*

$$X = \lambda \varepsilon_1 U^\ell, \quad Y = \mu \varepsilon_2 V^\ell, \quad Z = \nu \varepsilon_3 UV,$$

where U, V *are arbitrary integers from* $\mathbb{K}$, $\varepsilon_1, \varepsilon_2, \varepsilon_3$ *are units,* λ, μ, ν *are elements of* $\mathbb{K}$. *The last six numbers* $(\varepsilon_1, \varepsilon_2, \varepsilon_3, \lambda, \mu, \nu)$ *take only finitely many values and satisfy* $\lambda \mu \varepsilon_1 \varepsilon_2 = c \nu^\ell \varepsilon_3^\ell$.

Let us go back to Eq. (4.34). We consider it in the field $\mathbb{K} = \mathbb{Q}(\vartheta_i)$. We will give a sketch of the algorithm for the third case when the cubic polynomial $f(x)$

is irreducible. The transformation of the elliptic curve into Thue equations in the remaining two cases is carried out similarly. From Lemma 4.4.1, we get a relation of the form

$$x - \vartheta_i = m(r + s\vartheta_i + t\vartheta_i^2)^2, \tag{4.35}$$

where $r, s, t \in \mathbb{Z}$, and m takes finitely many values from $\mathbb{Q}(\vartheta_i)$. Indeed, ϑ_i is an algebraic integer of degree 3, so every element of $\mathbb{K}$ can be written in the form $\alpha + \beta\vartheta_i + \gamma\vartheta_i^2, \alpha, \beta, \gamma \in \mathbb{Q}$. Note that $\{1, \vartheta_i, \vartheta_i^2\}$ does not have to be a basis for $O_\mathbb{K}$. However, it can be shown that if $\frac{1}{d}(r + s\vartheta_i + t\vartheta_i^2) \in O_\mathbb{K}$ and $\gcd(d, r, s, t) = 1$, then d^2 divides the discriminant $\Delta(1, \vartheta_i, \vartheta_i^2) = (\vartheta_1 - \vartheta_2)^2(\vartheta_1 - \vartheta_3)^2(\vartheta_2 - \vartheta_3)^2$. Therefore, we have finitely many possibilities for d, which we can "transfer" to m.

The number m (each of finitely many of them) also can be written in the form $m = r_0 + s_0\vartheta_i + t_0\vartheta_i^2$, where $r_0, s_0, t_0 \in \mathbb{Q}$. Let us insert this in (4.35), multiply and express $\vartheta_i^3, \vartheta_i^4, \vartheta_i^5$ and ϑ_i^6 in terms of $1, \vartheta_i, \vartheta_i^2$. If we compare the coefficients of $1, \vartheta_i$ and ϑ_i^2 on both sides of the equation, we get three equations of the form

$$f_1(r, s, t) = 0, \quad f_2(r, s, t) = 1, \quad f_3(r, s, t) = x,$$

where f_1, f_2, f_3 are ternary quadratic forms with rational coefficients.

It is known that the solvability of the equation $f_1(r, s, t) = 0$ can be efficiently decided. Namely, such equations satisfy the local-global principle, i.e. if they have solutions over $\mathbb{R}$ and over the p-adic field $\mathbb{Q}_p$ for all prime numbers p, then they also have solutions over $\mathbb{Q}$. Even more explicitly, any such equation can be written in the form $ax^2 + by^2 + cz^2 = 0$, where a, b, c are integers such that abc is square-free, and a necessary and sufficient condition for the solvability of this equation in rationals is that the numbers a, b, c are not all positive nor all negative, and $-bc$ is a quadratic residue modulo a, $-ac$ is a quadratic residue modulo b, and $-ab$ quadratic residue modulo c (Legendre's theorem, for the proof see [116, Chapter 10.7] and [332, Chapter IV.2]).

Furthermore, assuming that a non-trivial solution exists, all solutions are given by

$$gr = q_1(u, v), \quad gs = q_2(u, v), \quad gt = q_3(u, v),$$

where q_1, q_2, q_3 are binary quadratic forms with integer coefficients, and g takes finitely many integer values. If we insert this in the equation $f_2(r, s, t) = 1$, we get finitely many equations of the form

$$h(u, v) = g^2, \tag{4.36}$$

where h is a homogeneous polynomial of the fourth degree with integer coefficients. The polynomial h is not the square of a second degree polynomial (otherwise, the starting curve would be of genus 0). Therefore, according to Thue's theorem

(Theorem 4.3.4), we conclude that Eq. (4.36) has finitely many solutions, and that is why the initial elliptic curve also has only finitely many integer points (which we can get from $f_3(r, s, t) = x$).

4.5   Algorithm for Solving Thue Equations

In the previous section, we saw that finding integer points on elliptic curves can be reduced to solving Thue equations. Thue equations also appear in other important problems in number theory such as the question of whether a given number field has a power integral basis (see [188]). Therefore, the existence of an algorithm for the systematic solution of Thue equations is of great interest. We will present one such algorithm—de Weger's algorithm from 1989 [78, Chapter 8].

We will divide the set of possible solutions into four classes:

- very small solutions: $|y| \leq Y_1$—can be found by the "brute force" check;
- small solutions: $Y_1 < |y| \leq Y_2$—correspond to the convergents of the continued fraction of ϑ_i;
- large solutions: $Y_2 < |y| \leq Y_3$—eliminated by LLL reduction;
- very large solutions: $|y| > Y_3$—it is proved that they do not exist using linear forms in logarithms.

Let $g(x) = F(x, 1) = a_0(x - \vartheta_1) \cdots (x - \vartheta_n)$. Then we get the equation $F(x, y) = a_0(x - \vartheta_1 y) \cdots (x - \vartheta_n y) = m$, which we will observe in the number field $\mathbb{K} = \mathbb{Q}(\vartheta_i)$ (all these fields are $\mathbb{Q}$-isomorphic). Let $\vartheta_1, \ldots, \vartheta_s$ be real, and $\vartheta_{s+1} = \overline{\vartheta_{s+t+1}}, \ldots, \vartheta_{s+t} = \overline{\vartheta_{s+2t}}$ conjugate complex roots of $g(x)$. We saw in Theorem 4.3.1 that the case $s = 0$ is very simple. Therefore, we will further assume that $s \geq 1$.

Let us introduce the notation $\beta_i = x - \vartheta_i y$, for $i = 1, \ldots, n$. Assume that $|a_0| = |m| = 1$. Then from $\beta_1 \beta_2 \cdots \beta_n = \pm 1$, we conclude that β_i is a unit (invertible element) in the ring $O_{\mathbb{K}}$. By Dirichlet's theorem on units (see [23, Chapter 12.2]), β_i can be written in the form

$$\pm \varepsilon_1^{a_1} \cdots \varepsilon_r^{a_r},$$

where $\varepsilon_1, \ldots, \varepsilon_r$ is a system of fundamental units and $r = s + t - 1$ (note that since g is irreducible and $s \geq 1$, the only roots of unity belonging to $\mathbb{K}$ are ± 1). In the general case, β_i will have the form

$$\pm \mu_i \varepsilon_1^{a_1} \cdots \varepsilon_r^{a_r},$$

where μ_i belongs to a finite set M of non-associated elements of $O_{\mathbb{K}}$ whose norm satisfies the condition $a_0 N(\mu_i) = m$.

Let $|\beta_{i_0}| = \min\{|\beta_i| : 1 \leq i \leq n\}$. It can be shown that $i_0 \in \{1, \dots, s\}$ and

$$|\beta_{i_0}| \leq c_1 |y|^{-(n-1)}, \tag{4.37}$$

$$|\beta_i| \geq c_2 |y|, \quad i \neq i_0, \tag{4.38}$$

where

$$c_1 = \frac{2^{n-1}|m|}{\min\{|g'(\vartheta_i)| : 1 \leq i \leq s\}},$$

$$c_2 = \frac{1}{2} \min\{|\vartheta_i - \vartheta_j| : 1 \leq i < j \leq n\}$$

(see the proof of Theorem 4.3.4).

Lemma 4.5.1 *If* $|y| > Y_1 = (4c_1)^{\frac{1}{n-2}}$, *then* $\frac{x}{y}$ *is a convergent of the continued fraction expansion of* ϑ_{i_0}.

Proof We have

$$\left| \frac{x}{y} - \vartheta_{i_0} \right| = |\beta_{i_0}| \cdot |y|^{-1} \leq c_1 \cdot |y|^{-n} \leq \frac{1}{4} Y_1^{n-2} \cdot |y|^{-n} < \frac{1}{4y^2},$$

so the statement follows from Legendre's theorem (see Theorem 4.1.5).   □

Let us now consider the connection between three different elements $\beta_i = x - \vartheta_i y$, $\beta_j = x - \vartheta_j y$, $\beta_k = x - \vartheta_k y$. By eliminating x and y from these three equations, we get *Siegel's identity*

$$\beta_i(\vartheta_j - \vartheta_k) + \beta_j(\vartheta_k - \vartheta_i) + \beta_k(\vartheta_i - \vartheta_j) = 0.$$

For β_i we will take β_{i_0}, while we will take ϑ_j and ϑ_k to be real (if $s \geq 3$), or $\vartheta_k = \overline{\vartheta_j}$ (if $s = 1$). We will give details only for the first, real case. Siegel's identity can be written in the form

$$\frac{\vartheta_{i_0} - \vartheta_j}{\vartheta_{i_0} - \vartheta_k} \cdot \frac{\beta_k}{\beta_j} - 1 = -\frac{\vartheta_k - \vartheta_j}{\vartheta_k - \vartheta_{i_0}} \cdot \frac{\beta_{i_0}}{\beta_j}. \tag{4.39}$$

Identity (4.39) will give us a connection with linear forms in logarithms. Let us define

$$\Lambda = \log\left(\frac{\vartheta_{i_0} - \vartheta_j}{\vartheta_{i_0} - \vartheta_k} \cdot \frac{\beta_k}{\beta_j} \right).$$

It is not difficult to see that the expression under the logarithm is positive.

Lemma 4.5.2 *Let*

$$c_3 = \max\left\{\left|\frac{\vartheta_{i_1} - \vartheta_{i_2}}{\vartheta_{i_1} - \vartheta_{i_3}}\right| : i_1 \neq i_2 \neq i_3 \neq i_1\right\},$$

$$Y_2 = \max\left\{Y_1, \left(\frac{2c_1c_3}{c_2}\right)^{\frac{1}{n}}\right\}.$$

If $|y| > Y_2$, *then*

$$|\Lambda| < \frac{1.39c_1c_3}{c_2}|y|^{-n}.$$

Proof The left-hand side of (4.39) is equal to $e^\Lambda - 1$. From the definition of c_3 and inequalities (4.37) and (4.38), we have

$$|e^\Lambda - 1| < c_3 \cdot \frac{c_1|y|^{-(n-1)}}{c_2|y|} = \frac{c_1c_3}{c_2}|y|^{-n} < \frac{1}{2}.$$

We will use the following simple fact: if $a \in \langle 0, 1 \rangle$ and $0 < |X| < a$, then

$$|\log(X + 1)| < \frac{-\log(1 - a)}{a} \cdot |X|. \qquad (4.40)$$

Indeed,

$$|\log(X + 1)| = \left|\sum_{i=1}^{\infty} \frac{(-1)^{i-1}X^i}{i}\right| < |X| \cdot \sum_{i=1}^{\infty} \frac{a^{i-1}}{i} = |X| \cdot \frac{-\log(1 - a)}{a}.$$

Therefore,

$$|\Lambda| \leq 2\log 2 \cdot |e^\Lambda - 1| \leq \frac{1.39c_1c_3}{c_2}|y|^{-n}.$$

$\square$

Recall that $\beta_i = \mu_i(\varepsilon_1^{(i)})^{a_1} \cdots (\varepsilon_r^{(i)})^{a_r}$. Then (4.39) becomes

$$\frac{\vartheta_{i_0} - \vartheta_j}{\vartheta_{i_0} - \vartheta_k} \cdot \frac{\mu_k}{\mu_j} \cdot \prod_{i=1}^{r}\left(\frac{\varepsilon_i^{(k)}}{\varepsilon_i^{(j)}}\right)^{a_i} - 1 = -\frac{\vartheta_k - \vartheta_j}{\vartheta_k - \vartheta_{i_0}} \cdot \frac{\mu_{i_0}}{\mu_j} \cdot \prod_{i=1}^{r}\left(\frac{\varepsilon_i^{(i_0)}}{\varepsilon_i^{(j)}}\right)^{a_i}.$$

In the real case, our linear form Λ is

$$\log\left|\frac{\vartheta_{i_0} - \vartheta_j}{\vartheta_{i_0} - \vartheta_k} \cdot \frac{\mu_k}{\mu_j}\right| + \sum_{i=1}^{r} a_i \cdot \log\left|\frac{\varepsilon_i^{(k)}}{\varepsilon_i^{(j)}}\right|.$$

Let $A = \max\{|a_i| : i = 1, \ldots, r\}$. We want to get an upper bound for A. If we manage to find a relatively small bound for A, then we will be able to examine all possibilities for $\mu \cdot \varepsilon_1^{a_1} \cdots \varepsilon_r^{a_r}$, check which of them has the form $x - \vartheta y$ and finally check if $F(x, y) = m$.

First, we will show how to obtain an upper bound for A in terms of y. For $I = \{i_1, \ldots, i_r\} \subset \{1, \ldots, s + t\}$, we define the matrix $U_I = (\log |\varepsilon_i^{(i_l)}|)_{1 \leq l, i \leq r}$. Note that from $\beta = \mu \cdot \varepsilon_1^{a_1} \cdots \varepsilon_r^{a_r}$, it follows that

$$
U_I \begin{bmatrix} a_1 \\ \vdots \\ a_r \end{bmatrix} = \begin{bmatrix} \log \left| \frac{\beta_{i_1}}{\mu_{i_1}} \right| \\ \vdots \\ \log \left| \frac{\beta_{i_r}}{\mu_{i_r}} \right| \end{bmatrix}.
$$

Analyzing this relation, the following inequality can be obtained

$$A < c_5 \log(c_4 |y|). \tag{4.41}$$

The constants c_4 and c_5 are defined as follows:

$$U_I^{-1} = [u_{il}],$$

$$N(U_I^{-1}) = \max\left\{ \sum_{l=1}^{r} |u_{il}| : i = 1, \ldots, r \right\},$$

$$\mu_- = \min\{|\mu_i| : \mu \in M, \; i = 1, \ldots, n\},$$

$$\mu_+ = \max\{|\mu_i| : \mu \in M, \; i = 1, \ldots, n\},$$

$$c_4 = \frac{\frac{1}{2} + \max\{|\vartheta_{i_1} - \vartheta_{i_2}| : 1 \leq i_1 < i_2 \leq n\}}{\mu_-},$$

$$c_5 = \min\{(n - 1) \cdot \min_I N(U_I^{-1}), \max_I N(U_I^{-1})\}.$$

Then it can be shown that (4.41) is valid as soon as $|y| > \max\{Y_1, 2|m|^{\frac{1}{n}}, \frac{\mu_+}{c_2}\}$.

Combining Lemma 4.5.2 and (4.41), we get the inequality

$$|\Lambda| < c_6 \cdot e^{-\frac{n}{c_5} A}, \tag{4.42}$$

where $c_6 = \frac{1.39 c_1 c_3 c_4^n}{c_2}$. On the other hand, from Baker's theory of linear forms in logarithms of algebraic numbers (for example, the Baker-Wüstholz theorem from Sect. 4.1.4 applied to the form Λ), we obtain an inequality of the form

$$|\Lambda| > e^{-c_7 (\log A + c_8)}. \tag{4.43}$$

By comparing the inequalities (4.42) and (4.43), we obtain a (very large) upper bound for A. Then we reduce that large bound by applying the LLL reduction to inequality (4.42) (see Sect. 4.6 or [78]).

In 1993, Tzanakis [343] described how one class of Thue equations of the fourth degree can be reduced to the problem of solving a system of Pellian equations. This is the case when $\mathbb{Q}(\theta_i)$ is the compositum of two real quadratic fields, i.e. $\mathbb{Q}(\theta_i) = \mathbb{Q}(\sqrt{d_1}, \sqrt{d_2})$, where $d_1, d_2 > 0$. The method is especially useful in solving Thue inequalities $|f(x, y)| \le m$ because from the continued fraction expansions of $\sqrt{d_1}$ and $\sqrt{d_2}$, by results from Diophantine approximations (e.g. Worley's theorem 4.1.5), we can eliminate some of those $|\mu| \le m$ for which the equation $f(x, y) = \mu$ does not have solutions (see, e.g. [131]).

In 1996, Bilu and Hanrot [37] proposed an algorithm for solving Thue equations which enables the solving of Thue equations of a fairly large degree in a reasonable time (if all the necessary data on the corresponding number field are available). As an illustration of their algorithm, they solved some Thue equations of degrees 19 and 33. The algorithm is implemented in MAGMA and in PARI through the function `thue`. To solve the equation $F(x, y) = m$, first, the polynomial $f(x) = F(x, 1)$ is initialized with `tnf = thueinit(f)`, and then the function `thue(tnf,m)` is called.

4.6   Application of Elliptic Logarithms

Previously described methods for finding integer points on elliptic curves do not use the Mordell-Weil group. We will now show how knowing the rank and the Mordell-Weil basis can be used to find integer points on an elliptic curve.

As we have already seen in Example 2.5.1, when the rank is equal to 0 (and we manage to prove it), by the Lutz-Nagell theorem, we can find all rational and then also all integer points on that elliptic curve.

In general, the so-called elliptic logarithms can be used to obtain an estimate $N \le N_0$ for $N = \max\{|n_1|, \ldots, |n_r|\}$ in the representation of an integer point in the form $P = n_1 P_1 + \cdots + n_r P_r + T$. Then this bound can be significantly reduced by the LLL algorithm. This method was proposed in 1994 by Gebel, Pethő, and Zimmer [192] and Stroeker and Tzanakis [340]. It is also described in [70, Chapter 8.7] and [345]. For this method to be applicable, it is necessary to know the rank and generators $P_1, \ldots, P_r$, which can be a very difficult problem.

Consider an elliptic curve given by the Weierstrass equation with integer coefficients

$$E : y^2 + a_1 xy + a_3 y = x^3 + a_2 x^2 + a_4 x + a_6.$$

It is isomorphic to a curve with the equation

$$E' : Y^2 = 4X^3 - g_2 X - g_3.$$

This is exactly the equation satisfied by the Weierstrass $\wp$-function and its derivative. We already said that function $\wp$ is doubly periodic. We can assume that its periods satisfy $\omega_1 \in \mathbb{R}$ and $\mathrm{Im}(\omega_1/\omega_2) > 0$. Let L be the lattice that corresponds to ω_1 and ω_2. We have an isomorphism $\phi : \mathbb{C}/L \to E$, given by

$$z \mapsto \begin{cases} (\wp(z) - \frac{b_2}{12}, (\wp'(z) - a_1 x - a_3)/2), & z \notin L, \\ O, & z \in L, \end{cases}$$

where $b_2 = a_1^2 + 4a_2$. The inverse function ψ is called the *elliptic logarithm*. It can be computed as

$$\psi(P) = \int_{\infty}^{x(P)+b_2/12} \frac{dt}{\sqrt{4t^3 - g_2 t - g_3}} \ (\mathrm{mod}\ L) \tag{4.44}$$

(using the arithmetic-geometric mean, see [69, Section 7.4] and [332, Section XIII.2]). The following property justifies the mention of logarithms in the name

$$\psi(P + Q) = \psi(P) + \psi(Q) \ (\mathrm{mod}\ L).$$

The principal determination of the elliptic logarithm is the unique one belonging to the parallelogram $\{x_1\omega_1 + x_2\omega_2,\ -1/2 < x_1, x_2 \le 1/2\}$.

Let P be an integer point on the curve E. The point P can be written in the form $P = n_1 P_1 + \cdots + n_r P_r + T$, where $T \in E(\mathbb{Q})_{\mathrm{tors}}$, and $\{P_1, \ldots, P_r\}$ is the Mordell-Weil basis of $E(\mathbb{Q})$. We would like to find an upper bound for the number $N = \max\{|n_1|, \ldots, |n_r|\}$, because after that we would have finitely many linear combinations to examine, and we could (in principle) find all integer points.

Integer points on the odd (compact) component $E^{gg}(\mathbb{Q})$ are easy to find, so in the following, we will assume that $P \in E^0(\mathbb{Q})$. We would like to transfer elements of the Mordell-Weil basis to the even component. For this purpose, we define: $m_i = 1$ if $P_i \in E^0(\mathbb{Q})$, and $m_i = 2$ otherwise. Then for all $i = 1, \ldots, r$ we have that $Q_i = m_i P_i \in E^0(\mathbb{Q})$. Furthermore, we define the numbers q_i, r_i by

$$n_i = m_i q_i + r_i, \quad 0 \le r_i < m_i.$$

If we now put $U = r_1 P_1 + \cdots r_r P_r$, then the integer point P can be written in the form

$$P = q_1 Q_1 + \cdots + q_r Q_r + T + U.$$

Since $P \in E^0(\mathbb{Q})$ and $Q_i \in E^0(\mathbb{Q})$, we must also have $T + U \in E^0(\mathbb{Q})$. We introduce the notation $Q_{r+1} = T + U$. We have only finitely many possibilities for Q_{r+1}, which we can effectively determine (if we know the torsion group and the Mordell-Weil basis). Let us put $H = \max\{|q_i| : i = 1, \ldots r\}$. It is clear that $H \le N$. If we find an upper bound for H, that will be enough to solve our problem.

Our starting point is the estimate from the following lemma. Here and until the end of this section, the constants $c_1, c_2, \ldots$ depend only on E and its Mordell-Weil basis.

Lemma 4.6.1 *For every integer point P, with the above notations, we have*

$$\frac{1}{x(P)} \le c_1 e^{-c_2 H^2}. \tag{4.45}$$

Proof Since the point P has integer coordinates, we have $h(P) = \log |x(P)|$. According to Proposition 2.6.7, there exists a constant c_3 such that $h(P) \ge \hat{h}(P) - c_3$, so $\log |x(P)| \ge \hat{h}(P) - c_3$. Let $R = (\langle P_i, P_j \rangle)$ be the height matrix of E. Then $\hat{h}(P) = nRn^\tau$, where $n = (n_1, \ldots, n_r)$. The matrix R is symmetric, so it can be represented in the form ODO^τ, where O is an orthogonal matrix and D is a diagonal matrix consisting of (real positive) eigenvalues of R. Let us put $c_2 = \min\{D_{i,i} : i = 1, \ldots r\}$, and $m = nO$. Then

$$\hat{h}(P) = nRn^\tau = mDm^\tau = \sum_{i=1}^{r} D_{i,i} m_i^2$$

$$\ge c_2 \sum_{i=1}^{r} m_i^2 = c_2 mm^\tau = c_2 nOO^\tau n^\tau = c_2 nn^\tau$$

$$= c_2 \sum_{i=1}^{r} n_i^2 \ge c_2 N^2.$$

If we put $c_1 = e^{c_3}$ and notice that $N \ge H$, we get inequality (4.45). $\qquad\square$

On the other hand, by examining the integral appearing in (4.44), one can prove that the inequality

$$|\psi(P)|^2 \le \frac{c_5}{|x(P)|} \tag{4.46}$$

holds for all points $P \in E^0(\mathbb{Q})$ and $|x(P) + b_2/12| > c_4$. Here one may take $c_4 = 2\max\{|e_1|, |e_2|, |e_3|\}$, where e_i are roots of the polynomial $f(X) = 4X^3 - g_2 X - g_3$, and $c_5 = 8 + \frac{|\omega_1^2 b_2|}{12}$ (see [70, Section 8.7] for the missing details).

By combining inequalities (4.45) and (4.46), we get that for all integer points $P \in E^0(\mathbb{Q})$ for which $|x(P) + b_2/12| > c_4$, it holds

$$|\psi(P)| \le \sqrt{c_5 c_1} e^{-c_2 H^2/2} = c_6 e^{-c_7 H^2}. \tag{4.47}$$

By the additivity modulo L of the elliptic logarithms and the choice of the principal determination, we have

$$\psi(P) = q_1\psi(Q_1) + \cdots + q_r\psi(Q_r) + \psi(Q_{r+1}) + m\omega_1,$$

where $|m| \le rN + 2$.

Now we can use David's result from 1995 [77] (it can be understood as an analogue of Baker's results on linear forms in logarithms of algebraic numbers from Sect. 4.1.4), which gives us an inequality of the form

$$|\psi(P)| > e^{-c_8(\log H + c_9)(\log\log H + c_{10})^{r+2}}. \tag{4.48}$$

Comparing (4.47) and (4.48), we get that $H \le H_0$, where H_0 is a (very large) constant (usually something like 10^{100}). However, by the LLL reduction, this huge upper bound can be significantly reduced. That gives a new bound $H \le H_1$, where H_1 is usually around 10 (H_1 has about the same magnitude as $\sqrt{\log H_0}$). Therefore, if the rank r is not too large (say, if $r \le 8$), we can test all the $(2H_1 + 1)^r$ candidates and find all integer points on the starting elliptic curve E.

Example 4.6.2 Let us determine all integer points on the elliptic curve

$$E \ : \ y^2 + y = x^3 - 7x + 6.$$

Using the program `mwrank` (in which the general 2-descent is implemented; it is applicable for curves that do not have points of order 2), we get that the rank of $E(\mathbb{Q})$ is equal to 3, and that a Mordell-Weil basis of E is given by $Q_1 = (1, 0)$, $Q_2 = (2, 0)$, $Q_3 = (0, 2)$ (this is the curve of rank 3 with the smallest conductor). Substituting $Y - 2y + 1$, $X - x$, we get the equation

$$Y^2 = 4X^3 - 28X + 25 = f(X).$$

The roots of the polynomial $f(X)$ are $e_1 = -3.0124$, $e_2 = 1.0658$, $e_3 = 1.9466$. Therefore, the integer points on $E^{gg}(\mathbb{Q})$ satisfy $-3 \le x(P) \le 1$, and it is easy to see that all such points are $(x, y) = (-3, 0)$, $(-2, 3)$, $(-1, 3)$, $(0, 2)$, $(1, 0)$ (and their opposite points).

We see that the points Q_1 and Q_3 are from $E^{gg}(\mathbb{Q})$. Following [70, Section 8.7.4], we will replace them by $Q_1' = -Q_1 - Q_2 - Q_3 = (4, 6)$, $Q_3' = -2Q_2 - 2Q_3 = (\frac{114}{49}, \frac{-720}{343})$ from $E^0(\mathbb{Q})$. So, we are looking for points P in the form $P = q_1Q_1' + q_2Q_2 + q_3Q_3' + U$. Since E has the trivial torsion group, we have $U = O$ or $-Q_2 - Q_3$, so since we are looking for points from $E^0(\mathbb{Q})$, we have that, in fact, $U = O$. By applying the method described above, we obtain the bound $H \le H_0 = 10^{60}$.

Now this very large upper bound can be significantly reduced by an application of the LLL reduction.

We will present de Weger's algorithm [78], which can be applied in many other similar situations. Let us consider an inequality

$$|\alpha_0 + x_1\alpha_1 + \cdots + x_n\alpha_n| < C_2 e^{-C_3 X^q}, \tag{4.49}$$

where α_i are given real or complex numbers, C_2 and C_3 are positive real constants, $q \in \mathbb{N}$, and $X = \max\{|x_1|, \ldots, |x_n|\}$. We are looking for solutions in the integers $x_1, \ldots, x_n$. Assume that it is known that $X \le X_0$, where X_0 is some (large) constant. We want to get a new upper bound of the shape $X^q \le c \log X_0$. We will consider the case when all α_i are real.

Let us choose the constant $C \approx X_0^n$. To the linear form $\alpha_0 + \sum_{i=1}^n x_i\alpha_i$, we assign the lattice L generated by the columns of the matrix

$$A = \begin{bmatrix} 1 & \cdots & 0 & 0 \\ \vdots & \ddots & 0 & 0 \\ 0 & \cdots & 1 & 0 \\ \lfloor C\alpha_1 \rceil & \cdots & \lfloor C\alpha_{n-1} \rceil & \lfloor C\alpha_n \rceil \end{bmatrix}.$$

Here $\lfloor \alpha \rceil$ denotes the nearest integer to the real number α. We chose the constant C to be approximately equal to X_0^n because then we can expect from the properties of the LLL-reduced basis that the shortest vector of the LLL-reduced basis will have the norm approximately X_0. We can also use the LLL algorithm to find a lower bound C_4 for the quantity

$$l(L, y) = \begin{cases} \min\{\|x - y\| : x \in L\}, & y \notin L, \\ \min\{\|x\| : x \in L, \ x \ne 0\}, & y \in L, \end{cases}$$

where $y = [0, \ldots, 0, -\lfloor C\alpha_0 \rceil]^\tau$.

Lemma 4.6.3 *Let $S = (n-1)X_0^2$ and $T = \frac{1+nX_0}{2}$. If $C_4^2 \ge T^2 + S$, then*

$$X^q \le \frac{1}{C_3}\left(\log(CC_2) - \log(\sqrt{C_4^2 - S} - T)\right),$$

or $x_1 = \cdots = x_{n-1} = 0$, $x_n = -\frac{\lfloor C\alpha_0 \rceil}{\lfloor C\alpha_n \rceil}$.

Proof Let $\varphi = \lfloor C\alpha_0 \rceil + \sum_{i=1}^n x_i \lfloor C\alpha_i \rceil$. Then

$$\left|\varphi - C\left(\alpha_0 + \sum_{i=1}^n x_i\alpha_i\right)\right| \le \frac{1}{2} + \sum_{i=1}^n \frac{X_0}{2} = T.$$

Therefore, $|\varphi| \le T + C \cdot C_2 e^{-C_3 X^q}$. Let $x = [x_1, \ldots, x_n]^\tau$ and $z = Ax$. Then we have $z - y = [x_1, \ldots, x_{n-1}, \varphi]^\tau$. Since $z \in L$, we have that either $z = y$ (and then

$$x_1 = \cdots = x_{n-1} = 0 \text{ and } x_n = -\frac{\lceil C\alpha_0 \rceil}{\lceil C\alpha_n \rceil}) \text{ or}$$

$$C_4^2 \le l(L, y)^2 \le \sum_{i=1}^{n-1} x_i^2 + \varphi^2 \le S + \left(T + CC_2 e^{-C_3 X^q}\right)^2.$$

By assumption, we have $C_4^2 \ge S$, so we get

$$e^{-C_3 X^q} \ge \frac{1}{CC_2}(\sqrt{C_4^2 - S} - T). \tag{4.50}$$

Using the assumption that $C_4^2 \ge T^2 + S$, from (4.50), by taking the logarithm, we get

$$X^q \le \frac{1}{C_3}\left(\log(CC_2) - \log(\sqrt{C_4^2 - S} - T)\right).$$

$\square$

Let us go back to our example. Inequality (4.49) has the form

$$|m\omega_1 + q_1\psi(Q_1') + q_2\psi(Q_2) + q_3\psi(Q_3')| \le 58.21 e^{-0.1614H^2}$$

(it was obtained under the condition $|x(P)| \ge 7$). Let us choose $C = 10^{250}$, and apply the procedure from Lemma 4.6.3. We get a new, significantly smaller, upper bound $H \le 51$. The process can be continued, and this bound can be further reduced. If we take $C = 10^9$, we get that $H \le H_1 = 11$.

Now it remains to find all integer points on $E^0(\mathbb{Q})$ that satisfy $|x(P)| \le 6$, which are

$$(-3, 0), \ (-2, 3), \ (-1, 3), \ (0, 2), \ (1, 0), \ (2, 0), \ (3, 3), \ (4, 6),$$

(and their opposites $-(x, y) = (x, -y - 1)$), and all integer points of the form $q_1 Q_1' + q_2 Q_2 + q_3 Q_3'$, $|q_i| \le 11$ (in total $23^3 = 12167$ possibilities), and these are (apart from those already listed)

$$(8, 21) = Q_1' - Q_2' + Q_3', \ (11, 35) = Q_1' + 2Q_2', \ (14, 51) = -2Q_1' - Q_3',$$

$$(21, 95) = -2Q_2', \ (37, 224) = Q_2' - Q_3', \ (52, 374) = Q_1' + Q_2' + Q_3',$$

$$(93, 896) = -2Q_1' - Q_2, \ (342, 6324) = -3Q_2' + Q_3,$$

$$(406, 8180) = -2Q_1' + 2Q_2' - Q_3', \ (816, 23309) = 3Q_1' + 2Q_2' + Q_3'$$

(and their opposites).

The described method that uses elliptic logarithms for finding integer points on an elliptic curve is implemented in software packages SageMath (function `integral_points`) and Magma (function `IntegralPoints`). Let us mention that in Magma, there is also the function `IntegralQuarticPoints`, based on the algorithm due to Tzanakis [344], which finds integer points on a genus 1 curve of the form $y^2 = ax^4 + bx^3 + cx^2 + dx + e$.

4.7   Baker-Davenport Theorem

In this section, we will present the result proved by Baker and Davenport in 1969 [24] on the uniqueness of the extension of the Diophantine triple $\{1, 3, 8\}$ to a Diophantine quadruple. First, we will use this example to illustrate the transformation of an elliptic equation to finitely many Thue equations and apply the previously mentioned implementation for solving Thue equations. And then, we will present the original Baker-Davenport proof (with minor modifications), which leads more directly to linear forms in logarithms and where we will also provide details that are "hidden" when using ready-made programs. In doing so, we will also solve a more general problem than the one considered by Baker and Davenport. Namely, we will determine all integer points on the elliptic curve

$$y^2 = (x + 1)(3x + 1)(8x + 1). \tag{4.51}$$

According to Conjecture 1.4.5, we expect that the integer points on (4.51) are $(-1, 0)$, $(0, \pm 1)$ and $(120, \pm 6479)$ (note that here $d_- = 0$). In particular, we expect that there will be no integer points except the point $(-1, 0)$ of order 2 and points that satisfy the condition that all three linear factors in the cubic polynomial in (4.51) are perfect squares. Let us first prove the last statement using the following lemma.

Lemma 4.7.1 *Let x, y, z be integers such that $xyz \neq 0$ and xyz is a perfect square. Then there are integers u, v, w and square-free integers α, β, γ such that*

$$x = \alpha\beta u^2, \quad y = \alpha\gamma v^2, \quad z = \beta\gamma w^2.$$

Proof Let $x = x'u^2$, $y = y'v^2$, $z = z'w^2$, where x', y', z' are square-free integers. Then $x'y'z'$ is a perfect square. Let $x' = (-1)^{\epsilon_1(-1)} \prod_p p^{\epsilon_1(p)}$, $y' = (-1)^{\epsilon_2(-1)} \prod_p p^{\epsilon_2(p)}$, $z' = (-1)^{\epsilon_3(-1)} \prod_p p^{\epsilon_3(p)}$ be the canonical decompositions of the numbers x', y', z' into prime factors. From the conditions that x', y', z' are square-free and $x'y'z'$ is a perfect square, it follows that $\epsilon_1(p) + \epsilon_2(p) + \epsilon_3(p) = 0$ or 2 for every prime number p, and also for $p = -1$. Let $P_1 = \{p : p \text{ prime or } -1, \epsilon_1(p) = \epsilon_2(p) = 1\}$, $P_2 = \{p : p \text{ prime or } -1, \epsilon_1(p) = \epsilon_3(p) = 1\}$, $P_3 = \{p : p \text{ prime or } -1, \epsilon_2(p) = \epsilon_3(p) = 1\}$, and put $\alpha = \prod_{p \in P_1} p$, $\beta = \prod_{p \in P_2} p$, $\gamma = \prod_{p \in P_3} p$. Then $x' = \alpha\beta$, $y' = \alpha\gamma$ and $z' = \beta\gamma$. □

Let integers x and y satisfy Eq. (4.51). If $y = 0$, then $x = -1$, so we can further assume that $y \neq 0$. Since the product $(x + 1)(3x + 1)(8x + 1)$ is positive, we have that $x \in (-1, -\frac{1}{3}) \cup (-\frac{1}{8}, \infty)$, so since x is an integer, we conclude that $x \geq 0$.

According to Lemma 4.7.1, there are integers u, v, w and square-free integers α, β, γ such that

$$x + 1 = \alpha\beta u^2, \quad 3x + 1 = \alpha\gamma v^2, \quad 8x + 1 = \beta\gamma w^2.$$

Since $x \geq 0$, the numbers α, β, γ have the same sign, so we can assume they are all positive. Since α divides $3(x + 1) - (3x + 1) = 2$, we conclude that $\alpha = 1$ or 2. Analogously, β divides $8(x + 1) - (8x + 1) = 7$, so $\beta = 1$ or 7, and γ divides $8(3x + 1) - 3(8x + 1) = 5$, so $\gamma = 1$ or 5. Now from $\beta\gamma w^2 = 8x + 1 \equiv 1 \pmod 8$, it follows that $\beta = \gamma = 1$, and then from $\alpha\gamma v^2 = 3x + 1 \equiv 1 \pmod 3$, it follows that $\alpha = 1$. Thus, we proved that all integer points on the elliptic curve (4.7.1), except for the torsion point $(-1, 0)$, satisfy the system of equations

$$x + 1 = u^2, \quad 3x + 1 = v^2, \quad 8x + 1 = w^2. \tag{4.52}$$

In the next step, we will transform the system (4.52) into a finite number of Thue equations. First, we will assign the appropriate ternary Diophantine equation to the system. Eliminating x from system (4.52), we get the system of simultaneous Pellian equations

$$3u^2 - v^2 = 2, \tag{4.53}$$

$$8u^2 - w^2 = 7. \tag{4.54}$$

Later we will show how this system can be solved by solving each of the individual Pellian equations and then finding the intersection of (infinite) sets of solutions using linear forms in logarithms and Baker-Davenport reductions (this was the original approach of Baker and Davenport). However, here we follow the approach via Thue equations. We multiply (4.53) by 7 and (4.54) by 2 and subtract the equations that we obtain. Thus we get the ternary Diophantine equation

$$5u^2 - 7v^2 + 2w^2 = 0. \tag{4.55}$$

We are looking for a general solution to the ternary Diophantine equation in the form

$$u = ru_0, \quad v = rv_0 + s, \quad w = rw_0 + t, \tag{4.56}$$

where (u_0, v_0, w_0) is one non-trivial solution. In our case, we can take $(u_0, v_0, w_0) = (1, 1, 1)$. Inserting it into (4.55), we obtain a linear equation in terms of r, whose solution is

$$r = \frac{7s^2 - 2t^2}{2(-7s + 2t)}.$$

By inserting in (4.56), we obtain all rational solutions of the equation. However, we are interested in integer solutions. By inserting and eliminating the denominators and common factors of s and t, we get

$$gu = 7s^2 - 2t^2,$$

$$gv = -7s^2 + 4st - 2t^2,$$

$$gw = 7s^2 - 14st + 2t^2,$$

for some integer g. It remains to determine which values g can take. We will establish this by writing this last system in the matrix form:

$$g \begin{pmatrix} x \\ y \\ z \end{pmatrix} = A \begin{pmatrix} s^2 \\ st \\ t^2 \end{pmatrix}.$$

From here, we have

$$g \operatorname{adj}(A) \begin{pmatrix} x \\ y \\ z \end{pmatrix} = \det A \begin{pmatrix} s^2 \\ st \\ t^2 \end{pmatrix}.$$

Here, $\operatorname{adj}(A)$ is the adjoint of the matrix A, i.e. the matrix that at the position (i, j) has the element $(-1)^{i+j} \det A_{ij}^{\tau}$, where A_{ij}^{τ} is the matrix obtained from the transposed matrix A^{τ} by dropping the ith row and jth column. In the previous equality, we used the formula $\operatorname{adj}(A)A = A \operatorname{adj}(A) = \det(A)I$, where I is the unit matrix. Since s and t are relatively prime (we divided them by their gcd), we conclude that $g \mid \det A$. In our case, this means that $g \mid 280$.

Let us now return to Eq. (4.53). We get

$$2g^2 = g^2(3u^2 - v^2) = 3(gu)^2 - (gv)^2 = 8t^4 + 16t^3s - 128t^2s^2 + 56ts^3 + 98s^4,$$

i.e. finitely many Thue equations

$$4t^4 + 8t^3s - 64t^2s^2 + 28ts^3 + 49s^4 = g^2,$$

where $g \in \{1, 2, 4, 5, 7, 8, 10, 14, 20, 28, 35, 40, 56, 70, 140, 280\}$.

Using the program `thue` from the software package PARI, we find that solutions with relatively prime t and s exist for $g = 2$ (which are $(t, s) = (\pm 1, 0)$, $(\pm 5, \pm 2)$), $g = 5$ (which are $(\pm 1, \pm 1)$, $(\pm 2, \mp 3)$, $g = 7$ (which are $(0, \pm 1)$, $(\pm 7, \pm 5)$) and $g = 70$ (which are $(\pm 7, \pm 2)$, $(\pm 21, \mp 4)$. When we insert these solutions in the expressions for u, v, w, we get $(u, v, w) = (\pm 1, \pm 1, \pm 1)$ and $(u, v, w) = (\pm 11, \pm 19, \pm 31)$ (for all possible combinations of signs). Finally, we get integer points on the curve (4.51): $(x, y) = (0, \pm 1)$ and $(x, y) = (120, \pm 6479)$.

All integer points on the curve (4.51) can also be obtained by applying the methods from Sect. 4.6, i.e. the programs in Magma listed at the end of that section. Recall that in Example 2.5.2, we showed that the rank of curve (4.51) is equal to 1, while in Example 2.6.6, we showed that the point $P = (0, 1)$ can be taken as a generator.

To apply the function `IntegralPoints`, we will take the equation

$$y^2 = (x + 1 \cdot 3)(x + 1 \cdot 8)(x + 3 \cdot 8) = x^3 + 35x^2 + 288x + 576$$

and discard those points whose coordinates are not divisible by 24. We get that the integer points on the modified curve are $(-24, 0)$, $(-18, \pm 30)$, $(-10, \pm 14)$, $(-8, 0)$, $(-3, 0)$, $(0, \pm 24)$, $(1, \pm 30)$, $(32, \pm 280)$, $(2880, \pm 155496)$, of which the condition of divisibility by 24 is satisfied only by $(-24, 0)$, $(0, \pm 24)$ and $(2880, \pm 155496)$ so, as before, we get points on (4.51) with x-coordinates -1, 0 and 120.

We can find the solution also by applying the function `IntegralQuartic Points`. Namely, from Eqs. (4.53) and (4.54), we get that $3u^2 - 2$ and $8u^2 - 7$ are squares, from where, by multiplying, we get the quartic $24u^4 - 37u^2 + 14 = z^2$. Now Magma for `IntegralQuarticPoints([24,0,-37,0,14])` gives us that all integer points on the quartic are $(\pm 1, \pm 1)$ and $(\pm 11, \pm 589)$, so from $x = u^2 - 1$, we get $x = 0$ and $x = 120$ (we already noticed that $x = -1$ is the only solution for $y = 0$, so we continued working with the assumption that $y \neq 0$).

It remains to present the announced Baker-Davenport proof of the uniqueness of the extension of the Diophantine triple $\{1, 3, 8\}$ to a quadruple with $d = 120$. As we have already argued, this result also finds all integer points on the elliptic curve $y^2 = (x + 1)(3x + 1)(8x + 1)$.

We have already mentioned that many problems on Diophantine equations can be transformed into inequalities for linear forms in logarithms of algebraic numbers. These inequalities usually have the form

$$|n_1 \log \alpha_1 + \cdots + n_k \log \alpha_k| < c_1 e^{-c_2 N},$$

where $n_1, \ldots, n_k \in \mathbb{Z}$, $N = \max\{|n_1|, \ldots, |n_k|\}$, and c_1, c_2 are positive constants. Then the results from Sect. 4.1.4 can be applied (e.g. the Baker-Wüstholz theorem). This gives an inequality of the form

$$c_3 \log N > c_2 N - \log c_1,$$

from which it follows that $N \leq N_0$, where N_0 is an explicit constant (usually quite large, e.g. 10^{20} or 10^{100}). Because N_0 is most likely too large (except when $k = 2$) to examine all possible remaining cases, the question arises whether this bound can be reduced. The answer is affirmative. Namely, two algorithms are known which (if certain technical conditions are met) replace the bound $N \leq N_0$ by $N \leq N_1$, where $N_1 \approx \log N_0$.

We have already encountered one of those two methods, the LLL reduction, and now we will present another ("older") method, the *Baker-Davenport reduction*. It was introduced by Baker and Davenport in the paper [24] from 1969 in which they proved that if d is a positive integer such that $\{1, 3, 8, d\}$ is a Diophantine quadruple, then d should be equal to 120, and thus at the same time they proved that the Diophantine quadruple $\{1, 3, 8, 120\}$, which was found by Fermat, cannot be extended to a Diophantine quintuple. Harold Davenport (1907–1969) was an English mathematician, and the doctoral supervisor of Alan Baker.

We will first show how the Baker-Wüstholz theorem (or some other result of a similar type) can be applied to the problem of extending the triple $\{1, 3, 8\}$.

Let d be a positive integer such that

$$1 \cdot d + 1 = x^2, \quad 3 \cdot d + 1 = y^2, \quad 8 \cdot d + 1 = z^2.$$

Then we have

$$y^2 - 3x^2 = -2, \tag{4.57}$$

$$z^2 - 8x^2 = -7, \tag{4.58}$$

so we got a system of Pellian equations with the common unknown x.

It is obvious that the fundamental solutions of the corresponding Pell's equations $y^2 - 3x^2 = 1$ and $z^2 - 8x^2 = 1$, are $2 + \sqrt{3}$ and $3 + \sqrt{8}$, respectively. Now from Theorem 4.1.8, we easily obtain that all solutions of Eq. (4.57) are given by

$$y + x\sqrt{3} = \pm(1 \pm \sqrt{3})(2 + \sqrt{3})^n, \tag{4.59}$$

and all solutions of Eq. (4.58) by

$$z + x\sqrt{8} = \pm(1 \pm \sqrt{8})(3 + \sqrt{8})^m. \tag{4.60}$$

We may assume that x is positive. Considering that $(1 - \sqrt{3})(2 + \sqrt{3}) = -(1 + \sqrt{3})$, in (4.59) the minus signs can be omitted (the class is ambiguous), so we get the following exponential equation

$$\frac{1}{2\sqrt{3}} \left((1 + \sqrt{3})(2 + \sqrt{3})^n - (1 - \sqrt{3})(2 - \sqrt{3})^n \right)$$

$$= \frac{1}{4\sqrt{2}} \left((2\sqrt{2} \pm 1)(3 + 2\sqrt{2})^m + (2\sqrt{2} \mp 1)(3 - 2\sqrt{2})^m \right), \tag{4.61}$$

i.e. an equation of the form $v_n = w_m^{+,-}$, where (v_n), $(w_m^{+,-})$ are binary recurrence sequences. We have $v_0 = w_0^{+,-} = 1$, from which we find $d = 0$ (trivial solution) and $v_2 = w_2^- = 11$, from which we deduce that $d = 11^2 - 1 = 120$. We want to prove that there are no other solutions.

It is easy to see that for $n > 2$, $w_n^{+,-} > v_n$, from which it follows that $m < n$.

Lemma 4.7.2 *Let* $m, n > 2$ *be integers satisfying (4.61). Then*

$$0 < |\Lambda| < 7.3 \cdot (2 + \sqrt{3})^{-2n}, \tag{4.62}$$

where

$$\Lambda = n \log(2 + \sqrt{3}) - m \log(3 + 2\sqrt{2}) + \log \frac{2\sqrt{2}(1 + \sqrt{3})}{\sqrt{3}(2\sqrt{2} \pm 1)}.$$

Proof It is obvious that

$$v_n > \frac{(1 + \sqrt{3})(2 + \sqrt{3})^n}{2\sqrt{3}}, \quad w_m^{+,-} < \frac{(2\sqrt{2} + 1)(3 + 2\sqrt{2})^m}{2\sqrt{2}},$$

so from $v_n = w_m^{+,-}$, it follows

$$(3 - 2\sqrt{2})^m < \frac{(2\sqrt{2} + 1)\sqrt{3}}{(\sqrt{3} + 1)\sqrt{2}}(2 - \sqrt{3})^n < 1.7163(2 - \sqrt{3})^n.$$

Now from (4.61), dividing by $\frac{2\sqrt{2} \pm 1}{2\sqrt{2}} \cdot (3 + 2\sqrt{2})^m$, we get

$$\left| \frac{2\sqrt{2}(1 + \sqrt{3})}{\sqrt{3}(2\sqrt{2} \pm 1)} \cdot \frac{(2 + \sqrt{3})^n}{(3 + 2\sqrt{2})^m} - 1 \right|$$

$$\leq \frac{2\sqrt{2} + 1}{2\sqrt{2} - 1} \cdot (3 - 2\sqrt{2})^{2m} + \frac{2\sqrt{2}(\sqrt{3} - 1)}{\sqrt{3}(2\sqrt{2} - 1)}(2 - \sqrt{3})^n (3 - 2\sqrt{2})^m$$

$$< 7.29(2 - \sqrt{3})^{2n}.$$

By applying inequality (4.40) to $X = \frac{2\sqrt{2}(1+\sqrt{3})}{\sqrt{3}(2\sqrt{2}\pm1)} \cdot \frac{(2+\sqrt{3})^n}{(3+2\sqrt{2})^m} - 1$ and $a = 0.0027$ ($\approx 7.29 \cdot (2 - \sqrt{3})^6$), we get the required inequality

$$|\Lambda| < 7.3 \cdot (2 - \sqrt{3})^{2n}.$$

It remains to prove that $|\Lambda| > 0$. If it were $\Lambda = 0$, then by squaring, we would get that

$$16(2 + \sqrt{3})^{2n+1} = 3(9 \pm 4\sqrt{2})(3 + 2\sqrt{2})^{2m},$$

which is a contradiction (because an equality of the form $a + b\sqrt{3} = c + d\sqrt{2}$, $a, b, c, d \in \mathbb{Q}$, is possible only if $b = d = 0$). $\qquad\square$

Now everything is ready to apply the Baker-Wüstholz theorem to the form Λ from Lemma 4.7.2. We have: $\alpha_1 = 2 + \sqrt{3}$, $\alpha_2 = 3 + 2\sqrt{2}$, $\alpha_3 = \frac{2(4\pm\sqrt{2})(3+\sqrt{3})}{21}$, $b_1 = n$, $b_2 = -m$, $b_3 = 1$, $d = 4$ (since $\mathbb{Q}(\alpha_1, \alpha_2, \alpha_3) = \mathbb{Q}(\sqrt{2}, \sqrt{3})$). The minimal polynomials over $\mathbb{Z}$ are

$$P_{\alpha_1}(x) = x^2 - 4x + 1,$$

$$P_{\alpha_2}(x) = x^2 - 6x + 1,$$

$$P_{\alpha_3}(x) = 441x^4 - 2016x^3 + 2880x^2 - 1536x + 256,$$

so the heights are $h(\alpha_1) = \frac{1}{2}\log(2 + \sqrt{3}) \approx 0.6585$, $h(\alpha_2) = \frac{1}{2}\log(3 + 2\sqrt{2}) \approx 0.8814$, $h(\alpha_3) = \frac{1}{4}\log\left(441 \cdot \frac{2(4+\sqrt{2})(3+\sqrt{3})}{21} \cdot \frac{2(4-\sqrt{2})(3+\sqrt{3})}{21}\right) \approx 1.7836$. Thus, the Baker-Wüstholz theorem gives us

$$\log(|\Lambda|) \geq -18 \cdot 4! \cdot 3^4 \cdot (32 \cdot 4)^5 \cdot \log 24 \cdot 0.6585 \cdot 0.8814 \cdot 1.7836 \log n$$

$$\geq -3.96 \cdot 10^{15} \log n.$$

If we compare this with (4.62), we get an inequality

$$3.96 \cdot 10^{15} \log n > 2.63n - 1.99,$$

which implies that $n < 6 \cdot 10^{16}$.

The following lemma from the paper [152] by Dujella and Pethő represents one of the variants of the Baker-Davenport reduction.

Lemma 4.7.3 *Let κ be a real number and N a positive integer. Let $\frac{p}{q}$ be a convergent of the continued fraction expansion of κ such that $q > 6N$ and let μ, A, B be real numbers such that $A > 0$ and $B > 1$. Let $\varepsilon = \|\mu q\| - N \cdot \|\kappa q\|$, where $\|\ \|$ denotes the distance to the nearest integer. If $\varepsilon > 0$, then the inequality*

$$0 < n\kappa - m + \mu < A \cdot B^{-n},$$

has no solution in the positive integers m and n such that

$$\frac{\log(\frac{Aq}{\varepsilon})}{\log B} \leq n \leq N.$$

Proof Suppose that $1 \leq n \leq N$. We have

$$n(\kappa q - p) + np - mq + \mu q < qAB^{-n}.$$

From here, we have

$$qAB^{-n} > |\mu q - (mq - np)| - n\|\kappa q\| \geq \|\mu q\| - N\|\kappa q\| = \varepsilon,$$

which implies that

$$n < \frac{\log(\frac{Aq}{\varepsilon})}{\log B}.$$

$\square$

Remark 4.7.4 The condition $q > 6N$ in the lemma is somewhat arbitrary. Namely, on the one hand, we want to be as sure as possible that the condition $\varepsilon > 0$ holds, and on the other hand, we want q to be as small as possible (so that the new bound is as small as possible). From the properties of convergents of continued fractions, we know that $\|\kappa q\| < \frac{1}{q}$ (see (4.8)), while about $\|\mu q\|$, in general, we only know that it is $\leq \frac{1}{2}$. Therefore, it is reasonable to assume that at least $q > 2N$, and $q > 6N$ is experimentally confirmed as a good choice.

Remark 4.7.5 If the condition $\varepsilon > 0$ is not satisfied, we can try to take the next convergent and check whether the condition will be met. Even if $\varepsilon < 0$, it is possible to get some information about n. Namely, if we denote $r = \lfloor \mu q + \frac{1}{2} \rfloor$, then

$$|np - mq + r| < qAB^{-n} + |\mu q - r| + n|\kappa q - p| \leq qAB^{-n} + \|\mu q\| + N\|\kappa q\|$$

$$< qAB^{-n} + \frac{1}{2} + \frac{1}{6}.$$

If $qAB^{-n} > \frac{1}{3}$, then $n < \frac{\log(3Aq)}{\log B}$. If $qAB^{-n} \leq \frac{1}{3}$, then $np - mq + r = 0$, which means that $np \equiv -r \pmod{q}$. This congruence has a unique solution $n \equiv n_0 \pmod{q}$, so from $n \leq N < q$, it follows that $n = n_0$

Theorem 4.7.6 *If $\{1, 3, 8, d\}$ is a Diophantine quadruple, then $d = 120$.*

Proof Let us apply the reduction from Lemma 4.7.3 to the form Λ from Lemma 4.7.2 with $N = 6 \cdot 10^{16}$. Here $\kappa = \frac{\log(2+\sqrt{3})}{\log(3+2\sqrt{2})}$, $\mu_{1,2} = \frac{\log\left(\frac{2\sqrt{2}(1+\sqrt{3})}{\sqrt{3}(2\sqrt{2}\pm1)}\right)}{\log(3+2\sqrt{2})}$, $A = \frac{7.3}{\log(3+2\sqrt{2})}$, $B = (2 + \sqrt{3})^2$. The continued fraction expansion of κ is

$$[0; 1, 2, 1, 20, 1, 5, 3, 8, 5, 1, 2, 1, 1, 1, 1, 4, 3, 3, 3, 1, 6, 3, 1, 2, 22,$$

$$1, 2, 8, 2, 1, 2, 6, 3, 20, 2, 10, 3, \ldots],$$

so the first convergent of κ that satisfies the condition $q > 6N$ is

$$\frac{p}{q} = \frac{742265900639684111}{993522360732597120}.$$

We will see that for μ_1, we need to take the next convergent, so let us first consider what is obtained for μ_2. We have $\|\mu_2 q\| \approx 0.24492$, $\|\kappa q\| \cdot N \approx 0.01878$, so $\varepsilon \approx 0.22614 > 0$. By applying Lemma 4.7.3, we get that $n \leq 16$. Now we can repeat the reduction with $N = 16$. The corresponding convergent is now $\frac{p'}{q'} = \frac{387}{518}$ and the reduction gives $n \leq 3$.

As we already said, applying Lemma 4.7.3 for μ_1 and p, q given above, we get a negative ε ($\|\mu_1 q\| \approx 0.007626$, $\|\kappa q\| \cdot N \approx 0.01878$). Therefore, we take the next convergent

$$\frac{P}{Q} = \frac{22975706401873544392}{30752966607888933649}.$$

We get $\|\mu_1 Q\| \approx 0.2989$, $\|\kappa Q\| \cdot N \approx 0.002254$, so the corresponding $\varepsilon \approx 0.29665 > 0$, and we can apply the reduction, which gives us $n \leq 17$. If we repeat the reduction for $N = 17$, we get that the corresponding convergent is $\frac{p'}{q'} = \frac{387}{518}$ and the reduction gives us $n \leq 4$.

It is easy to check that the equations $v_n = w_m^{+,-}$ do not have solutions for $n = 3, 4$. Thus we have completed the proof of the claim that if $\{1, 3, 8, d\}$ is a Diophantine quadruple, then it has to be $d = 120$. $\qquad\square$

4.8 Infinite Families of Elliptic Curves

In the previous section, we presented several methods with the help of which, in principle, all integer points on an elliptic curve induced by a Diophantine triple can be determined. Several problems with all these methods appear when we want to determine all integer points on a parametric family of elliptic curves induced by Diophantine triples. To find out what the values of α, β, γ are possible in the system

$$ax + 1 = \alpha\beta u^2, \quad bx + 1 = \alpha\gamma v^2, \quad cx + 1 = \beta\gamma w^2,$$

we should know something about divisors of $b - a$, $c - a$ and $b - c$. These numbers and their factors also appear in the corresponding Thue equations as candidates for the numbers g that should be factors of the determinant of a certain system of equations, and also when computing the rank of the corresponding elliptic curve, which is necessary information for the application of the elliptic logarithm method.

Already for the simplest parametric family of Diophantine triples $\{k - 1, k + 1, 4k\}$, which we encountered in the work of Diophantus, this includes divisors of the numbers $3k - 1$ and $3k + 1$, about which we can say almost nothing for an arbitrary k. If we could somehow conclude (at least under some additional assumptions) that $\alpha = \beta = \gamma = 1$ is the only possibility (or that apart from that, we have some other explicit additional possibilities), then we could try to solve the corresponding systems using the method of linear forms in logarithms. It will, like in the case of the triple $\{1, 3, 8\}$, give us an upper bound for the indices m and n in the sequences of solutions of the corresponding Pellian equations. However, these bounds will depend on a, b, c, i.e. on the parameter k in the case of a parametric family of triples.

To solve the problem, at least for sufficiently large values of the parameter k, we need some method for obtaining a lower bound for the solutions. For this purpose, the "congruence method" is usually used, either in the rather general form introduced in [152] by Dujella and Pethő, or in a more special form that uses properties of a specific family of triples.

If, by comparing the upper and lower bounds, we obtain a contradiction for sufficiently large values of the parameter k, then the problem will be solved for $k \geq k_0$. If k_0 is not too large, we can try to solve the remaining concrete equations by one of the methods described in the previous section.

Additional assumptions that enable the conclusion that $\alpha = \beta = \gamma = 1$ is the only possibility, usually include the assumption that the rank of the elliptic curve is the "smallest possible", i.e. equal to the "generic rank" of the observed family of elliptic curves (the rank over $\mathbb{Q}(k)$ in the case when a, b, c are polynomials in parameter k).

Suppose we want to obtain the same conclusion without assumptions on the rank. In that case, as soon as the elements of the triple (and then also the coefficients of the curve) become large (and when elements of the triple are given by elements of some recurrence sequence this will happen very quickly), the problem of finding fundamental solutions of Pell's equation arises. Namely, in such cases, the standard method by continued fractions is not efficient enough, so the so-called compact representation of quadratic irrationalities should be used (see [223, 294]).

The procedure just described was carried out for several parametric families of Diophantine triples:

- for the triple $\{k - 1, k + 1, 4k\}$ under the assumption that the rank of the curve is equal to 1 (Dujella [101] in 2000) or $k \leq 5000$ (Najman in 2010 [294]); also if $k = 3n^2 + 2n - 2$ or $k = \frac{1}{2}(3m^2 + 5m)$ and the rank is equal to 2 [101];
- for the triples $\{k - 1, k + 1, 16k^3 - 4k\}$ and $\{k - 1, k + 1, 64k^5 - 48k^3 + 8k\}$ under the assumption that the rank of the curve is equal to 2 or $k \leq 10000$, with possible exception $k = 6300$ for the first triple (Najman [293] in 2009);
- for the triple $\{F_{2k}, F_{2k+2}, F_{2k+4}\}$ under the assumption that the rank of the curve is equal to 1 (Dujella [103] in 2001) or $k \leq 50$ [294];
- for triples of the form $\{1, 3, c\}$ under the assumption that the rank of the curve is equal to 2 (Dujella and Pethő [153] in 2000) or $c = c_k$, $k \leq 100$ (for $k = 37$ under the assumption that ERH holds) (Jacobson and Williams [222] in 2002).

In all the mentioned cases, it was obtained that the integer points on the curves are exactly those predicted by Conjecture 1.4.5 ($x \in \{0, -1, d_+, d_-\}$).

In what follows, we will present the mentioned result for triples of the form $\{1, 3, c\}$, with the assumption that the rank of the curve is equal to 2. However, we will prove some auxiliary results in full generality, i.e. for an arbitrary Diophantine triple $\{a, b, c\}$.

4.8.1   System of Generalized Pellian Equations

Let $\{a, b, c\}$ be an (integer) Diophantine triple such that

$$ab + 1 = r^2, \quad ac + 1 = s^2, \quad bc + 1 = t^2,$$

which we want to extend to a quadruple $\{a, b, c, d\}$. Then the positive integer d must satisfy the system of equations

$$ad + 1 = x^2, \quad bd + 1 = y^2, \quad cd + 1 = z^2.$$

Eliminating d from this system, we obtain the system of generalized Pellian equations

$$az^2 - cx^2 = a - c, \tag{4.63}$$

$$bz^2 - cy^2 = b - c. \tag{4.64}$$

Sometimes also the third equation

$$ay^2 - bx^2 = a - b, \tag{4.65}$$

which is of course a direct consequence of (4.63) and (4.64), will be useful. We will often assume that $a < b < c$.

We now describe the set of solutions to Eqs. (4.63) and (4.64) (analogous results hold for (4.65)). Let us first note that $s + \sqrt{ac}$ and $t + \sqrt{bc}$ are non-trivial units with norm 1 in the rings $\mathbb{Z}[\sqrt{ac}]$ and $\mathbb{Z}[\sqrt{bc}]$. Indeed, $s^2 - ac \cdot 1^2 = 1$ and $t^2 - bc \cdot 1^2 = 1$.

The following lemma from [104] is a direct generalization of Theorem 4.1.8.

Lemma 4.8.1 *There are positive integers i_0, j_0 and integers $z_0^{(i)}$, $x_0^{(i)}$, $z_1^{(j)}$, $y_1^{(j)}$, $i = 1, \ldots, i_0$, $j = 1, \ldots, j_0$, such that:*

(i) $(z_0^{(i)}, x_0^{(i)})$ and $(z_1^{(j)}, y_1^{(j)})$ are solutions of Eqs. (4.63) and (4.64), respectively.

(ii) $z_0^{(i)}$, $x_0^{(i)}$, $z_1^{(j)}$, $y_1^{(j)}$ *satisfy the following inequalities:*

$$1 \le x_0^{(i)} \le \sqrt{\frac{a(c-a)}{2(s-1)}} < \sqrt{\frac{s+1}{2}} < 0.841 \sqrt[4]{ac}, \tag{4.66}$$

$$1 \le |z_0^{(i)}| \le \sqrt{\frac{(s-1)(c-a)}{2a}} < \sqrt{\frac{c\sqrt{c}}{2\sqrt{a}}} < 0.421\,c, \tag{4.67}$$

$$1 \le y_1^{(j)} \le \sqrt{\frac{b(c-b)}{2(t-1)}} < \sqrt{\frac{t+1}{2}} < 0.783 \sqrt[4]{bc}, \tag{4.68}$$

$$1 \le |z_1^{(j)}| \le \sqrt{\frac{(t-1)(c-b)}{2b}} < \sqrt{\frac{c\sqrt{c}}{2\sqrt{b}}} < 0.32\,c. \tag{4.69}$$

(iii) *If (z, x) and (z, y) are solutions in positive integers of Eqs. (4.63) and (4.64), respectively, then there exist $i \in \{1, \ldots, i_0\}$, $j \in \{1, \ldots, j_0\}$ and integers $m, n \ge 0$ so that*

$$z\sqrt{a} + x\sqrt{c} = (z_0^{(i)}\sqrt{a} + x_0^{(i)}\sqrt{c})(s + \sqrt{ac})^m, \tag{4.70}$$

$$z\sqrt{b} + y\sqrt{c} = (z_1^{(j)}\sqrt{b} + y_1^{(j)}\sqrt{c})(t + \sqrt{bc})^n. \tag{4.71}$$

Proof It is clear that it is sufficient to prove the statement of the lemma for Eq. (4.63). Let (z, x) be a solution of Eq. (4.63) in positive integers. Let us consider all pairs (z^*, x^*) of integers of the form

$$z^*\sqrt{a} + x^*\sqrt{c} = (z\sqrt{a} + x\sqrt{c})(s + \sqrt{ac})^m, \quad m \in \mathbb{Z}.$$

Obviously, (z^*, x^*) satisfies (4.63) and x^* is a positive integer. Among all pairs (z^*, x^*), let us choose a pair with the property that x^* is minimal, and denote that pair by (z_0, x_0). Let us define the integers z' and x' by

$$z'\sqrt{a} + x'\sqrt{c} = (z_0\sqrt{a} + x_0\sqrt{c})(s - \varepsilon\sqrt{ac}),$$

where $\varepsilon = 1$ if $z_0 > 0$, and $\varepsilon = -1$ if $z_0 < 0$. From the minimality of x_0, we conclude that $x' = sx_0 - \varepsilon a z_0 \ge x_0$, and that implies $a|z_0| \le (s-1)x_0$. Squaring this inequality, we get

$$acx_0^2 + a(a-c) \le (ac + 2 - 2s)x_0^2$$

and finally

$$x_0^2 \leq \frac{a(c-a)}{2(s-1)}.$$

Now we have

$$z_0^2 = \frac{1}{a}(cx_0^2 + a - c) \leq \frac{1}{a}\left(\frac{ac(c-a)}{2(s-1)} + a - c\right) = \frac{(s-1)(c-a)}{2a}. \tag{4.72}$$

Note that $a \geq 1$, $b \geq 3$, and $c \geq 8$ (and so $s \geq 3$ and $t \geq 5$). So, we proved that there is a solution (z_0, x_0) of (4.63) that satisfies inequalities (4.66) and (4.67) (and therefore belongs to a finite set of solutions) and the integer m so that

$$z\sqrt{a} + x\sqrt{c} = (z_0\sqrt{a} + x_0\sqrt{c})(s + \sqrt{ac})^m.$$

It remains to prove that $m \geq 0$. Assume that $m < 0$. Then $(s + \sqrt{ac})^m = (s - \sqrt{ac})^{|m|} = \alpha - \beta\sqrt{ac}$, where α and β are positive integers that satisfy $\alpha^2 - ac\beta^2 = 1$. We have $z = \alpha z_0 - \beta c x_0$, so from the condition $z > 0$, we get $z_0^2 > \beta^2 c(c-a) \geq c(c-a)$, which is obviously in contradiction with (4.72). $\qquad\square$

From (4.70), we conclude that $z = v_m^{(i)}$ for some index i and integer $m \geq 0$, where

$$v_0^{(i)} = z_0^{(i)}, \quad v_1^{(i)} = sz_0^{(i)} + cx_0^{(i)}, \quad v_{m+2}^{(i)} = 2sv_{m+1}^{(i)} - v_m^{(i)}, \tag{4.73}$$

while from (4.71), we conclude that $z = w_n^{(j)}$ for some index j and integer $n \geq 0$, where

$$w_0^{(j)} = z_1^{(j)}, \quad w_1^{(j)} = tz_1^{(j)} + cy_1^{(j)}, \quad w_{n+2}^{(j)} = 2tw_{n+1}^{(j)} - w_n^{(j)}. \tag{4.74}$$

So, we have reduced our problem to solving equations of the form $v_m^{(i)} = w_n^{(j)}$, i.e. searching for the intersection of binary recurrence sequences. Solving recurrences (4.73) and (4.74), we get

$$v_m^{(i)} = \frac{1}{2\sqrt{a}}((z_0^{(i)}\sqrt{a} + x_0^{(i)}\sqrt{c})(s + \sqrt{ac})^m + (z_0^{(i)}\sqrt{a} - x_0^{(i)}\sqrt{c})(s - \sqrt{ac})^m),$$

$$\tag{4.75}$$

$$w_n^{(j)} = \frac{1}{2\sqrt{b}}((z_1^{(j)}\sqrt{b} + y_1^{(j)}\sqrt{c})(t + \sqrt{bc})^n + (z_1^{(j)}\sqrt{b} - y_1^{(j)}\sqrt{c})(t - \sqrt{bc})^n).$$

$$\tag{4.76}$$

Now we will, in a similar way as we did previously in the particular case of the Diophantine triple $\{1, 3, 8\}$, transform the equation $v_m^{(i)} = w_n^{(j)}$ into an inequality for linear forms in logarithms.

Lemma 4.8.2 *Assume that $c > 4b$. If $v_m^{(i)} = w_n^{(j)}$ and $m, n \neq 0$, then*

$$0 < m \log(s + \sqrt{ac}) - n \log(t + \sqrt{bc}) + \log \frac{\sqrt{b}(x_0^{(i)} \sqrt{c} + z_0^{(i)} \sqrt{a})}{\sqrt{a}(y_1^{(j)} \sqrt{c} + z_1^{(j)} \sqrt{b})}$$

$$< \frac{8}{3} ac(s + \sqrt{ac})^{-2m}.$$

Proof Let us put

$$P = \frac{1}{\sqrt{a}}(z_0^{(i)} \sqrt{a} + x_0^{(i)} \sqrt{c})(s + \sqrt{ac})^m, \quad Q = \frac{1}{\sqrt{b}}(z_1^{(j)} \sqrt{b} + y_1^{(j)} \sqrt{c})(t + \sqrt{bc})^n.$$

Then we have

$$P^{-1} = \frac{\sqrt{a}(x_0^{(i)} \sqrt{c} - z_0^{(i)} \sqrt{a})}{c - a}(s - \sqrt{ac})^m,$$

$$Q^{-1} = \frac{\sqrt{b}(y_1^{(j)} \sqrt{c} - z_1^{(j)} \sqrt{b})}{c - b}(t - \sqrt{bc})^n.$$

Therefore, the equation $v_m = w_n$ becomes $P - \frac{c-a}{a} P^{-1} = Q - \frac{c-b}{b} Q^{-1}$. If $m, n \geq 1$, then we have

$$P \geq \frac{1}{\sqrt{a}}(x_0^{(i)} \sqrt{c} - |z_0^{(i)}|\sqrt{a}) \cdot 2\sqrt{ac} = \frac{2\sqrt{c}(c - a)}{x_0^{(i)} \sqrt{c} + |z_0^{(i)}|\sqrt{a}} > \frac{\frac{3}{2}c\sqrt{c}}{c\sqrt{a}} = \frac{3}{2}\sqrt{\frac{c}{a}} > 1,$$

$$Q \geq \frac{2\sqrt{c}(c - b)}{y_1^{(j)} \sqrt{c} + |z_1^{(j)}|\sqrt{b}} > \frac{\frac{3}{2}c\sqrt{c}}{2\sqrt{c}\sqrt[4]{bc}} > 1.$$

Now from

$$P - Q = \left(\frac{c}{a} - 1\right)P^{-1} - \left(\frac{c}{b} - 1\right)Q^{-1} > \left(\frac{c}{a} - 1\right)(P^{-1} - Q^{-1})$$

$$= \left(\frac{c}{a} - 1\right)(Q - P)P^{-1}Q^{-1},$$

we conclude that $P > Q$. Furthermore, we have $\frac{P-Q}{P} < (\frac{c-a}{a})P^{-2}$. Since $P > \frac{3}{2}\sqrt{\frac{c}{a}}$, we have $\frac{c-a}{a}P^{-2} < \frac{4}{9} < \frac{1}{2}$, and finally we get

$$0 < \log \frac{P}{Q} = -\log\left(1 - \frac{P-Q}{P}\right)$$

$$< \frac{2(c-a)}{a}P^{-2} = \frac{2(c-a)}{a} \cdot \frac{a}{(z_0^{(i)}\sqrt{a} + x_0^{(i)}\sqrt{c})^2}(s + \sqrt{ac})^{-2m}.$$

Now the statement of the lemma follows from

$$\frac{2(c-a)}{a} \cdot \frac{a}{(z_0^{(i)}\sqrt{a} + x_0^{(i)}\sqrt{c})^2} = \frac{2(z_0^{(i)}\sqrt{a} - x_0^{(i)}\sqrt{c})^2}{c-a} \leq \frac{2(|z_0^{(i)}|\sqrt{a} + x_0^{(i)}\sqrt{c})^2}{c-a}$$

$$\leq \frac{2ac^2}{\frac{3}{4}c} = \frac{8}{3}ac.$$

$$\square$$

We can now apply the Baker-Wüstholz theorem to the linear form from Lemma 4.8.2. We have: $l = 3$, $d = 4$, $B = m$ (see Lemma 4.10.4 below),

$$\alpha_1 = s + \sqrt{ac}, \qquad \alpha_2 = t + \sqrt{bc},$$

$$\alpha_3 = \frac{\sqrt{b}(x_0^{(i)}\sqrt{c} + z_0^{(i)}\sqrt{a})}{\sqrt{a}(y_1^{(j)}\sqrt{c} + z_1^{(j)}\sqrt{b})}.$$

Furthermore,

$$h'(\alpha_1) = \frac{1}{2}\log\alpha_1 < \frac{1}{2}\log c, \quad h'(\alpha_2) = \frac{1}{2}\log\alpha_2 < \frac{1}{2}\log c,$$

The conjugates of α_3 are

$$\frac{\sqrt{b}(z_0^{(i)}\sqrt{a} \pm x_0^{(i)}\sqrt{c})}{\sqrt{a}(z_1^{(j)}\sqrt{b} \pm y_1^{(j)}\sqrt{c})},$$

and the leading coefficient of the minimal polynomial of α_3 divides $a^2(c-b)^2$. Under assumptions $c > 4b$ and $m \geq 2$, we obtain

$$
\begin{aligned}
h'(\alpha_3) &< \frac{1}{4} \log\left(a^2(c-b)^2 \cdot 2\sqrt[4]{\frac{b^2 c}{a}} \cdot \frac{16}{3}\sqrt[4]{\frac{b^3 c^2}{a}} \cdot \sqrt{\frac{b}{a}} \cdot \frac{8}{3}\sqrt[4]{\frac{b^3 c}{a^2}} \right) \\
&< \frac{1}{4} \log\left(\frac{256}{9} a^{1/2} b^2 c^3 \right) < 1.375 \log c
\end{aligned}
$$

and

$$
\log \frac{8}{3} ac(s + \sqrt{ac})^{-2m} < -\frac{1}{2} m \log c
$$

(see [107] for the missing details and slightly stronger inequalities). Thus, we get

$$
\frac{1}{2} m \log c < 3.822 \cdot 10^{15} \cdot \frac{1}{2} \log c \cdot \frac{1}{2} \log c \cdot 1.375 \log c \cdot \log m
$$

and

$$
\frac{m}{\log m} < 2.628 \cdot 10^{15} \log^2 c \, . \tag{4.77}
$$

4.8.2   Congruence Method

The idea of the congruence method consists of replacing the equation $v_m = w_n$ with a congruence $v_m \equiv w_n \pmod{M}$ for a suitably chosen modulus M. If a congruence of the form $f(m) \equiv g(n) \pmod{M'}$ is obtained from it by certain admissible transformations, where, assuming that m, n are "small", we have $0 < f(m), g(n) < M'$, then we can replace the congruence with the equation $f(m) = g(n)$. If this equation leads us to a contradiction, we will have the desired conclusion that m, n cannot be "small".

The method will be illustrated in the example of the triple $\{1, 3, c\}$, although it will be clear that the method can be applied much more generally. Omitted technical details in the proofs can be found in [152], where the congruence method was introduced in this form.

We consider Diophantine triples of the form $\{1, 3, c\}$. This means that $c + 1 = s^2$, $3c + 1 = t^2$, from which we deduce that $t^2 - 3s^2 = -2$. This is exactly Eq. (4.57), for which we have seen that all solutions are given by

$$
s = \frac{1}{2\sqrt{3}}\left((1 + \sqrt{3})(2 + \sqrt{3})^k - (1 - \sqrt{3})(2 - \sqrt{3})^k \right).
$$

By inserting this in $c = s^2 - 1$, we obtain that $c = c_k$, where

$$c_k = \frac{1}{6}\left((2 + \sqrt{3})(7 + 4\sqrt{3})^k + (2 - \sqrt{3})(7 - 4\sqrt{3})^k - 4\right).$$

The sequence (c_k) satisfies the recurrence

$$c_k = 14c_{k-1} - c_{k-2} + 8.$$

The first few elements of the sequence (c_k) are $c_0 = 0$, $c_1 = 8$, $c_2 = 120$, $c_3 = 1680$, $c_4 = 23408$. It is not difficult to verify that for the triple $\{1, 3, c_k\}$, we have $d_+ = c_{k+1}$ and $d_- = c_{k-1}$. So, we want to prove the following theorem.

Theorem 4.8.3 *Let k be a positive integer. If d is an integer with the property that there are integers x, y, z such that*

$$d + 1 = x^2, \quad 3d + 1 = y^2, \quad c_k d + 1 = z^2, \tag{4.78}$$

then $d \in \{0, c_{k-1}, c_{k+1}\}$.

Assume that k is the smallest positive integer for which the statement of Theorem 4.8.3 does not hold. We want to obtain a (preferably small) upper bound for k and then check the assertion of the theorem for all values of k less than that bound. By Theorem 4.7.6, we know that $k \geq 2$ and thus $c_k \geq 120$.

So we have to solve the equation $v_m^{(i)} = w_n^{(j)}$, with $a = 1$ and $b = 3$. First, we will show that, in our case, the initial members of these sequences can be determined by observing the corresponding congruences.

The following lemma is easily proved by induction, using (4.73) and (4.74), and it is valid for all Diophantine triples $\{a, b, c\}$.

Lemma 4.8.4

$$v_{2m}^{(i)} \equiv z_0^{(i)} \pmod{2c}, \quad v_{2m+1}^{(i)} \equiv sz_0^{(i)} + cx_0^{(i)} \pmod{2c},$$

$$w_{2n}^{(j)} \equiv z_1^{(j)} \pmod{2c}, \quad w_{2n+1}^{(j)} \equiv tz_1^{(j)} + cy_1^{(j)} \pmod{2c}.$$

Lemma 4.8.5

1) *If the equation $v_{2m}^{(i)} = w_{2n}^{(j)}$ has a solution, then $z_0^{(i)} = z_1^{(j)} = \pm 1$.*
2) *If the equation $v_{2m+1}^{(i)} = w_{2n}^{(j)}$ has a solution, then $z_0^{(i)} = \pm 1$ and $z_1^{(j)} = sz_0^{(i)} = \pm s$.*
3) *If the equation $v_{2m}^{(i)} = w_{2n+1}^{(j)}$ has a solution, then $z_1^{(j)} = \pm 1$ and $z_0^{(i)} = tz_1^{(j)} = \pm t$.*
4) *If the equation $v_{2m+1}^{(i)} = w_{2n+1}^{(j)}$ has a solution, then $z_0^{(i)} = \pm t$ and $z_1^{(j)} = \pm s$.*

Proof We will prove only statement 1). The proofs of the other statements of the lemma are analogous, although somewhat more technically demanding.

From Lemma 4.8.4, we have $z_0^{(i)} \equiv z_1^{(j)} \pmod{2c}$, so (4.67) and (4.69) imply that $z_0^{(i)} = z_1^{(j)}$. Let $d_0 = ((z_1^{(j)})^2 - 1)/c_k$. Then d_0 satisfies system (4.78). Let us compare d_0 with c_{k-1}. On the one hand, we have that $c_{k-1} \geq c_k/15$, and on the other hand

$$d_0 < \frac{1}{c_k} \cdot \frac{c_k\sqrt{c_k}}{2\sqrt{3}} = \frac{\sqrt{c_k}}{2\sqrt{3}} < 0.027 c_k < c_k/15 \, .$$

Therefore, $d_0 < c_{k-1}$, so from the minimality of k, it follows that $d_0 = 0$. Therefore, $|z_1^{(j)}| = 1$, so $z_0^{(i)} = z_1^{(j)} = 1$ or $z_0^{(i)} = z_1^{(j)} = -1$.   □

Now we will consider the sequences $(v^{(i)} \bmod c_k^2)$ and $(w^{(j)} \bmod c_k^2)$, taking into account that from Lemma 4.8.5, we know what their initial terms are. (We will omit the labels (i) and (j) since we have shown that the initial terms are almost unique and we will often write c instead of c_k.)

Lemma 4.8.6

1) $v_{2m} \equiv z_0 + 2c(m^2 z_0 + msx_0) \pmod{c^2}$
2) $v_{2m+1} \equiv sz_0 + c(2m(m+1)sz_0 + (2m+1)x_0) \pmod{c^2}$
3) $w_{2n} \equiv z_1 + 2c(3n^2 z_1 + nty_1) \pmod{c^2}$
4) $w_{2n+1} \equiv tz_1 + c(6n(n+1)tz_1 + (2n+1)y_1) \pmod{c^2}$

Proof We will prove the statement 1) by induction. The proofs of the other statements of the lemma are analogous. We use the fact that the sequences (v_{2m}) and (v_{2m+1}) satisfy the recurrence

$$a_{m+2} = 2(2c+1)a_{m+1} - a_m \, ,$$

while the sequences (w_{2n}) and (w_{2n+1}) satisfy the recurrence

$$b_{n+2} = 2(6c+1)b_{n+1} - b_n \, .$$

We have $v_0 = z_0$ and $v_2 = 2s^2 z_0 + 2scx_0 - z_0 = z_0 + 2c(z_0 + sx_0)$. Assume that the statement 1) holds for $m-1$ and m. Then we have

$$v_{2m+2} = (4c+2)v_{2m} - v_{2m-2}$$

$$\equiv 4cz_0 + 2z_0 + 4c(m^2 z_0 + msx_0) - z_0 - 2c((m-1)^2 z_0 + (m-1)sx_0)$$

$$= z_0 + 2c(z_0(2 + 2m^2 - m^2 + 2m - 1) + sx_0(2m - m + 1))$$

$$= z_0 + 2c((m+1)^2 z_0 + (m+1)sx_0) \pmod{c^2} \, .$$

□

Corollary 4.8.7 *The equations $v_{2m} = w_{2n+1}$ and $v_{2m+1} = w_{2n}$ have no solutions.*

Proof If $v_{2m} = w_{2n+1}$, then Lemmas 4.8.5 and 4.8.6 imply

$$\pm 2m^2 t + 4ms \equiv \pm 6n(n+1)t + (2n+1)\,(\mathrm{mod}\ c)\,.$$

However, this is a contradiction with the fact that c is even.

If $v_{2m+1} = w_{2n}$, then Lemmas 4.8.5 and 4.8.6 imply

$$\pm 2m(m+1)s + (2m+1) \equiv \pm 6n^2 s + 4nt\,(\mathrm{mod}\ c)$$

so we get a contradiction again for the same reason. $\square$

For the remaining two equations, we know that some small solutions exist. Namely, $v_0 = w_0 = \pm 1$ corresponds to $d = 0$, while $v_1 = w_1 = cr \pm st$ corresponds to $d = d_+ = c_{k+1}$ and $d = d_- = c_{k-1}$. The congruence method gives us the following conclusions for these two equations.

Lemma 4.8.8

1) *If $v_{2m} = w_{2n}$ and $n \neq 0$, then $n > 0.105\sqrt{c}$.*
2) *If $v_{2m+1} = w_{2n+1}$ and $n \neq 0$, then $n > 0.156\sqrt[4]{c}$.*

Proof We prove again only the statement (1). The proof of the statement (2) is analogous. Note that $k \geq 2$ and thus $c \geq 120$, $s \geq 11$, $t \geq 19$.

From Lemma 4.8.2, we have

$$2m \log(s + \sqrt{c}) - 2n \log(t + \sqrt{3c}) < 0\,,$$

and so

$$\frac{m}{n} < \frac{\log(t + \sqrt{3c})}{\log(s + \sqrt{c})} = \frac{\log \sqrt{3}}{\log(s + \sqrt{c})} + \frac{\log(\sqrt{c + \frac{1}{3}} + \sqrt{c})}{\log(\sqrt{c+1} + \sqrt{c})} < 1.178\,.$$

On the other hand, Lemma 4.8.6 implies

$$\pm 2m^2 + 2ms \equiv \pm 6n^2 + 2nt\,(\mathrm{mod}\ c)\,.$$

Assume that $n \leq 0.105\sqrt{c}$. Then $m < 0.124\sqrt{c}$. We have

$$2|\pm m^2 + ms| \leq 2c(0.124^2 + 0.124 \cdot 1.005) < \frac{c}{3}\,,$$

$$2|\pm 3n^2 + nt| \leq 2c(3 \cdot 0.105^2 + 0.105 \cdot 1.735) < \frac{c}{2}\,.$$

Therefore $\pm m^2 + ms = \pm 3n^2 + nt$. However,

$$0.876ms \leq \pm m^2 + ms \leq 1.124ms\,,$$

$$0.685nt \leq \pm 3n^2 + nt \leq 1.315nt\,.$$

Moreover, $1.727 \le t/s < \sqrt{3}$. Therefore, for the plus sign, we get

$$\frac{ms}{nt} \ge 0.889 \quad \Rightarrow \quad \frac{m}{n} \ge 1.535,$$

while for the minus sign, we get

$$\frac{ms}{nt} \ge 0.685 \quad \Rightarrow \quad \frac{m}{n} \ge 1.182.$$

So, in both cases, we got a contradiction with the previously shown $\frac{m}{n} < 1.178$. $\square$

Proposition 4.8.9 *If k is the smallest positive integer for which the statement of Theorem 4.8.3 does not hold, then $k \le 79$.*

Proof We will prove the statement of the proposition by comparing the upper bound for the indices m, n, obtained from the Baker-Wüstholz theorem, with the lower bound obtained by the congruence method.

If $v_{2m} = w_{2n}$, with $n \ne 0$, then from inequality (4.77), we have

$$\frac{2m}{\log(2m)} < 2.628 \cdot 10^{15} \log^2 c.$$

On the other hand, from Lemma 4.8.8, we have that $m > n > 0.105\sqrt{c}$. So, we get the following inequality for m

$$m < 1.314 \cdot 10^{15} \log(2m) \log^2(91m^2),$$

which implies that $m < 7 \cdot 10^{20}$. Hence $c < 5 \cdot 10^{43}$, and finally, from

$$\frac{1}{6}(2 + \sqrt{3})(7 + 4\sqrt{3})^k < 5 \cdot 10^{43},$$

it follows that $k \le 39$.

Analogously, if $v_{2m+1} = w_{2n+1}$, with $n \ne 0$, then from inequality (4.77), we have

$$\frac{2m + 1}{\log(2m + 1)} < 2.628 \cdot 10^{15} \log^2 c.$$

On the other hand, from Lemma 4.8.8, we have that $m > n > 0.156\sqrt[4]{c}$, so we get the following inequality for m

$$m < 1.314 \cdot 10^{15} \log(2m + 1) \log^2(1689m^4).$$

Hence, $m < 3 \cdot 10^{21}$ and $c < 2 \cdot 10^{89}$. Finally from

$$\frac{1}{6}(2 + \sqrt{3})(7 + 4\sqrt{3})^k < 2 \cdot 10^{89},$$

it follows that $k \leq 79$. $\square$

To prove Theorem 4.8.3, it remains to examine the cases $k \leq 79$, which, as we said before, can be done quite analogously to the proof by Baker and Davenport of the case $k = 1$, so that with the help of the Baker-Davenport reduction, we reduce the upper bounds for m, n, which we have already obtained by applying the Baker-Wüstholz theorem.

4.8.3 *Integer Points Under Assumption of Minimal Rank*

After we found all the integer solutions of the system of equations

$$x + 1 = \Box, \quad 3x + 1 = \Box, \quad c_k x + 1 = \Box$$

in the previous subsection, we ask whether it is possible to find all integer points on the elliptic curve induced by the Diophantine triple $\{1, 3, c_k\}$, i.e. on the elliptic curve

$$E_k: \quad y^2 = (x + 1)(3x + 1)(c_k x + 1).$$

We introduce the notation: $c_k + 1 = s_k^2$, $3c_k + 1 = t_k^2$. We will present the proof of the following result from the paper [153] by Dujella and Pethő.

Theorem 4.8.10 *Let k be a positive integer. If $\operatorname{rank}(E_k(\mathbb{Q})) = 2$, then all integer points on the curve E_k are given by*

$$(x, y) \in \{(-1, 0), (0, \pm 1), (c_{k-1}, \pm s_{k-1} t_{k-1}(2c_k - s_k t_k)),$$

$$(c_{k+1}, \pm s_{k+1} t_{k+1}(2c_k + s_k t_k))\}.$$

Let us recall that on the curve E_k we have three rational points of order 2: $A_k = (-1, 0)$, $B_k = (-\frac{1}{3}, 0)$, $C_k = (-\frac{1}{c_k}, 0)$ (of which only A_k is an integer point), and two additional rational points $P_k = (0, 1)$ and $R_k = (\frac{s_k t_k + 2 s_k + 2 t_k + 1}{3 c_k}, \frac{(s_k + t_k)(s_k + 2)(t_k + 2)}{3 c_k})$, for which we can expect that they will be independent points of infinite order (except in the case $k = 1$ when the triple $\{1, 3, 8\}$ is regular).

Let us first determine the torsion group of E_k. According to Theorem 3.8.9, the only possibilities are $\mathbb{Z}/2\mathbb{Z} \times \mathbb{Z}/2\mathbb{Z}$ and $\mathbb{Z}/2\mathbb{Z} \times \mathbb{Z}/6\mathbb{Z}$. We will eliminate the second possibility with the help of Lemma 3.8.12.

Proposition 4.8.11 $E_k(\mathbb{Q})_{\text{tors}} \cong \mathbb{Z}/2\mathbb{Z} \times \mathbb{Z}/2\mathbb{Z}$

Proof As we have already said, in order to prove the proposition, we only need to eliminate the possibility that the torsion group of E_k is $\mathbb{Z}/2\mathbb{Z} \times \mathbb{Z}/6\mathbb{Z}$. We will do this by applying Lemma 3.8.12. First, we need to bring the curve E_k into the form $y^2 = x(x + M)(x + N)$. By using standard transformations, we get the equivalent equation $y^2 = x(x + c_k - 3)(x + 3c_k - 3)$, so $M = c_k - 3$, $N = 3c_k - 3$. Thus, E_k will have the torsion group $\mathbb{Z}/2\mathbb{Z} \times \mathbb{Z}/6\mathbb{Z}$ if and only if there exist integers α, β, δ such that $\frac{\alpha}{\beta} \notin \{-2, -1, -\frac{1}{2}, 0, 1\}$ and

$$c_k - 3 = \delta^2(\alpha^4 + 2\alpha^3\beta), \quad 3c_k - 3 = \delta^2(2\alpha\beta^3 + \beta^4).$$

By adding these two conditions, we get

$$4c_k - 6 = \delta^2((\alpha^2 + \alpha\beta + \beta^2)^2 - 3\alpha^2\beta^2). \tag{4.79}$$

Since c_k is even, the left-hand side of (4.79) is $\equiv 2 \pmod 8$. This implies that δ is odd, and also that α and β are both odd. But then the right-hand side of (4.79) is $\equiv 6 \pmod 8$, which is a contradiction. $\qquad\square$

From Lemma 3.8.8, we know that $A_k, B_k, C_k \notin 2E_k(\mathbb{Q})$. Now we want to prove an analogous statement for other relevant combinations of points P_k, R_k, A_k, B_k, C_k. In doing so, as usual, we will work with the equivalent curve

$$E'_k: \quad y^2 = (x + 3)(x + c_k)(x + 3c_k)$$

and corresponding points on it obtained by multiplying the coordinates by $3c_k$ so that we can directly apply Theorem 2.4.9.

Lemma 4.8.12 $P'_k, P'_k + A'_k, P'_k + B'_k, P'_k + C'_k \notin 2E'_k(\mathbb{Q})$

Proof We have

$$P'_k + A'_k = (-c_k - 2, -2c_k + 2),$$
$$P'_k + B'_k = (-3c_k + 6, 6c_k - 18),$$
$$P'_k + C'_k = (c_k^2 - 4c_k, -c_k^3 + 4c_k^2 - 3c_k).$$

It follows from Theorem 2.4.9 that $P'_k, P'_k + A'_k, P'_k + B'_k, P'_k + C'_k \notin 2E'_k(\mathbb{Q})$, because for each of these points, at least one of the numbers $x + 3, x + c_k, x + 3c_k$ is obviously not a square, e.g. $0 + 3 = 3$, $-c_k - 2 + c_k = -2$, $-3c_k + 6 + 3c_k = 6$, $c_k^2 - 4c_k + 3c_k = c_k(c_k - 1)$. $\qquad\square$

Lemma 4.8.13 R'_k, $R'_k + A'_k$, $R'_k + B'_k$, $R'_k + C'_k \notin 2E'_k(\mathbb{Q})$

Proof We have

$$R'_k = (s_k t_k + 2s_k + 2t_k + 1, (t_k + s_k)(s_k + 2)(t_k + 2)),$$

$$R'_k + A'_k = (2s_k - 2t_k - s_k t_k + 1, (s_k - t_k)(s_k + 2)(t_k - 2)),$$

$$R'_k + B'_k = (2t_k - 2s_k - s_k t_k + 1, (t_k - s_k)(s_k - 2)(t_k + 2)),$$

$$R'_k + C'_k = (s_k t_k - 2s_k - 2t_k + 1, (t_k + s_k)(2 - s_k)(t_k - 2)).$$

Since $2s_k - 2t_k - s_k t_k + 4 = (s_k + 2)(2 - t_k) < 0$ and $2t_k - 2s_k - s_k t_k + 4 = (t_k + 2)(2 - s_k) < 0$, we conclude that $R'_k + A'_k$, $R'_k + B'_k \notin 2E'_k(\mathbb{Q})$.

If $R'_k \in 2E'_k(\mathbb{Q})$, then $(t_k + s_k)(t_k + 2) = \square$ and $(t_k + s_k)(s_k + 2) = \square$, where $\square$ denotes a perfect square. Let $\delta = \gcd(t_k + s_k, t_k + 2, s_k + 2)$. Then δ divides $(t_k + 2) + (s_k + 2) - (t_k + s_k) = 4$, so since s_k and t_k are odd, we conclude that $\delta = 1$. Thus, we have

$$t_k + s_k = \square, \quad t_k + 2 = \square, \quad s_k + 2 = \square.$$

Let $s_k + 2 = z^2$. Then from $t_k^2 = 3s_k^2 - 2$, we get that

$$3z^4 - 12z^2 + 10 = \square. \tag{4.80}$$

Using the function `IntegralQuarticPoints` in Magma, we get that all integer solutions of (4.80) are given by $z = \pm 1$, corresponding to $c_k = 0$.

If $R'_k + C'_k \in 2E'_k(\mathbb{Q})$, then $(t_k + s_k)(t_k - 2) = \square$ and $(t_k + s_k)(s_k - 2) = \square$, so as above, we get that

$$t_k + s_k = \square, \quad t_k - 2 = \square, \quad s_k - 2 = \square.$$

Now let $s_k - 2 = z^2$. Then from $t_k^2 = 3s_k^2 - 2$, we get that

$$3z^4 + 12z^2 + 10 = \square. \tag{4.81}$$

Using the function `IntegralQuarticPoints` in Magma, we get that all integer solutions of (4.81) are given by $z = \pm 1, \pm 3$, corresponding to $c_k = 8$ and $c_k = 120$. But for $c_k = 8$, $t_k - 2 = 5 - 2 = 3$ is not a square, while for $c_k = 120$, $t_k - 2 = 19 - 2 = 17$ is not a square. $\qquad\square$

Lemma 4.8.14 *If* $k \geq 2$*, then* $R'_k + P'_k$, $R'_k + P'_k + A'_k$, $R'_k + P'_k + B'_k$, $R'_k + P'_k + C'_k \notin 2E'_k(\mathbb{Q})$.

Proof We use Theorem 2.4.9 again.

If $R'_k + P'_k + A'_k \in 2E'_k(\mathbb{Q})$, then $0 > c_k(s_k + 2)(s_k - t_k) = \square$, while if $R'_k + P'_k + B'_k \in 2E'_k(\mathbb{Q})$, then $0 > c_k(s_k - 2)(s_k - t_k) = \square$. Thus, $R'_k + P'_k + A'_k$, $R'_k + P'_k + B'_k \notin 2E'_k(\mathbb{Q})$.

If $R'_k + P'_k \in 2E'_k(\mathbb{Q})$, then

$$3c_k(t_k + s_k)(t_k + 2) = \square, \quad c_k(t_k + s_k)(s_k + 2) = \square, \quad 3(s_k + 2)(t_k + 2) = \square. \qquad (4.82)$$

By inserting $2c_k = (t_k + s_k)(t_k - s_k)$ in (4.82), we obtain

$$(t_k - s_k)(t_k + 2) = 6\square, \quad (t_k - s_k)(s_k + 2) = 2\square, \quad (s_k + 2)(t_k + 2) = 3\square.$$

Let $\delta = \gcd(s_k + 2, t_k + 2)$. It follows from $t_k^2 - 3s_k^2 = -2$ that $\delta \mid 6$. Since $t_k + 2$ is odd, we have that $\delta \in \{1, 3\}$. Therefore,

$$s_k + 2 = \square \quad \text{or} \quad s_k + 2 = 3\square. \qquad (4.83)$$

In the proof of Lemma 4.8.13, we have already shown that $s_k + 2 = \square$ implies $c_k = 0$, while from the condition $s_k + 2 = 3z^2$, we get $27z^4 - 36z^2 + 10 = \square$, which implies $z = \pm 1$ and again $c_k = 0$.

If $R'_k + P'_k + C'_k \in 2E'_k(\mathbb{Q})$, then

$$3c_k(t_k + s_k)(t_k - 2) = \square, \quad c_k(t_k + s_k)(s_k - 2) = \square, \quad 3(s_k - 2)(t_k - 2) = \square.$$

As above, we get

$$(t_k - s_k)(t_k - 2) = 6\square, \quad (t_k - s_k)(s_k - 2) = 2\square,$$
$$(s_k - 2)(t_k - 2) = 3\square, \qquad (4.84)$$

and we conclude that

$$s_k - 2 = \square \quad \text{or} \quad s_k - 2 = 3\square.$$

In the proof of Lemma 4.8.13, we have already shown that $s_k - 2 = \square$ implies $c_k = 8$ or $c_k = 120$, while from $s_k - 2 = 3z^2$, we get $27z^4 + 36z^2 + 10 = \square$, which has no integer solutions. We excluded the case $c_k = 8$ by assuming that $k \geq 2$ (otherwise, we would have $(s_k, t_k) = (3, 5)$, which satisfies (4.84)), while for $c_k = 120$, we have $(s_2, t_2) = (11, 19)$, which does not satisfy the condition $(s_2 - 2)(t_2 - 2) = 3\square$. $\qquad \square$

Proposition 4.8.15 *If $k \geq 2$, then the points P'_k and R'_k generate a subgroup of rank 2 in the group $E'_k(\mathbb{Q})/E'_k(\mathbb{Q})_{\text{tors}}$.*

Proof We have to prove that $mP'_k + nR'_k \in E'_k(\mathbb{Q})_{\text{tors}}$, for $m, n \in \mathbb{Z}$, implies that $m = n = 0$.

Assume that $mP'_k + nR'_k = T \in E'_k(\mathbb{Q})_{\text{tors}} = \{O, A_k, B_k, C_k\}$. If m and n are not both even, then $T \equiv P'_k, R'_k$ or $P'_k + R'_k \pmod{2E'_k(\mathbb{Q})}$, which is not possible according to Lemmas 4.8.12, 4.8.13, and 4.8.14. Therefore, m and n are both even, say $m = 2m_1, n = 2n_1$, and since $A_k, B_k, C_k \notin 2E'_k(\mathbb{Q})$, we conclude that

$$2m_1 P'_k + 2n_1 P'_k = O.$$

So, we got that $m_1 P'_k + n_1 R'_k \in E'_k(\mathbb{Q})_{\text{tors}}$. By the same argumentation as above, we get that m_1 and n_1 are both even, and by continuing this procedure, we conclude that $m = n = 0$. $\qquad\square$

Proof of Theorem 4.8.10 Let U and V be generators of $E'_k(\mathbb{Q})/E'_k(\mathbb{Q})_{\text{tors}}$ and $X \in E'_k(\mathbb{Q})$. Then there exist integers m, n and a torsion point T such that $X = mU + nV + T$. Also, the points P'_k and R'_k can be written in the form $P'_k = m_P U + n_P V + T_P$, $R'_k = m_R U + n_R V + T_R$. Let $\mathcal{U} = \{O, U, V, U + V\}$. Then there exist $U_1, U_2 \in \mathcal{U}$, $T_1, T_2 \in E'_k(\mathbb{Q})_{\text{tors}}$ so that $P'_k \equiv U_1 + T_1 \pmod{2E'_k(\mathbb{Q})}$, $R'_k \equiv U_2 + T_2 \pmod{2E'_k(\mathbb{Q})}$. Let $U_3 \in \mathcal{U}$ be such that $U_3 \equiv U_1 + U_2 \pmod{2E'_k(\mathbb{Q})}$. Then $P'_k + R'_k \equiv U_3 + (T_1 + T_2) \pmod{2E'_k(\mathbb{Q})}$. Now Lemmas 4.8.12, 4.8.13, and 4.8.14 imply that $U_1, U_2, U_3 \neq O$. Therefore $\{U_1, U_2, U_3\} = \{U, V, U + V\}$. Hence, $X \equiv X_1 \pmod{2E'_k(\mathbb{Q})}$, where

$$\begin{aligned} X_1 \in S = \{ & O, A'_k, B'_k, C'_k, P'_k, P'_k + A'_k, P'_k + B'_k, P'_k + C'_k, R'_k, R'_k + A'_k, R'_k + B'_k, \\ & R'_k + C'_k, R'_k + P'_k, R'_k + P'_k + A'_k, R'_k + P'_k + B'_k, R'_k + P'_k + C'_k \}. \end{aligned}$$

According to Theorem 2.4.10, the functions φ_a, φ_b and φ_c defined in the proof of Corollary 3.1.1 for $\{a, b, c\} = \{1, 3, c_k\}$ are group homomorphisms.

This fact, together with Theorem 4.8.3, implies that it is sufficient to prove that for all $X_1 \in S \setminus \{P'_k\}$, $X_1 = (3c_k u, 3c_k v)$, the system

$$x + 1 = \alpha\square, \quad 3x + 1 = \beta\square, \quad c_k x + 1 = \gamma\square \qquad (4.85)$$

has no integer solutions, where $\square$ denotes the square of a rational number, and α, β, γ are defined by $u + 1 = \alpha, 3u + 1 = \beta, c_k u + 1 = \gamma$ if all these numbers are $\neq 0$, and if $u + 1 = 0$, then we put $\alpha = \beta\gamma$ (so that $\alpha\beta\gamma = \square$). (Instead of O, we can insert any other element of $2E'_k(\mathbb{Q})$, for example $2R'_k = S'_k = (1, 2s_k t_k)$.) Let us note that for $X_1 = P'_k$ we get the system $x + 1 = \square, 3x + 1 = \square, c_k x + 1 = \square$, which is completely solved in Theorem 4.8.3.

For $X_1 \in \{A'_k, B'_k, P'_k + A'_k, P'_k + B'_k, R'_k + A'_k, R'_k + B'_k, R'_k + P'_k + A'_k, R'_k + P'_k + B'_k\}$, precisely two of the numbers α, β, γ are negative, and therefore system (4.85) has no integer solutions.

So we need to consider the remaining seven cases. We will denote the square-free part of an integer g by g'.

(1) $X_1 = O$

We have

$$x + 1 = 3c_k\square, \quad 3x + 1 = c_k\square, \quad c_k x + 1 = 3\square. \tag{4.86}$$

From the second equation in (4.86), we see that $3 \nmid c'_k$ and therefore, the first and second equations imply that c'_k divides $3x + 1$ and $x + 1$. Therefore, $c'_k \mid 3(x + 1) - (3x + 1) = 2$ and we conclude that $c'_k = 1$ or 2. Thus,

$$c_k = \square, \quad \text{or} \quad c_k = 2\square.$$

However, $c_k = s_k^2 - 1 = \square$ is obviously impossible, while $c_k = 2w^2$ leads to the system of Pell's equations

$$s_k^2 - 2w^2 = 1, \quad t_k^2 - 6w^2 = 1,$$

from which we obtain that $(2w^2 + 1)(6w^2 + 1) = 12w^4 + 8w^2 + 1 = \square$. Function `IntegralQuarticPoints` show us that the only solution in positive integers is $w = 2$, which corresponds to $c_k = c_1 = 8$ and this contradicts our assumption that $k \geq 2$. (Let us note that for $c_k = c_1 = 8$, the system also has no solution because, in this case, the first and third equations in (4.86) imply that $3 \mid 7$.)

(2) $X_1 = C'_k$

We have

$$x + 1 = c_k(c_k - 1)\sqcup, \quad 3x + 1 = c_k(c_k - 3)\square, \quad c_k x + 1 = (c_k - 1)(c_k - 3)\square.$$

If $3 \nmid c_k$, then, as in **(1)**, we get $c'_k = 1$ or 2, and $c_k = \square$ or $2\square$, which we have shown to be impossible.

If $c_k = 3e_k$, then e'_k divides $3x + 1$ and $3x + 3$, so $e'_k = 1$ or 2. Therefore,

$$c_k = 3\square, \quad \text{or} \quad c_k = 6\square.$$

The equation $c_k = 3\square$ is impossible because it implies $t_k^2 - 1 = 9\square$, while $c_k = 6w^2$ leads to the system of Pell's equations

$$s_k^2 - 6w^2 = 1, \quad t_k^2 - 18w^2 = 1,$$

i.e. to the conditions $(6w^2 + 1)(18w^2 + 1) = 108w^4 + 24w^2 + 1 = \square$, for which the function `IntegralQuarticPoints` shows that there are no solutions in positive integers.

(3) $X_1 = P'_k + C'_k$

We have

$$x + 1 = 3(c_k - 1)\square, \quad 3x + 1 = (c_k - 3)\square, \quad c_k x + 1 = 3(c_k - 1)(c_k - 3)\square.$$

Since $c_k = s_k^2 - 1$, we see that $c_k \not\equiv 1 \pmod 3$, so $x \equiv -1 \pmod 3$. From the second equation, we have that $(c_k - 3)'$ is not divisible by 3, so that the third equation gives $c_k x + 1 \equiv 0 \pmod 3$. This implies that $c_k \equiv 1 \pmod 3$, a contradiction.

(4) $X_1 = R_k'$

We have

$$x + 1 = 6(t_k - s_k)(t_k + 2)\square, \quad 3x + 1 = 2(t_k - s_k)(s_k + 2)\square,$$
$$c_k x + 1 = 3(s_k + 2)(t_k + 2)\square.$$

From the relation $t_k^2 - 3s_k^2 = -2$, it follows that $\gcd(t_k - s_k, s_k + 2) = \gcd(t_k - s_k, t_k + 2) = 1$ or 3.

If $3 \nmid t_k - s_k$, then $(2(t_k - s_k))'$ divides $x + 1$ and $3x + 1$, and therefore $(2(t_k - s_k))' = 1$ or 2. Thus,

$$t_k - s_k = 2\square \quad \text{or} \quad t_k - s_k = \square.$$

Let us denote $t_k - s_k = 2a_k$. Then we have

$$a_0 = 0, \quad a_1 = 1, \quad a_{k+2} = 4a_{k+1} - a_k, \ k \geq 0.$$

It is easy to see that a_k represents the sequence of y-components of the solutions of Pell's equation $x^2 - 3y^2 = 1$. If we combine this with the obtained condition that $a_k = w^2$ or $a_k = 2w^2$, we get the conditions that $3w^4 + 1 = \square$ or $12w^4 + 1 = \square$, the first of which gives $w = 1$ or $w = 2$, while the second has no solution in positive integers. We conclude that $k = 2$. Now we get $120x + 1 = 91\square$, which is impossible modulo 4.

If $t_k - s_k = 3z_k$, then $(2z_k)'$ divides $x + 1$ and $9x + 3$. Therefore, $(2z_k)'$ divides 6, which implies that $a_k = \square, 2\square, 3\square$ or $6\square$, and as above, we get that this is possible only for $k = 2$. But for $k = 2$, $t_k - s_k = 8 \not\equiv 0 \pmod 3$.

(5) $X_1 = R_k' + C_k'$

We have

$$x + 1 = 6(t_k - s_k)(t_k - 2)\square, \quad 3x + 1 = 2(t_k - s_k)(s_k - 2)\square,$$
$$c_k x + 1 = 3(s_k - 2)(t_k - 2)\square.$$

This case is completely analogous to the case **(4)**.

(6) $X_1 = R_k' + P_k'$

We have

$$x + 1 = (t_k + s_k)(t_k + 2)\square, \quad 3x + 1 = (t_k + s_k)(s_k + 2)\square,$$
$$c_k x + 1 = (s_k + 2)(t_k + 2)\square.$$

As in the case **(4)**, we get that if $3 \nmid t_k + s_k$, then $(t_k + s_k)'$ divides 2, and if $t_k + s_k = 3z_k$, then z_k' divides 6. Since $t_k + s_k = 2a_{k+1}$, we have that $a_{k+1} = \square, 2\square, 3\square$ or $6\square$, which we have seen is impossible for $k \geq 2$.

(7) $X_1 = R_k' + P_k' + C_k'$

We have

$$x + 1 = (t_k + s_k)(t_k - 2)\square, \quad 3x + 1 = (t_k + s_k)(s_k - 2)\square,$$
$$c_k x + 1 = (s_k - 2)(t_k - 2)\square.$$

This case is completely analogous to the case **(6)**.

$\square$

Remark 4.8.16 The question arises how realistic is the condition that the rank of the curve E_k is equal to 2. Since the coefficients of E_k grow exponentially, it is possible to calculate the rank of E_k only for relatively small values of k. For example, $c_{50} \approx 9.739822684 \cdot 10^{56}$, $c_{100} = 1.525126277 \cdot 10^{115}$. Nevertheless, with the help of the function $\verb|ellrank|$ in PARI, it was obtained that for $k \leq 79$, the rank is equal to 2 for the following 28 values of k:

$$2, 5, 6, 12, 14, 17, 22, 24, 26, 27, 29, 36, 38, 41, 43, 47, 50, 53, 57,$$
$$61, 62, 64, 66, 67, 68, 70, 72, 79.$$

Furthermore, for other 16 values of k, the rank is 2 or 4 (assuming the parity conjecture holds), and for another value of k ($k = 42$), the rank is 2 or 4 or 6. These data may suggest that about 50% of curves E_k could have rank 2. That would be in line with the general expectation that within families of elliptic curves, curves mostly have almost minimal possible rank, while even and odd rank curves appear equally, unless there is some good reason for different behaviour.

For a given positive integer k, we may try to find all integer points on E_k by solving several systems of Pellian equations. This approach does not use any information or assumption on the rank of E_k. Dujella and Pethő showed in [153] that it suffices to consider systems of the form

$$d_1 x_1^2 - d_2 x_2^2 = j_1, \quad d_3 x_1^2 - d_2 x_3^2 = j_2, \tag{4.87}$$

where d_1 is a square-free divisor of $c_k - 1$ which is divisible by 3, d_2 is a square-free divisor of $c_k - 3$ which is not divisible by 3, $(d_3, j_1, j_2) = (c_k, 2, 3(c_k - 1)/d_1)$ or $(d_3, j_1, j_2) = (2c_k, 1, 3(c_k - 1)/d_1)$. It can be shown that many such systems are unsolvable by considering the solvability of the equations in (4.87) modulo appropriate prime powers. In the remaining cases, to show the unsolvability of a Pellian equation, we need information on the fundamental solution of the corresponding Pell's equation. With this approach, it was shown in [153] that for $k \leq 40$, with the possible exceptions of $k = 23$ and 37, all integer points on E_k satisfy $x \in \{-1, 0, c_{k-1}, c_{k+1}\}$. By using an efficient algorithm for modular arithmetic in quadratic fields, Jacobson and Williams [222] were able to extend this result to all $k \leq 100$, with the possible exception of $k = 37$, for which the result holds under the assumption of the extended Riemann hypothesis (ERH).

4.9   Fibonacci Numbers and Hoggatt-Bergum Conjecture

4.9.1   Hoggatt-Bergum Conjecture

In the famous Fermat's example of a Diophantine quadruple, the set $\{1, 3, 8, 120\}$, the first three elements are members of the Fibonacci sequence, defined by

$$F_0 = 0, \quad F_1 = 1, \quad F_{n+2} = F_{n+1} + F_n \quad \text{for } n \geq 0.$$

This fact motivated several authors, starting with the paper by Hoggatt and Bergum [217] from 1977, to consider connections between Diophantine m-tuples and Fibonacci numbers. In fact, we may say that this was the most active area of research related with Diophantine m-tuples in the late 1970s and 1980s. In this context, apart from the Fibonacci numbers, also other binary recurrence sequences appeared, in particular, the Lucas numbers defined by

$$L_0 = 2, \quad L_1 = 1, \quad L_{n+2} = L_{n+1} + L_n \quad \text{for } n \geq 0$$

and the Pell numbers defined by

$$P_0 = 0, \quad P_1 = 1, \quad P_{n+2} = 2P_{n+1} + P_n \quad \text{for } n \geq 0.$$

Several books have been devoted to Fibonacci numbers and especially their connection to number theory, here we mention [216, 242, 348, 353].

In 1977, Hoggatt and Bergum [217] proved that for any positive integer k, the set

$$\{F_{2k}, \ F_{2k+2}, \ F_{2k+4}, \ 4F_{2k+1}F_{2k+2}F_{2k+3}\} \tag{4.88}$$

is a Diophantine quadruple. In fact, quadruple (4.88) has Euler's form

$$\{a, \ b, \ a + b + 2r, \ 4r(r + a)(r + b)\}.$$

The proof is based on Cassini's identity

$$F_{n-1}F_{n+1} - F_n^2 = (-1)^n, \tag{4.89}$$

which can be easily proved by induction and can be viewed as a special case of (4.7) because for the convergents of the continued fraction expansion of the number

$(1 + \sqrt{5})/2 = [1; 1, 1, \ldots]$, it holds $p_n = F_{n+2}$ and $q_n = F_{n+1}$. By inserting $n = 2k + 1$ in (4.89), we get

$$F_{2k} F_{2k+2} + 1 = F_{2k+1}^2,$$

and by taking $a = F_{2k}$, $b = F_{2k+2}$, $r = F_{2k+1}$, we obtain $a + b + 2r = (F_{2k} + F_{2k+1}) + (F_{2k+1} + F_{2k+2}) = F_{2k+2} + F_{2k+3} = F_{2k+4}$ and $4r(r + a)(r + b) = 4F_{2k+1}(F_{2k} + F_{2k+1})(F_{2k+2} + F_{2k+1}) = 4F_{2k+1}F_{2k+2}F_{2k+3}$.

Of course, the conditions for a Diophantine quadruple can also be checked directly:

$$F_{2k} \cdot F_{2k+2} + 1 = F_{2k+1}^2,$$

$$F_{2k} \cdot F_{2k+4} + 1 = F_{2k+2}^2,$$

$$F_{2k} \cdot 4F_{2k+1}F_{2k+2}F_{2k+3} + 1 = (2F_{2k+1}F_{2k+2} - 1)^2,$$

$$F_{2k+2} \cdot F_{2k+4} + 1 = F_{2k+3}^2,$$

$$F_{2k+2} \cdot 4F_{2k+1}F_{2k+2}F_{2k+3} + 1 = (2F_{2k+2}^2 + 1)^2,$$

$$F_{2k+4} \cdot 4F_{2k+1}F_{2k+2}F_{2k+3} + 1 = (2F_{2k+2}F_{2k+3} + 1)^2.$$

Hoggatt and Bergum conjectured that the fourth element of set (4.88) is unique. This conjecture can be viewed as a special case of Conjecture 1.4.3. In 1999, the Hoggatt-Bergum conjecture was settled by Dujella [98].

Theorem 4.9.1 *Let k be a positive integer. If the set $\{F_{2k}, F_{2k+2}, F_{2k+4}, d\}$ is a Diophantine quadruple, then $d = 4F_{2k+1}F_{2k+2}F_{2k+3}$.*

Note that for $k = 1$, (4.88) becomes the Fermat set $\{1, 3, 8, 120\}$. Thus the case $k = 1$ of Theorem 4.9.1 is exactly the result of Baker and Davenport [24] discussed in Sect. 4.7. We will outline the proof of Theorem 4.9.1.

Assume that d is a positive integer such that $\{F_{2k}, F_{2k+2}, F_{2k+4}, d\}$ is a Diophantine quadruple. This time we will eliminate d from

$$F_{2k}d + 1 = x^2, \quad F_{2k+2}d + 1 = y^2, \quad F_{2k+4}d + 1 = z^2$$

in a slightly different way than in Sect. 4.8.1, to take advantage of the fact that in this case, for $a = F_{2k}$ and $b = F_{2k+2}$, we have $a < b < 4a$. We obtain the following system of generalized Pellian equations

$$F_{2k}y^2 - F_{2k+2}x^2 = F_{2k} - F_{2k+2}, \tag{4.90}$$

$$F_{2k}z^2 - F_{2k+4}x^2 = F_{2k} - F_{2k+4}. \tag{4.91}$$

We know that all solutions of a single generalized Pellian equation are contained in the union of finitely many binary recurrence sequences. Therefore, our problem reduces to finitely many equations of the form $v_m = w_n$, where $\{v_m\}$ and $\{w_n\}$ are binary recurrence sequences. However, in the case when $a < b < 4a$, we can be more precise, since we have at most two such sequences, by the following result due to Jones [224].

Lemma 4.9.2 *If $\{a, b, c\}$ is a Diophantine triple and $a < b < 4a$, then $c = c_k$ or $c = \bar{c}_k$ for some $k \geq 0$, where the sequences (c_k) and $(\bar{c}_k)$ are given by*

$$c_0 = 0, \quad c_1 = a + b + 2r, \quad c_{k+2} = (4ab + 2)c_{k+1} - c_k + 2(a + b), \quad (4.92)$$

$$\bar{c}_0 = 0, \quad \bar{c}_1 = a + b - 2r, \quad \bar{c}_{k+2} = (4ab + 2)\bar{c}_{k+1} - \bar{c}_k + 2(a + b). \quad (4.93)$$

If $b = a + 2$, then $\bar{c}_{k+1} = c_k$ for $k \geq 0$.

Proof Let $ab + 1 = r^2$, $ac + 1 = s^2$, $bc + 1 = t^2$. Then we have

$$at^2 - bs^2 = a - b. \quad (4.94)$$

By Lemma 4.8.1 applied to Eq. (4.94), we have $s = s_k^{(i)}$, where

$$s_1^{(i)} = rs_0^{(i)} + at_0^{(i)}, \quad s_{k+2}^{(i)} = 2rs_{k+1}^{(i)} - s_k^{(i)} \quad (4.95)$$

and

$$1 \leq s_0^{(i)} < \sqrt{\frac{r + 1}{2}} \leq \sqrt{\frac{2a + 1}{2}}, \quad (4.96)$$

since from $a < b < 4a$, we have $r \leq 2a$.

From $ac + 1 = s^2$, we have $(s_k^{(i)})^2 \equiv 1 \pmod{a}$. However, from (4.95), we easily get by induction that $s_{2k}^{(i)} \equiv s_0^{(i)} \pmod{a}$, while $s_{2k+1}^{(i)} \equiv rs_0^{(i)} \pmod{a}$, and thus $(s_k^{(i)})^2 \equiv (s_0^{(i)})^2 \pmod{a}$ for any $k \geq 0$. Therefore, $(s_0^{(i)})^2 \equiv 1 \pmod{a}$. On the other hand, from (4.96), we have $1 \leq (s_0^{(i)})^2 \leq a + \frac{1}{2} < a + 1$. Hence, we conclude that $(s_0^{(i)})^2 = 1$, i.e. $s_0^{(i)} = 1$, $t_0^{(i)} = \pm 1$ and $s_1^{(i)} = r \pm a$. From these two sequences for s_k, we obtain two sequences for c_k as claimed (see [224] for the missing details). $\qquad\square$

Note that if we allow negative indices k, then $\bar{c}_k = c_{-k}$, so the two sequences of solutions in Lemma 4.9.2 can be replaced by one two-sided sequence $(c_k)_{k \in \mathbb{Z}}$.

From Lemma 4.9.2, applied to $a = F_{2k}$, $b = F_{2k+2}$, $r = F_{2k+1}$, we see that the solutions (in x) of Eq. (4.90) belong to the two-sided sequence $(v_m)_{m \in \mathbb{Z}}$, defined by

$$v_0 = 1, \quad v_1 = F_{2k+2}, \quad v_{m+2} = 2F_{2k+1}v_{m+1} - v_m, \quad m \in \mathbb{Z}.$$

For Eq. (4.91), Lemma 4.8.1 gives that the fundamental solutions satisfy $0 < x_0^{(i)} < \sqrt{\frac{3F_{2k}}{2}} < F_{2k}$. For the solutions $x = w_n^{(i)}$ of (4.91), similarly as in the proof of Lemma 4.9.2, we have $w_{2n}^{(i)} \equiv x_0^{(i)} \pmod{F_{2k}}$, $w_{2n+1}^{(i)} \equiv F_{2k+2}x_0^{(i)} \equiv (F_{2k+2} - 2F_{2k})x_0^{(i)} \equiv F_{2k-1}x_0^{(i)} \pmod{F_{2k}}$. By comparing this with $v_{2m} \equiv 1 \pmod{F_{2k}}$, $v_{2m+1} \equiv F_{2k+2} \equiv F_{2k-1} \pmod{F_{2k}}$, we conclude (with some additional work for small values of k, see [98] for the details) that $x_0^{(i)} = 1$ and $z_0^{(i)} = \pm 1$. Thus, if the equation $v_m = w_n^{(i)}$ has a solution in integers m and n, then the solutions (in x) of Eq. (4.91) belong to the two-sided sequence $(w_n)_{n \in \mathbb{Z}}$, defined by

$$w_0 = 1, \quad w_1 = F_{2k} + F_{2k+2}, \quad w_{n+2} = 2F_{2k+2}w_{n+1} - w_n, \quad n \in \mathbb{Z}.$$

We claim that the only solutions of the equation $v_m = w_n$, $m, n \in \mathbb{Z}$, are $v_0 = w_0 = 1$ and $v_2 = w_{-2} = 2F_{2k+1}F_{2k+2} - 1$. These solutions correspond to $d = 0$ and $d = 4F_{2k+1}F_{2k+2}F_{2k+3}$.

In the proof of this statement, we derive a lower bound and an upper bound for solutions. These bounds have the property that the lower bound is greater than the upper bound for sufficiently large k. These bounds are obtained by two different transformations of the original exponential equation $v_m = w_n$.

For obtaining a lower bound, we use a variant of the congruence method. The method from Sect. 4.8.2 is a general method, but here we use its variant based on some additional properties of studied equations. We again start with the trivial observation that the equality $v_m = w_n$ implies the congruence $v_m \equiv w_n \pmod{M}$ for all positive integers M. We would like to choose the modulus M in such a way that:

(1) $\{v_m \bmod M\}$ is a polynomial sequence in m;
(2) $\{w_n \bmod M\}$ has a small period.

Both goals are achieved for $M = 2F_{2k}F_{2k+2}$. Indeed, we have

$$v_{2m} \equiv 2mF_{2k+2}^2 - (2m - 1)\,(\mathrm{mod}\ 2F_{2k}F_{2k+2}),$$

$$v_{2m+1} \equiv F_{2k+2} + 2(m - 1)F_{2k}\,(\mathrm{mod}\ 2F_{2k}F_{2k+2}),$$

$$w_{4n} \equiv 1\,(\mathrm{mod}\ 2F_{2k}F_{2k+2}), \quad w_{4n+1} \equiv F_{2k} + F_{2k+2}\,(\mathrm{mod}\ 2F_{2k}F_{2k+2}),$$

$$w_{4n+2} \equiv 2F_{2k+2}^2 - 1\,(\mathrm{mod}\ 2F_{2k}F_{2k+2}), \quad w_{4n+3} \equiv F_{2k+1}\,(\mathrm{mod}\ 2F_{2k}F_{2k+2}).$$

It is easy to see that these congruences imply that from $v_m = w_n$, it follows $m \equiv 0$ or $2 \pmod{2F_{2k+2}}$. This gives a very good lower bound for solutions. Indeed, if $d \neq 0$ and $d \neq 4F_{2k+1}F_{2k+2}F_{2k+3}$, then $m \neq 0, 2$, and therefore

$$|m| \geq 2F_{2k+2} - 2. \tag{4.97}$$

An upper bound can be obtained using Baker's theory. Namely, we can transform our exponential equation $v_m = w_n$ into a logarithmic inequality, and then we may apply results on linear forms in logarithms of algebraic numbers. In [98], the theorem of Baker and Wüstholz [25] (Theorem 4.1.13) was used to obtain the following upper bound

$$\frac{|m|}{\log |m|} < 6.423 \cdot 10^{15} \log^2 F_{2k},$$

which contradicts (4.97) for large k. Indeed, this finishes the proof for $k \geq 49$. In the remaining 48 cases, the corresponding systems of Pellian equations can be solved by a version of the Baker-Davenport reduction procedure (Lemma 4.7.3).

As a simple consequence of Theorem 4.9.1, we obtain the result that if $\{F_{2k}, F_{2k+2}, F_{2k+4}, d\}$ is a Diophantine quadruple, then d cannot be a Fibonacci number. This answers the question posed by Jones [224]. The statement follows directly from Theorem 4.9.1 and the inequalities

$$F_{6k+5} < 4F_{2k+1}F_{2k+2}F_{2k+3} < F_{6k+6},$$

proved by Jones, which follow easily from Binet's formula

$$F_k = \frac{1}{\sqrt{5}} \left(\left(\frac{1+\sqrt{5}}{2} \right)^k - \left(\frac{1-\sqrt{5}}{2} \right)^k \right).$$

Motivated by the Hoggatt-Bergum set, several authors considered the question how large Diophantine tuples consisting of Fibonacci numbers can be. He, Luca, and Togbé [209] proved that if $\{F_{2n}, F_{2n+2}, F_k\}$ is a Diophantine triple, then $k = 2n+4$ or $k = 2n - 2$ (when $n > 1$), except when $n = 2$, in which case $k = 1$ is also possible. Fujita and Luca [260] proved that there are only finitely many Diophantine quadruples consisting of Fibonacci numbers, and in [261] they proved that there are no such quadruples. Rihane, Luca, and Togbé [308] proved the analogous result for Pell numbers.

There are several papers devoted to various generalizations of the Hoggatt-Bergum quadruple. Let us briefly mention some of them.

In [285], Morgado showed that the product of any two distinct elements of the set

$$\{F_k, F_{k+2r}, F_{k+4r}, 4F_{k+r}F_{k+2r}F_{k+3r}\}$$

increased by $F_a^2 F_b^2$ or $-F_a^2 F_b^2$, for suitable positive integers a and b, is a perfect square. Similar results are valid for more general binary recurrence sequences (see [218, 287, 288, 322, 346, 347]).

In [90], the following generalization of the Hoggatt-Bergum result was obtained. Let the sequences $h_k = h_k(A)$ and $g_k = g_k(A)$ be defined by

$$h_0 = 0, \quad h_1 = 1, \quad h_k = Ah_{k-1} + h_{k-2}, \quad k \geq 2;$$

$$g_0 = 0, \quad g_1 = 1, \quad g_k = Ag_{k-1} - g_{k-2}, \quad k \geq 2,$$

where A is a positive integer ($A \geq 2$ in the case of the sequence (g_k)). Then the sets

$$\{h_{2k}, h_{2k+2}, 2h_{2k} + (A + 2)h_{2k+1}, 4h_{2k+1}((A + 2)h_{2k+1}^2 + 2h_{2k}h_{2k+1} - 1))\} \tag{4.98}$$

and

$$\{g_k, g_{k+2}, (A \pm 2)g_{k+1}, 4g_{k+1}((A \pm 2)g_{k+1}^2 \mp 1)\} \tag{4.99}$$

are $D(1)$-quadruples. Since $h_k(1) = F_k$ and $g_k(3) = F_{2k}$, these quadruples are generalizations of the Hoggatt-Bergum quadruple. Moreover, since $g_k(2) = k$, we may consider (4.99) as a common generalization of the Hoggatt-Bergum quadruple (4.88) and a shift of well-known polynomial quadruple $\{k - 1, k + 1, 4k, 16k^3 - 4k\}$ (studied, e.g. in [93]).

If we concentrate our attention on the first three elements of the Hoggatt-Bergum quadruple, we see that the sequence of Fibonacci numbers with even subscripts (F_{2k}) has one interesting property. If we choose three successive elements of this sequence, the product of any two of them increased by 1 is a perfect square. In [82], Deshpande and Dujella characterized all non-degenerate binary recurrence sequences (G_k) of the form $G_k = AG_{k-1} - G_{k-2}$, which possess the property that there exists an integer n such that $G_k G_{k+1} + n$ and $G_k G_{k+2} + n$ are perfect squares for all $k \geq 0$. Their result is that this assumption implies $A = 3$ and $n = G_0^2 - 3G_0 G_1 + G_1^2$. In other words, $G_k = G_1 F_{2k} - G_0 F_{2k-2}$. Therefore, we see that this property is very closely related to Fibonacci numbers of even subscripts.

4.9.2 Regular Triples and Integer Points on Elliptic Curves

We have seen that all integer solutions of the system

$$F_{2k}d + 1 = \square, \quad F_{2k+2}d + 1 = \square, \quad F_{2k+4}d + 1 = \square, \tag{4.100}$$

are given by $d = 0$ and $d = 4F_{2k+1}F_{2k+2}F_{2k+3}$. Multiplying three conditions from (4.100), we obtain the elliptic curve

$$E_k : \quad y^2 = (F_{2k}x + 1)(F_{2k+2}x + 1)(F_{2k+4}x + 1). \tag{4.101}$$

Since we can solve the system (4.100) completely, we may try to find all integer points on elliptic curve (4.101) and verify Conjecture 1.4.5, at least under certain additional assumptions. That was done in [103] under the assumption that the rank is "smallest possible". In fact, in [103], the following more general result concerning integer points on elliptic curves induced by regular Diophantine triples was proved.

Theorem 4.9.3 *Let a, b and r be positive integers such that $a < b$ and $ab + 1 = r^2$ and let $c = a + b + 2r$. Assume that there are no perfect squares among the numbers a, $2a$, b, $2b$, c, $2c$. If $rank(E(\mathbb{Q})) = 1$, then all integer points (x, y) on the elliptic curve*

$$E : \quad y^2 = (ax + 1)(bx + 1)(cx + 1)$$

satisfy the system

$$ax + 1 = \square, \quad bx + 1 = \square, \quad cx + 1 = \square. \tag{4.102}$$

Proof As before (see, e.g. Sect. 4.8.3), we will work with the equivalent elliptic curve in the Weierstrass form

$$E' : \quad y^2 = (x + bc)(x + ac)(x + ab),$$

with obvious points $A' = (-bc, 0)$, $B' = (-ac, 0)$, $C' = (-ab, 0)$ and $P' = (0, abc)$ (there is also the point $S' = (1, r(r + a)(r + b))$, but for a regular triple we have $S' = -2P'$).

Let $E'(\mathbb{Q})/E'(\mathbb{Q})_{\text{tors}} = \langle U \rangle$, i.e. U is a free generator. If $X \in E'(\mathbb{Q})$, then we can represent X in the form $X = mU + T$, where $m \in \mathbb{Z}$ and $T \in E'(\mathbb{Q})_{\text{tors}}$. We also have $P' = nU + T_1$ for an integer n and a torsion point T_1. Since, by Corollary 3.8.10, $E'(\mathbb{Q})_{\text{tors}} \cong \mathbb{Z}/2\mathbb{Z} \times \mathbb{Z}/2\mathbb{Z}$ or $\mathbb{Z}/2\mathbb{Z} \times \mathbb{Z}/6\mathbb{Z}$, we have $T_1 \equiv O$, A', B' or C' (mod $2E'(\mathbb{Q})$).

We claim that n is odd. Indeed, it follows from Theorem 2.4.9 that P', $P' + A'$, $P' + B' \notin 2E'(\mathbb{Q})$ since ab, $a(a - b)$ and $b(b - c)$ are not squares. Assume that $P' + C' \in 2E'(\mathbb{Q})$. Then by Theorem 2.4.9, we have

$$c(c - a) = c(b + 2r) = \square, \qquad c(c - b) = c(a + 2r) = \square. \tag{4.103}$$

Assume that c, $c - a$ and $c - b$ are all squares multiplied by an integer δ. Then $a \equiv c \equiv b \equiv 2r \equiv 0 \,(\text{mod } \delta)$ and from $2ab + 2 = 2r^2$ we find that $\delta = 1$ or $\delta = 2$, i.e. c or $2c$ is a square, a contradiction. Thus $P' + T_1 \notin 2E'(\mathbb{Q})$, which implies that n is odd.

Since n is odd, we have $X \equiv X_1 \,(\text{mod } 2E'(\mathbb{Q}))$, where

$$X_1 \in S = \{O,\ A',\ B',\ C',\ P',\ P' + A',\ P' + B',\ P' + C'\}.$$

Now, since the functions φ_a, φ_b and φ_c, defined in the proof of Corollary 3.1.1, are homomorphisms, in order to find all integer points on E, it suffices to solve in integers all systems of the form

$$ax + 1 = \alpha\square, \quad bx + 1 = \beta\square, \quad cx + 1 = \gamma\square, \tag{4.104}$$

where for $X_1 = (abcu, abcv) \in S$, the numbers α, β, γ are defined by $\alpha = au + 1$, $\beta = bu + 1$, $\gamma = cu + 1$ if all of these three expressions are non-zero, and if e.g. $au + 1 = 0$, then we define $\alpha = \beta\gamma$. Here $\square$ denotes the square of a rational number.

Since for $X_1 = P'$ system (4.104) is equivalent to system (4.102), we have to prove that for $X_1 \in S \setminus \{P'\}$, system (4.104) has no integer solutions.

For $X_1 \in \{A', B', P' + A', P' + B'\}$, exactly two among the numbers α, β, γ are negative and therefore, system (4.104) has no integer solution.

Let e' denote the square-free part of an integer e and let $e'' = \min\{|e'|, |2e'|\}$.

If $X_1 = O$, then system (4.104) becomes

$$ax + 1 = bc\square, \quad bx + 1 = ac\square, \quad cx + 1 = ab\square.$$

First, we will prove that $\gcd(a', b') = 1$ or 2. Assume that a prime p divides a' and b'. Then from $ax + 1 = bc\square$, we conclude that $p \mid c'$, and from $c = a + b + 2r$ that $p \mid 2r$. Now from $2ab + 2 = 2r^2$, it follows that $p = 2$. Analogously we can prove that $\gcd(a', c') = 1$ or 2 and $\gcd(b', c') = 1$ or 2. Since a'' divides $bx + 1$ and $cx + 1$, we conclude that a'' divides $c - b = a + 2r$. Therefore $a'' \mid 2r$. Analogously we find that $b'' \mid 2r$ and $c'' \mid 2s$. However, now the equalities $2ab + 2 = 2r^2$ and $2ac + 2 = 2s^2$ imply $a'', b'', c'' \in \{1, 2\}$. Thus a is either a square of twice a square (and the same holds for b and c), a contradiction.

If $X_1 = C'$, then system (4.104) becomes

$$ax + 1 = c(c - a)\square, \quad bx + 1 = c(c - b)\square, \quad cx + 1 = (c - a)(c - b)\square.$$

Assume that $p \mid c'$ and $p \mid (c - a)'$ for a prime p. Then from $cx + 1 = (c - a)(c - b)\square$, we conclude that $p \mid (c - b)'$. Hence p divides a, b and c and therefore $p \mid 2r$, and we obtain that $p = 2$, as before. Hence we proved that $\gcd(c', (c - a)') = 1$ or 2, and in the same manner, we can prove that $\gcd(c', (c - b)') = 1$ or 2 and $\gcd((c - a)', (c - b)') = 1$ or 2. Since c'' divides $b - a = c - 2s$, we find, as above, that c is either a square or twice a square.

If $X_1 = P' + C'$, then system (4.104) becomes

$$ax + 1 = b(c - a)\square, \quad bx + 1 = a(c - b)\square, \quad cx + 1 = ab(c - a)(c - b)\square.$$

As before, we can prove that $\gcd(a', (c - b)')$, $\gcd(b', (c - a)')$, $\gcd(a', b')$ are all equal to 1 or 2, and since a'' divides $c - b$, we conclude that a is either a square or twice a square, (similarly, we can prove that b is either a square or twice a square), a contradiction. $\qquad\square$

Corollary 4.9.4 *Let $k \geq 2$ be an integer. If $\mathrm{rank}(E_k(\mathbb{Q})) = 1$, then all integer points on the elliptic curve E_k given by (4.101) are*

$$\begin{aligned}
(x, y) \in \{(0, \pm 1),\ (4F_{2k+1}F_{2k+2}F_{2k+3}, \\
\pm(2F_{2k+1}F_{2k+2} - 1)(2F_{2k+2}^2 + 1)(2F_{2k+2}F_{2k+3} + 1)\}.
\end{aligned} \tag{4.105}$$

Proof The statement follows directly from Theorems 4.9.1 and 4.9.3, unless at least one of the numbers F_{2k}, F_{2k+2}, F_{2k+4} is a square or twice a square. By the results of Ljunggren [255] and Cohn [71] (see also [116, Chapter 10.6]), this is the case if and only if $2 \leq k \leq 6$ (the only squares in the Fibonacci sequence are $F_1 = F_2 = 1$ and $F_{12} = 144$, while only double squares are $F_3 = 2$ and $F_6 = 8$).

However, from the proof of Theorem 4.9.3, it follows that if F_{2k+4} is neither a square nor twice a square, then we should have that F_{2k} and F_{2k+2} are both either a square or twice a square. This observation eliminates all cases except $k = 4$.

If $k = 4$, we have to solve system (4.104) for $X_1 = C'$. In this case, (4.104) becomes

$$21x + 1 = 123\square, \quad 55x + 1 = 89\square, \quad 144x + 1 = 123 \cdot 89\square. \tag{4.106}$$

But the first equation in (4.106) is clearly impossible modulo 3. $\qquad\square$

Remark 4.9.5 Similarly as in Remark 4.8.16, we may ask how realistic is the condition that the rank of the curve E_k is equal to 1. Here also the coefficients grow exponentially, so we can calculate the rank of E_k only for relatively small values of k. With the function `ellrank` in PARI, it is obtained that for $k \leq 100$, the rank is equal to 1 for the following 20 values of k:

$$1, 2, 6, 10, 12, 16, 20, 22, 30, 32, 34, 40, 46, 48, 54, 58, 66, 72, 80, 90.$$

Furthermore, for other 5 values of k (38, 52, 60, 92, 94), the rank is 1 or 3 (assuming that the parity conjecture holds). In this family, we can observe the usual behaviour of rank parities; there are 52 curves with even rank (mostly conditionally or unconditionally equal to 2), and 48 curves with odd rank. However, among the curves with odd rank, there are almost equal numbers of curves with expected rank 1 (25 curves) as curves with expected rank 3 (23 curves). So, in contrast with the situation in Remark 4.8.16, here we do not have enough data to support a speculation that 50% of curves E_k might have rank 1, although at least a conjecture that there are infinitely many such curves seems plausible.

In [103], it was also proved that for $2 \leq k \leq 50$, all integer points on E_k are given by (4.105), without any assumption on the rank (by considering systems of Pellian equations). For $k = 1$ there is one additional integer point, the torsion point $(-1, 0)$.

4.9.3 Diophantine Quadruples for Squares of Fibonacci Numbers

Sets with the property $D(\ell^2)$, where ℓ is a positive integer, were studied in [88]. It was shown that for any $D(\ell^2)$-pair $\{a, b\}$, such that ab is not a perfect square, there exist infinitely many $D(\ell^2)$-quadruples of the form $\{a, b, c, d\}$. This is the generalization of a well-known fact for $\ell = 1$, proved by Euler (see Sect. 1.3). The proof of this result is based on the construction of three double sequences $x_{n,m}$, $y_{n,m}$ and $z_{n,m}$ which are defined as follows. Let $ab + \ell^2 = r^2$ and let (s, t) be the fundamental solution of Pell's equation $S^2 - abT^2 = 1$. Define

$$y_{0,0} = \ell, \quad z_{0,0} = \ell, \quad y_{1,0} = r + a, \quad z_{1,0} = r + b,$$
$$y_{n+1,0} = \tfrac{2r}{\ell} y_{n,0} - y_{n-1,0}, \quad z_{n+1,0} = \tfrac{2r}{\ell} z_{n,0} - z_{n-1,0}, \quad n \in \mathbb{Z},$$
$$y_{n,1} = s y_{n,0} + at z_{n,0}, \quad z_{n,1} = bt y_{n,0} + s z_{n,0}, \quad n \in \mathbb{Z},$$
$$y_{n,m+1} = 2s y_{n,m} - y_{n,m-1}, \quad z_{n,m+1} = 2s z_{n,m} - z_{n,m-1}, \quad n, m \in \mathbb{Z}.$$

The desired quadruples have the form $\{a, b, x_{n,m}, x_{n+1,m}\}$, where

$$x_{n,m} = (y_{n,m}^2 - \ell^2)/a = (z_{n,m}^2 - \ell^2)/b.$$

In [88] and [95], the following result was proved.

Theorem 4.9.6 *The sets $\{a, b, x_{0,m}, x_{-1,m}\}$, $m \notin \{-1, 0, 1\}$, and $\{a, b, x_{0,m}, x_{1,m}\}$, $m \notin \{-2, -1, 0\}$, are Diophantine quadruples with the property $D(\ell^2)$.*

The proof uses the relations

$$a y_{n,m}^2 - b z_{n,m}^2 = \ell^2(a - b), \quad x_{n,m} x_{n+1,m} + \ell^2 = \left(\frac{y_{n,m} y_{n+1,m} - \ell r}{a}\right)^2$$

proved by induction, and the fact that $x_{-1,m}$, $x_{0,m}$ and $x_{1,m}$ are integers for all $m \in \mathbb{Z}$.

If $\ell - 1$, then $s = r$, $t = 1$, and we obtain $x_{n,m} = x_{n+m,0}$. Moreover, for $k \geq 0$, we have $x_{k,0} = c_k$ and $x_{-k,0} = \bar{c}_k$ in the notation from Lemma 4.9.2.

Hence, if we have a pair of identities of the form

$$ab + \ell^2 = r^2 \quad \text{and} \quad s^2 - abt^2 = 1,$$

then we can construct the sequence $(x_{m,n})$ and, by Theorem 4.9.6, obtain infinitely many Diophantine quadruples with the property $D(\ell^2)$. There are several pairs of identities for Fibonacci and Lucas numbers, which have the above form. For example, starting with the identities

$$4F_{k-1} F_{k+1} + F_k^2 = L_k^2,$$

$$4F_{k-1} F_k^2 F_{k+1} + 1 = (F_k^2 + F_{k-1} F_{k+1})^2,$$

we obtain $D(F_k^2)$-quadruples

$$\{2F_{k-1}, \ 2F_{k+1}, \ 2F_k^3 F_{k+1} F_{k+2}, \ 2F_{k+1} F_{k+2} F_{k+3}(2F_{k+1}^2 - F_k^2)\},$$

$$\{F_{k-1}, \ 4F_{k+1}, \ F_{k-2} F_{k-1} F_{k+1}(2F_k^2 - F_{k-1}^2), \ F_k^3 F_{k+2} F_{k+3}\},$$

$$\{4F_{k-1}, \ F_{k+1}, \ F_{k-2} F_{2k-2} F_{2k-1}, \ F_k^3 L_k L_{k+1}\}$$

(see [89]).

In general, $x_{2,0}$ need not be an integer. Indeed, $x_{2,0} = \frac{4r(r+a)(r+b)}{\ell^2}$. Nevertheless, if $x_{2,0}$ is a positive integer, then $\{a, b, x_{1,0}, x_{2,0}\}$ is a $D(\ell^2)$-quadruple. With this method, some formulas for quadruples with the properties $D(1)$, $D(4)$, $D(9)$ and $D(64)$, in terms of Fibonacci and Lucas numbers, were obtained in [88]. This idea was also applied in [89] to the Morgado identity [286]:

$$F_{k-3} F_{k-2} F_{k-1} F_{k+1} F_{k+2} F_{k+3} + L_k^2 = (F_k(2F_{k-1} F_{k+1} - F_k^2))^2.$$

Among other similar examples, the following $D(L_k^2)$-quadruple was obtained:

$$\{F_{k-3} F_{k-2} F_{k+1}, \ F_{k-1} F_{k+2} F_{k+3}, \ F_k L_k^2, \ 4F_{k-1}^2 F_k F_{k+1}^2 (2F_{k-1} F_{k+1} - F_k^2)\}.$$

In [90], the results mentioned in this subsection were generalized to the sequences (h_k) and (g_k), defined in Sect. 4.9.1.

4.10   An Absolute Bound for the Size of Diophantine Tuples

The result of Baker and Davenport that Fermat's set $\{1, 3, 8, 120\}$ cannot be extended to a Diophantine quintuple motivated the conjecture that there does not exist a Diophantine quintuple. However, obtaining any absolute upper bound for the size of Diophantine tuples was a non-trivial task. Namely, although the number of integer points on an elliptic curve

$$y^2 = (ax + 1)(bx + 1)(cx + 1) \tag{4.107}$$

is finite, so it is clear that there does not exist an infinite set of positive integers with the property of Diophantus, known bounds for the size [20, 206] and for the number [316] of solutions of (4.107) depend on parameters a, b, c and accordingly they do not immediately yield an absolute bound for the size of such set.

The first absolute upper bound was given by Dujella [104] in 2001, where it was proved that there does not exist a Diophantine 9-tuple and that there are only finitely many Diophantine 8-tuples. In this section, we sketch results and methods from that paper. In the next section, we will discuss further improvements which led to the proof of the non-existence of Diophantine quintuples by He, Togbé, and Ziegler in [211]. The approach of [104] is less technically involved, but already contains most of the ingredients used in later improvements. Furthermore, it is more suitable for generalizations, such as results of Adžaga [2] and Cipu and Fujita [66] that in the ring of integers in an imaginary quadratic field, there is no Diophantine 25-tuple.

The main idea is to prove the uniqueness of the extension of a Diophantine triple $\{a, b, c\}$, $a < b < c$, to a Diophantine quadruple $\{a, b, c, d\}$ with $d > c$ for sufficiently large class of triples; more precisely, for the triples satisfying some gap conditions such as $b > 4a$ and $c > \max(b^{13}, 10^{20})$ or $c > \max(b^5, 10^{1029})$.

The assumption that c is large compared with a and b, and also large in the absolute value, gives us roughly that $m/n \approx 1$. Indeed, from $v_m = w_n$, we get

$$\frac{m}{n} \approx \frac{\log(t + \sqrt{bc})}{\log(s + \sqrt{ac})} \approx \frac{\log b + \log c}{\log a + \log c} \approx 1.$$

On the other hand, from the congruence method, we may expect to obtain (under the assumption that m and n are small compared with c) something like $am^2 \approx bn^2$, i.e. $m/n \approx \sqrt{b/a}$. Thus, under the assumption that $b > 4a$, these two approximations for m/n are far enough to obtain a contradiction. Furthermore, a (very) large gap between b and c means that the numbers $\frac{s}{a}\sqrt{\frac{a}{c}} = \sqrt{1 + \frac{1}{ac}}$ and $\frac{t}{b}\sqrt{\frac{b}{c}} = \sqrt{1 + \frac{1}{bc}}$ are very close to 1, which will allow us to apply the theorem of Bennett (Theorem 4.1.20), and obtain better upper bounds for solutions than those obtained by linear forms in logarithms.

The first part of the following lemma shows that the assumption $b > 4a$ is not a big loss of generality if our goal is to get an upper bound for the size of Diophantine tuples.

Lemma 4.10.1

(1) If $\{a, b, c\}$ is a Diophantine triple and $a < b < c$, then $c > 4a$.
(2) If $\{a, b, c, d\}$ is a Diophantine quadruple and $a < b < c < d$, then $d \geq 4bc$.

Proof

(1) From Remark 3.8.6, we have $c \geq a + b + 2r > 4a$.
(2) If we apply Remark 3.8.6 to the triples $\{a, c, d\}$ and $\{b, c, d\}$, we obtain that $d = a + c + 2s$ or $d \geq 4ac$, and $d = b + c + 2t$ or $d \geq 4bc$. However,

$$a + c + 2s < b + c + 2t < 4c \leq 4ac$$

implies that we must have $d \geq 4bc$.

$\square$

On the other hand, if $b < 4a$, then by Lemma 4.9.2, we have two sequences c_k and $\bar{c}_k$ of extensions of the Diophantine pair $\{a, b\}$ to a Diophantine triple. The following lemma, based on [213], will be useful.

Lemma 4.10.2 *If $c_i c_{i+j} + 1$ is a perfect square and $i, j > 0$, then $j = 1$ or $j \geq 2i + 3$. If $\bar{c}_i \bar{c}_{i+j} + 1$ is a perfect square and $i > 1$, $j > 0$, then $j = 1$ or $j \geq 2i + 1$.*

Proof We give the details of the proof for the sequence (c_k). The claim for $(\bar{c}_k)$ follows the same lines, but for the shifted sequence $(\bar{c}_{k+1})$.

We have $ac_i + 1 = s_i^2$, $bc_i + 1 = t_i^2$. Note that s_i and t_i can be given in terms of the solution (u_i, v_i) of Pell's equation $u^2 - (r^2 - 1)v^2 = 1$, where $ab + 1 = r^2$, as

$$s_i = u_i s_0 + a v_i t_0, \quad t_i = b v_i s_0 + u_i t_0, \tag{4.108}$$

where for the sequence (c_k) we have $s_0 = 1$, $t_0 = 1$, while for the sequence $(\bar{c}_k)$ we have $s_0 = 1$, $t_0 = -1$. More generally, we have

$$s_{i+j} = u_j s_i + a v_j t_i, \quad t_{i+j} = b v_j s_i + u_j t_i. \tag{4.109}$$

From (4.109), we obtain (see [213] for details)

$$c_i c_{i+j} + 1 = (u_j c_i + v_j s_i t_i)^2 + 1 - v_j^2. \tag{4.110}$$

Suppose that $c_i c_{i+j} + 1$ is a square and that $j > 1$ (and thus $v_j > 1$). Set $L = u_j c_i + v_j s_i t_i$. Then from (4.110), we obtain $L^2 + 1 - v_j^2 \leq (L - 1)^2$, which implies

$$v_j^2 \geq 2L > 2u_j s_i t_i. \tag{4.111}$$

Let $\alpha = r + \sqrt{ab}$. Then

$$u_i = \frac{1}{2}\left(\alpha^i + \alpha^{-i}\right), \quad v_i = \frac{1}{2\sqrt{ab}}\left(\alpha^i - \alpha^{-i}\right).$$

Inserting this in (4.108), we obtain

$$s_i t_i = \frac{a + b}{4\sqrt{ab}}\left(\alpha^{2i} - \alpha^{-2i}\right) + \frac{1}{2}\left(\alpha^{2i} + \alpha^{-2i}\right).$$

Using this and the obvious inequalities $a + b > 2\sqrt{ab}$, $\alpha > 2\sqrt{ab}$, we obtain

$$\frac{2v_j s_i t_i}{v_j^2} > 4\sqrt{ab}\,\frac{\alpha^{2i}}{\alpha^j - \alpha^{-j}} > 4ab\alpha^{2i-j} > (2\sqrt{ab})^{2i-j+2} \geq 1,$$

if $2i - j + 2 \geq 0$, i.e. $j \leq 2i + 2$. This contradicts (4.111) and proves that $j \geq 2i + 3$.

The same arguments can be applied for the shifted sequence $\bar{c}_{k+1}$ since $\bar{s}_1 = r - a \geq 1$, $\bar{t}_1 = b - r \geq 1$, and they lead to a contradiction provided $j \leq 2(i - 2) + 2 = 2i$. Hence, we have $j \geq 2i + 1$ in that case. $\qquad\square$

As in Sect. 4.8, we consider the equation $v_m^{(i)} = w_n^{(j)}$, where the sequences $(v_m^{(i)})$ and $(w_n^{(j)})$ are given by (4.73) and (4.74), i.e. by (4.75) and (4.76). The following lemma is a generalization of Lemma 4.8.5.

Lemma 4.10.3

(1) If the equation $v_{2m}^{(i)} = w_{2n}^{(j)}$ has a solution, then $z_0^{(i)} = z_1^{(j)}$.

(2) If the equation $v_{2m+1}^{(i)} = w_{2n}^{(j)}$ has a solution, then $z_0^{(i)} \cdot z_1^{(j)} < 0$ and $cx_0^{(i)} - s|z_0^{(i)}| = |z_1^{(j)}|$. In particular, if $b > 4a$ and $c > 100a$, then this equation has no solution.

(3) If the equation $v_{2m}^{(i)} = w_{2n+1}^{(j)}$ has a solution, then $z_0^{(i)} \cdot z_1^{(j)} < 0$ and $cy_1^{(j)} - t|z_1^{(j)}| = |z_0^{(i)}|$.

(4) If the equation $v_{2m+1}^{(i)} = w_{2n+1}^{(j)}$ has a solution, then $z_0^{(i)} \cdot z_1^{(j)} > 0$ and $cx_0^{(i)} - s|z_0^{(i)}| = cy_1^{(j)} - t|z_1^{(j)}|$.

Proof

(1) From Lemmas 4.8.1 and 4.8.4, we have $|z_0^{(i)} - z_1^{(j)}| < c$ and $z_0^{(i)} \equiv z_1^{(j)}$ (mod $2c$), which implies $z_0^{(i)} = z_1^{(j)}$.

(2) Observe that

$$cx_0^{(i)} - s|z_0^{(i)}| = \frac{c^2 - ac - (z_0^{(i)})^2}{cx_0^{(i)} + s|z_0^{(i)}|} < \frac{c^2 - s^2}{c + s} < c,$$

and since $c > 4a$ (by Lemma 4.10.1), we also have $c^2 - ac - (z_0^{(i)})^2 \geq \frac{1}{2}c^2 > 0$. Hence,

$$0 < cx_0^{(i)} - s|z_0^{(i)}| < c.$$

By Lemma 4.8.4, we have $sz_0^{(i)} + cx_0^{(i)} \equiv z_1^{(j)}$ (mod $2c$). Thus we conclude that if $z_0^{(i)} > 0$, then $z_1^{(j)} = sz_0^{(i)} - cx_0^{(i)}$, and if $z_0^{(i)} < 0$, then $z_1^{(j)} = sz_0^{(i)} + cx_0^{(i)}$. Assume now that $b > 4a$ and $c > 100a$. Then

$$c^2 - ac - (z_0^{(i)})^2 \geq c^2 - \frac{c^2}{100} - \frac{c\sqrt{c}}{2\sqrt{a}} \geq 0.94c^2,$$

$$cx_0^{(i)} + s|z_0^{(i)}| < 2cx_0^{(i)} < 2c\sqrt{\frac{s+1}{2}} < 1.5c\sqrt[4]{ac}.$$

Thus, $cx_0^{(i)} - s|z_0^{(i)}| \geq 0.62\sqrt{\frac{c\sqrt{c}}{\sqrt{a}}}$. On the other hand, $|z_1^{(j)}| \leq \sqrt{\frac{c\sqrt{c}}{2\sqrt{b}}} \leq 0.5\sqrt{\frac{c\sqrt{c}}{\sqrt{a}}}$.

(3) As in (2), we find that

$$0 < cy_1^{(j)} - t|z_1^{(j)}| < c,$$

which implies that if $z_1^{(j)} > 0$, then $z_0^{(i)} = tz_1^{(j)} - cy_1^{(j)}$, and if $z_1^{(j)} < 0$, then $z_0^{(i)} = tz_1^{(j)} + cy_1^{(j)}$.

4) We have already proved that

$$0 < cx_0^{(i)} - s|z_0^{(i)}| < c, \quad 0 < cy_1^{(j)} - t|z_1^{(j)}| < c.$$

From Lemma 4.8.4, we have

$$sz_0^{(i)} \pm cx_0^{(i)} \equiv tz_1^{(j)} \pm cy_1^{(j)} \pmod{2c}.$$

Hence, we have two possibilities: if $z_0^{(i)} > 0$, $z_1^{(j)} > 0$, then $sz_0^{(i)} - cx_0^{(i)} = tz_1^{(j)} - cy_1^{(j)}$, and if $z_0^{(i)} < 0$, $z_1^{(j)} < 0$, then $sz_0^{(i)} + cx_0^{(i)} = tz_1^{(j)} + cy_1^{(j)}$.

$\square$

Lemma 4.10.4 *Assume that $b > 4a$ and $c > 100b$. If $v_m = w_n$ and $n \geq 3$, then $m \geq n$.*

Proof From Lemma 4.8.2 we have

$$\frac{m}{n} > \frac{\log(t + \sqrt{bc})}{\log(s + \sqrt{ac})} - \frac{\log \gamma}{n \log(s + \sqrt{ac})},$$

where $\gamma = \frac{\sqrt{b}(x_0\sqrt{c}+z_0\sqrt{a})}{\sqrt{a}(y_1\sqrt{c}+z_1\sqrt{b})}$. Accordingly, it suffices to prove

$$\frac{\log\left(\frac{t+\sqrt{bc}}{s+\sqrt{ac}}\right)}{\log(s + \sqrt{ac})} - \frac{\log \gamma}{n \log(s + \sqrt{ac})} > -\frac{1}{n},$$

and this is equivalent to

$$\left(\frac{t + \sqrt{bc}}{s + \sqrt{ac}}\right)^n > \frac{\gamma}{s + \sqrt{ac}}.$$

However,

$$\frac{\gamma}{s + \sqrt{ac}} \leq \frac{\sqrt{b} \cdot 1.45\sqrt{c}\sqrt[4]{ac} \cdot 1.45\sqrt{c}\sqrt[4]{bc}}{\sqrt{a} \cdot 0.99c \cdot 2\sqrt{ac}} \leq 2.124\left(\frac{b}{a}\right)^{3/4}$$

and

$$\left(\frac{t + \sqrt{bc}}{s + \sqrt{ac}}\right)^3 > \left(\frac{2\sqrt{bc}}{2.002\sqrt{ac}}\right)^3 > 0.997\left(\frac{b}{a}\right)^{3/2}.$$

Thus our condition becomes $(\frac{b}{a})^{3/4} > 2.131$ and this is clearly satisfied if $\frac{b}{a} > 4$. $\qquad\square$

Lemma 4.10.5 *Assume that* $c > \max(b^5, 10^{100})$ *or* $c > \max(b^9, 10^{30})$ *or* $c > \max(b^{13}, 10^{20})$. *If* $v_m = w_n$ *and* $n \geq 2$, *then* $m \leq \frac{3}{2}n$.

Proof From Lemma 4.8.2 we obtain

$$\frac{m}{n} < \frac{\log(t + \sqrt{bc})}{\log(s + \sqrt{ac})} + \frac{\log \gamma}{n \log(s + \sqrt{ac})} + \frac{\frac{8}{3}ac(s + \sqrt{ac})^{-2m}}{n \log(s + \sqrt{ac})}. \tag{4.112}$$

We will estimate the three summands on the right-hand side of (4.112) separately and give details only for the case $c > \max(b^5, 10^{100})$. We have

$$\frac{\log(t + \sqrt{bc})}{\log(s + \sqrt{ac})} = 1 + \frac{\log\left(\frac{t+\sqrt{bc}}{s+\sqrt{ac}}\right)}{\log(s + \sqrt{ac})} < 1 + \frac{\log\sqrt{\frac{b}{a}}}{\log(s + \sqrt{ac})} < 1 + \frac{1}{5}.$$

Assume that $n \geq 5$. From

$$\gamma \leq \frac{\sqrt{b}(x_0\sqrt{c} + |z_0|\sqrt{a})}{\sqrt{a}(y_1\sqrt{c} - |z_1|\sqrt{b})} = \frac{\sqrt{b}(x_0\sqrt{c} + |z_0|\sqrt{a})(y_1\sqrt{c} + |z_1|\sqrt{b})}{\sqrt{a}(c - b)}$$

$$\leq \frac{\sqrt{b} \cdot 2\sqrt{c}\sqrt[4]{ac} \cdot 2\sqrt{c}\sqrt[4]{bc}}{0.9c\sqrt{a}} < c^{0.657},$$

we obtain

$$\frac{\log \gamma}{n \log(s + \sqrt{ac})} < \frac{0.657}{0.5n} < 0.263.$$

Finally,

$$\frac{\frac{8}{3}ac(s + \sqrt{ac})^{-2m}}{n \log(s + \sqrt{ac})} < \frac{\frac{8}{3}ac}{5 \cdot 4ac \log(2\sqrt{ac})} < 0.002.$$

If we combine these three estimates, we obtain the desired result. In the remaining cases ($n \leq 4$), the above estimates give that if $n = 1$, $n = 2$, $n = 3$, $n = 4$ then $m \leq 2$, $m \leq 3$, $m \leq 4$, $m \leq 6$, respectively. $\qquad\square$

4.10.1 Lower Bounds for Solutions

The following the lemma is a straightforward generalization of Lemma 4.8.6.

Lemma 4.10.6

(1) $v_{2m} \equiv z_0 + 2c(az_0m^2 + sx_0m) \pmod{8c^2}$
(2) $v_{2m+1} \equiv sz_0 + c(2asz_0m(m+1) + x_0(2m+1)) \pmod{4c^2}$
(3) $w_{2n} \equiv z_1 + 2c(bz_1n^2 + ty_1n) \pmod{8c^2}$
(4) $w_{2n+1} \equiv tz_1 + c(2btz_1n(n+1) + y_1(2n+1)) \pmod{4c^2}$

In this subsection we will apply congruence relations from Lemma 4.10.6 to obtain lower bounds for the indices m and n satisfying the equation $v_m = w_n$ with $n \neq 0, 1$. As Lemma 4.10.3 suggests, we will consider the equations $v_{2m} = w_{2n}$, $v_{2m} = w_{2n+1}$ and $v_{2m+1} = w_{2n+1}$ separately.

Lemma 4.10.7 *Assume that $b > 4a$.*

(1) *If $v_{2m} = w_{2n}$, $n \neq 0$, and $c > \max(b^{13}, 10^{20})$, then $m \geq n > c^{0.086}$.*
(2) *If $v_{2m} = w_{2n}$, $n \neq 0$, and $c > \max(b^{9}, 10^{30})$, then $m \geq n > c^{0.077}$.*
(3) *If $v_{2m} = w_{2n}$, $n \neq 0$, and $c > \max(b^{5}, 10^{100})$, then $m \geq n > c^{0.048}$.*

Proof Lemmas 4.10.3 and 4.10.6 imply

$$az_0m^2 + sx_0m \equiv bz_0n^2 + ty_1n \pmod{4c}. \tag{4.113}$$

Let $c > \max(b^{13}, 10^{20})$ and assume that $n \leq c^{0.086}$. Then by Lemma 4.10.5 we have $m < c^{0.095}$. We will estimate all four summands in (4.113):

$$|az_0m^2| \leq a\sqrt{\frac{c\sqrt{c}}{2\sqrt{a}}}m^2 \leq c^{\frac{1}{13}\cdot\frac{3}{4}+\frac{3}{4}+2\cdot 0.095} < c,$$

$$sx_0m \leq s\sqrt{\frac{s+1}{2}}m \leq \sqrt[4]{a^3c^3}m < c,$$

$$|bz_0n^2| = |bz_1n^2| \leq b\sqrt{\frac{c\sqrt{c}}{2\sqrt{b}}}n^2 \leq c^{\frac{1}{13}\cdot\frac{3}{4}+\frac{3}{4}+2\cdot 0.086} < c,$$

$$ty_1n \leq \sqrt[4]{b^3c^3}n < c.$$

These estimates imply that in (4.113) we can replace $\equiv$ by $=$, i.e.

$$az_0m^2 + sx_0m = bz_0n^2 + ty_1n. \tag{4.114}$$

The same arguments can be used if $c > \max(b^{9}, 10^{30})$, $n \leq c^{0.077}$ and if $c > \max(b^{5}, 10^{100})$, $n \leq c^{0.048}$.

Let us consider the case when $|z_0| = 1$. Then $x_0 = y_1 = 1$ and (4.114) becomes

$$\pm am^2 + sm = \pm bn^2 + tn.$$

However, $\max(am, bn) < c^{0.2} < 0.001\sqrt{c}$. Hence, $|\pm am^2 + sm| \le 1.001sm$ and $|\pm bn^2 + tn| \ge 0.999tn$. Therefore,

$$\frac{m}{n} \ge \frac{0.999}{1.001} \cdot \frac{t}{s} \ge \frac{0.999}{1.001^2} \cdot \sqrt{\frac{b}{a}} > 1.99 \,,$$

which contradicts Lemma 4.10.5.

Hence, we may assume that $|z_0| > 1$. Then $x_0 \ge 2$ and from $az_0^2 = cx_0^2 - c + a$, we obtain that $z_0^2 \ge \frac{3c}{a}$. This implies

$$0 \le \frac{sx_0}{a|z_0|} - 1 = \frac{x_0^2 + ac - a^2}{a|z_0|(sx_0 + a|z_0|)} \le \frac{ac}{2a^2 z_0^2} \le \frac{1}{6}$$

and

$$0 \le \frac{ty_1}{b|z_1|} - 1 \le \frac{bc}{2b^2 z_1^2} \le \frac{a}{6b} \le \frac{1}{24} \,.$$

If $z_0 > 1$, then by (4.114), we have $az_0 m(m + \frac{7}{6}) \ge bz_0 n(n + 1)$, which for $m \ge 2$ implies $\frac{m}{n} \ge 1.59$ and for $m = n = 1$ gives $b \le 1.1a$, which are both impossible.

If $z_0 < -1$, then by (4.114), we have $a|z_0|m(m - 1) \ge b|z_0|n(n - \frac{25}{24})$, which for $n \ge 3$ implies $\frac{m}{n} \ge 1.61$, while for $m = n = 2$ gives $b < 1.1a$ and for $n = 2$, $m = 3$ gives $b < 3.14a$. The only remaining case is $m = n = 1$. However, in this case we have

$$|z_0|(b - a)(sx_0 + ty_1) = c(b - a) + y_1^2 - x_0^2 \,. \tag{4.115}$$

Since $ax_0^2 = by_1^2 - b + a > \frac{3}{4}by_1^2$, we have $x_0^2 > 3y_1^2$ and the right-hand side of (4.115) is less that $c(b - a)$. It follows that $|z_0|(sx_0 + ty_1) < c$, which is impossible since $|z_0|(sx_0 + ty_1) \ge \sqrt{\frac{3c}{a}} \cdot 2\sqrt{ac} > c$. $\qquad\square$

Lemma 4.10.8 *Assume that $b > 4a$.*

(1) If $v_{2m} = w_{2n+1}$, $n \ne 0$, and $c > \max(b^{13}, 10^{20})$, then $m \ge n > c^{0.0765}$.
(2) If $v_{2m} = w_{2n+1}$, $n \ne 0$, and $c > \max(b^9, 10^{30})$, then $m \ge n > c^{0.062}$.
(3) If $v_{2m} = w_{2n+1}$, $n \ne 0$, and $c > \max(b^5, 10^{100})$, then $m \ge n > c^{0.023}$.

Proof From Lemmas 4.10.3 and 4.10.6, we have

$$as z_0 m^2 + s x_0 m \equiv b z_0 n(n+1) + y_1 n' \pmod{2c},\tag{4.116}$$

where $n' = n$ if $z_0 > 0$ and $n' = n+1$ if $z_0 < 0$.

Assume that $c > \max(b^{13}, 10^{20})$ and $n \le c^{0.0765}$. Then $n+1 < c^{0.0772}$ and $m+1 < \frac{3}{2}(n+1) < c^{0.0861}$. As in the proof of Lemma 4.10.7, we have

$$|a z_0 m^2| < c, \qquad s x_0 m < c,$$

and also

$$|b z_0 n(n+1)| \le b \sqrt{\frac{c\sqrt{c}}{2\sqrt{a}}}(n+1)^2 \le c^{\frac{1}{13}+\frac{3}{4}+2\cdot 0.0772} < c,$$

$$y_1 n' \le \sqrt[4]{bc}(n+1) < c.$$

Hence, we have an equality in (4.116),

$$as z_0 m^2 + s x_0 m = b z_0 n(n+1) + y_1 n',$$

and the same is true for $c > \max(b^9, 10^{30})$, $n \le c^{0.062}$ and for $c > \max(b^5, 10^{100})$, $n \le c^{0.023}$.

Note that $|z_0| = 1$ is impossible since it implies

$$c^2 y_1^2 = t^2 z_1^2 + 2t z_1 + 1 = bc z_1^2 + z_1^2 + 2t z_1^2 + 1$$

and

$$(z_1 + t)^2 = c(c y_1^2 - b z_1^2 + b) = c^2,$$

while $z_1 = \pm c - t$ is impossible since $|\pm c - t| > \sqrt{\frac{c\sqrt{c}}{2\sqrt{b}}}$.

Therefore, we have $|z_0| \ge \sqrt{\frac{3c}{a}}$ and

$$1 \le \frac{s x_0}{a|z_0|} \le \frac{7}{6}, \qquad 0 < \frac{y_1}{b|z_0|} \le \frac{\sqrt[4]{bc}}{b\sqrt{\frac{3c}{a}}} \le \frac{1}{\sqrt[4]{c}} < 0.001.$$

If $z_0 > 0$, from (4.116), we obtain $a z_0 m(m + \frac{7}{6}) \ge b z_0 n(n+1)$. For $m = 2$, $n = 1$, we get $\frac{b}{a} \le 3.17$, and for $m \ge 3$ and $n \ge 2$ from $m(m + \frac{7}{6}) \le 1.39 m^2$ and $n(n+1) \ge 0.96(n+\frac{1}{2})^2$, we obtain $\frac{2m}{2n+1} > 1.66$, which contradicts Lemma 4.10.5.

If $z_0 < 0$, from (4.116), we obtain $az_0m^2 \geq bz_0(n+1)(n-0.001)$, which implies $\frac{2m}{2n+1} > 1.88$. $\qquad\qquad\square$

Lemma 4.10.9 *Assume that $b > 4a$.*

(1) If $v_{2m+1} = w_{2n+1}$, $n \neq 0$, and $c > \max(b^{13}, 10^{20})$, then $m \geq n > c^{0.0765}$.
(2) If $v_{2m+1} = w_{2n+1}$, $n \neq 0$, and $c > \max(b^9, 10^{30})$, then $m \geq n > c^{0.062}$.
(3) If $v_{2m+1} = w_{2n+1}$, $n \neq 0$, and $c > \max(b^5, 10^{100})$, then $m \geq n > c^{0.023}$.

Proof Since $sz_0 \pm cx_0 = tz_1 \pm cy_1$, Lemma 4.10.6 implies

$$asz_0m(m+1) + x_0m' \equiv btz_1n(n+1) + y_1n' \,(\mathrm{mod}\ 2c), \qquad (4.117)$$

where $m' = m$, $n' = n$ if $z_0 < 0$ and $m' = m+1$, $n' = n+1$ if $z_0 > 0$. Noticing that $tz_1 \equiv sz_0 \,(\mathrm{mod}\ c)$ and multiplying (4.117) by s and t, respectively, we obtain

$$az_0m(m+1) + sx_0m' \equiv bz_0n(n+1) + sy_1n' \,(\mathrm{mod}\ 2c), \qquad (4.118)$$

$$az_1m(m+1) + tx_0m' \equiv bz_1n(n+1) + ty_1n' \,(\mathrm{mod}\ 2c). \qquad (4.119)$$

Assume that $c > \max(b^{13}, 10^{20})$ and $n \leq c^{0.0765}$. Then $m + 1 < \frac{3}{2}(n+1) < c^{0.0861}$. This implies that all summands in (4.118) and (4.119) are less than c. Thus we actually have equalities in (4.118) and (4.119):

$$az_0m(m+1) + sx_0m' = bz_0n(n+1) + sy_1n', \qquad (4.120)$$

$$az_1m(m+1) + tx_0m' = bz_1n(n+1) + ty_1n'. \qquad (4.121)$$

We obtain the same conclusion if $c > \max(b^9, 10^{30})$, $n \leq c^{0.062}$ and if $c > \max(b^5, 10^{100})$, $n \leq c^{0.023}$.

Suppose now that $sz_1 \neq tz_0$. Then (4.120) and (4.121) imply

$$am(m+1) = bn(n+1), \qquad (4.122)$$

$$tm' = sn'. \qquad (4.123)$$

However, (4.122) implies $\frac{m}{n} \geq \sqrt{\frac{b}{a}} > 2$. On the other hand, for $n \geq 2$, by Lemma 4.10.5, we have

$$\frac{m}{n} \leq \frac{2m+1}{2n+1} \cdot \frac{5}{4} \leq \frac{3}{2} \cdot \frac{5}{4} = \frac{15}{8} < 2,$$

and for $n = 1$, we have $m = 1$ and $\frac{m}{n} = 1 < 2$.

Let us now consider the case $sz_1 = tz_0$. It is clear that $|z_0| \neq 1$ and $|z_1| \neq 1$. Hence, $|z_0| \geq \sqrt{\frac{3c}{a}}$, $|z_1| \geq \sqrt{\frac{3c}{b}}$ and we have

$$1 \leq \frac{sx_0}{a|z_0|} \leq \frac{7}{6}, \quad 1 \leq \frac{sy_1}{b|z_0|} = \frac{ty_1}{b|z_1|} \leq \frac{7}{6}.$$

If $z_0 > 0$, from (4.120), we obtain $az_0(m+1)(m+\frac{7}{6}) \geq bz_0(n+1)^2$, which implies $\frac{2m+1}{2n+1} \geq \frac{m+1}{n+1} > 1.58$.

If $z_0 < 0$, we obtain $az_0 m^2 \geq bz_0 n(n - \frac{1}{6})$, which for $m = n = 1$ gives $\frac{b}{a} \leq 1.2$ and for $n \geq 2$ implies $\frac{m}{n} \geq 1.91$. However, we have already seen that $\frac{m}{n} \leq \frac{15}{8} = 1.875$. $\square$

Proposition 4.10.10 *Let $\{a, b, c, d\}$, $a < b < c < d$, be a Diophantine quadruple.*

(1) If $c > \max(b^{13}, 10^{20})$, then $d < c^{26}$.
(2) If $c > \max(b^9, 10^{30})$, then $d < c^{62}$.

Proof We give details for the proof of (1). The proof of (2) is completely analogous.

Let $ad + 1 = x^2$, $bd + 1 = y^2$, $cd + 1 = z^2$. We will apply Theorem 4.1.20 with $a_0 = 0$, $a_1 = a$, $a_2 = b$, $N = abc$, $M = b$, $q = abz$, $p_1 = sbx$, $p_2 = tay$. Since $abc > b^9$, the condition $N > M^9$ is satisfied. For the quantity γ from Theorem 4.1.20, we obtain $\gamma = \frac{b^2(b-a)^2}{2b-a}$ if $b \geq 2a$, and $\gamma = \frac{b^2 a^2}{a+b}$ if $a < b < 2a$. Considering γ as a function of a, we get in both cases that

$$\frac{b^3}{6} \leq \gamma < \frac{b^3}{2}.$$

Furthermore, we have

$$\lambda = 1 + \frac{\log(32.04 abc\gamma)}{\log(1.68 c^2(b-a)^{-2})} = 2 - \lambda_1,$$

where

$$\lambda_1 = \frac{\log\left(\frac{1.68c}{32.04ab(b-a)^{-2}\gamma}\right)}{\log(1.68 c^2(b-a)^{-2})}.$$

Theorem 4.1.20 and Lemma 4.1.21 imply

$$\frac{c}{2az^2} > (130abc\gamma)^{-1}(abz)^{\lambda_1 - 2}.$$

This implies

$$z^{\lambda_1} < 65a^2b^3c^2\gamma$$

and

$$\log z < \frac{\log\left(65a^2b^3c^2\gamma\right)\log\left(1.68c^2(b-a)^{-2}\right)}{\log\left(\frac{1.68c}{32.04ab(b-a)^{-2}\gamma}\right)}. \tag{4.124}$$

We will estimate the right-hand side of (4.124). We have

$$65a^2b^3c^2\gamma < 32.5a^2b^6c^2 < 32.5b^8c^2 < c^{2.691},$$

$$1.68c^2(b-a)^{-2} < c^2,$$

$$\frac{1.68c}{32.04ab(b-a)^2\gamma} > \frac{1.68c}{16.02ab^6} > 0.106cb^{-7} > c^{0.412}.$$

Putting these estimates together, we obtain

$$\log z < \frac{2 \cdot 2.691\log^2 c}{0.412\log c} < 13.07\log c.$$

Hence,

$$z < c^{13.07} \tag{4.125}$$

and

$$d = \frac{z^2 - 1}{c} < c^{25.14}.$$

$$\square$$

4.10.2   Special Cases of the Unique Extension Conjecture

Theorem 4.10.11 *If $\{a, b, c, d\}$ is a Diophantine quadruple such that $b > 4a$, $c > \max(b^{13}, 10^{20})$ or $c > \max(b^9, 10^{30})$, and $d > c$, then $d = a+b+c+2abc+2rst$.*

Proof Let $ad + 1 = x^2$, $bd + 1 = y^2$, $cd + 1 = z^2$. Then there exist integers $m, n \geq 0$ such that

$$z = v_m = w_n, \tag{4.126}$$

where the sequences (v_m) and (w_n) are defined by (4.73) and (4.74).

Assume that $n > 1$. Let $c > \max(b^{13}, 10^{20})$. Then Lemmas 4.10.7, 4.10.8, and 4.10.9 imply $n > 2 \cdot c^{0.0765}$. On the other hand, since

$$y_1\sqrt{c} - |z_1|\sqrt{b} = \frac{c - b}{|z_1|\sqrt{b} + y_1\sqrt{c}} > \frac{0.45\sqrt{c}}{\sqrt[4]{bc}} > 3\sqrt{b},$$

it follows from (4.76) that

$$z = w_n > (t + \sqrt{bc})^n > (bc)^{n/2} > c^{n/2}.$$

Thus from (4.125), we conclude that $n \le 26$. If we compare this with $n > 2 \cdot c^{0.0765}$, we obtain $c < 4 \cdot 10^{14}$, which is a contradiction. Similarly, for $c > \max(b^9, 10^{30})$, we obtain $n \le 62$ and $n > 2 \cdot c^{0.062}$, which lead to $c < 2 \cdot 10^{24}$, a contradiction.

Hence, we have $n \le 1$. Let us consider three possible cases separately.

(1) m and n are even. It follows that $m = n = 0$. Then $z = z_0 < c$ and $d = \frac{z^2-1}{c} < c$.

(2) m is even and n is odd. It follows that $n = 1$ and $m = 0$ or 2. If $m = 0$, then $z = z_0 < c$ and $d < c$. If $m = 2$, then $z_0 < 0$ and Lemma 4.10.3 implies

$$z_0 = tz_1 - cy_1. \tag{4.127}$$

Since $v_2 = z_0 + 2c(az_0 + sx_0)$, $v_2 = w_1$ implies

$$az_0 + sx_0 = y_1. \tag{4.128}$$

Combining conditions (4.127) and (4.128) with $az_0^2 - cx_0^2 = a - c$ and $bz_1^2 - cy_1^2 = b - c$, we obtain that

$$z_1 = s, \quad y_1 = r, \quad z_0 = st - cr, \quad x_0 = rs - at.$$

However,

$$|z_0| = \frac{c^2 - ac - bc - 1}{cr + st} > \frac{0.9c^2}{2c\sqrt{ab}} > \sqrt{\frac{c\sqrt{c}}{2\sqrt{a}}},$$

a contradiction.

(3) m and n are odd. It follows that $m = n = 1$.

If $z_0 < 0$, then $z = v_1 = sz_0 + cx_0 < c$ and $d < c$.

If $z_0 > 0$, then

$$sz_0 - cx_0 = tz_1 - cy_1, \tag{4.129}$$

by Lemma 4.10.3. On the other hand, $z = v_1 = w_1$ implies

$$sz_0 + cx_0 = tz_1 + cy_1. \tag{4.130}$$

Clearly, (4.129) and (4.130) imply $x_0 = y_1$ and $sz_0 = tz_1$. If we put this in $az_0^2 - cx_0^2 = a - c$ and $bz_1^2 - cy_1^2 = b - c$, we obtain

$$(b - a)s^2 = (bz_1^2 - az_0^2)s^2 = z_1^2(abc + b - abc - a) = z_1^2(b - a).$$

This implies that $z_1 = s$, $z_0 = t$, $x_0 = y_1 = r$. Hence, $z = v_1 = st + cr$ and

$$d = \frac{z^2 - 1}{c} = \frac{abc^2 + ac + bc + 1 + 2rst + abc^2 + c^2 - 1}{c}$$

$$= a + b + c + 2abc + 2rst.$$

$\square$

Theorem 4.10.12 *If $\{a, b, c, d\}$ is a Diophantine quadruple such that $b > 4a$, $c > \max(b^5, 10^{1029})$ and $d > c$, then $d = a + b + c + 2abc + 2rst$.*

Proof We have already deduced in the proof of Theorem 4.10.11 that $m \leq 1$ and $d > c$ imply $d = a + b + c + 2abc + 2rst$. Thus we may assume that $m \geq 2$. Nevertheless, we proved in Lemmas 4.10.7, 4.10.8, and 4.10.9 that under the assumptions of the theorem, $m > 2 \cdot c^{0.023}$. If we put this in (4.77), we obtain

$$\frac{m}{\log^3 m} < 4.968 \cdot 10^{18},$$

which implies $m < 9 \cdot 10^{23}$ and finally $c < 10^{1029}$. $\square$

4.10.3   *Proofs of Absolute Upper Bounds*

Theorem 4.10.13 *There does not exist a Diophantine 9-tuple.*

Proof Assume that $\{a_1, a_2, \ldots, a_9\}$ is a Diophantine 9-tuple and $a_1 < a_2 < \cdots < a_9$. We will consider two cases depending on whether $a_2 < 4a_1$ or $a_2 > 4a_1$.

Let us first consider the case $a_2 > 4a_1$. Then by Lemma 4.10.1, we have the following estimates

$$a_3 > a_2, \quad a_4 \geq 4a_3a_2 > 4a_2^2, \quad a_5 \geq 4a_4a_3 > 4^2a_2^3,$$

$$a_6 \geq 4a_5a_4 > 4^4a_2^5, \quad a_7 \geq 4a_6a_5 > 4^7a_2^8, \quad a_8 \geq 4a_7a_6 > 4^{12}a_2^{13}.$$

We will apply Theorem 4.10.11 to the quadruple $\{a_1, a_2, a_8, a_9\}$. It is clear that the conditions $a_2 > 4a_1$ and $a_8 > a_2^{13}$ are satisfied. We must check the condition $a_8 > 10^{20}$. But if $a_2 \geq 10$, then $a_8 > 4^{12} \cdot 10^{13} > 10^{20}$, and the only Diophantine pair $\{a_1, a_2\}$ which satisfies $4a_1 < a_2 \leq 10$ is the pair $\{1, 8\}$. However, in this case $a_3 \geq 1 + 8 + 6 = 15$, and we have $a_8 > 4^{12}a_3^8 a_2^5 > 10^{20}$.

Thus, the conditions of Theorem 4.10.11 are satisfied and we have the conclusion that

$$a_9 = a_1 + a_2 + a_8 + 2a_1 a_2 a_8 + 2\sqrt{(a_1 a_2 + 1)(a_1 a_8 + 1)(a_2 a_8 + 1)}.$$

This implies that $a_9 \leq 2(a_1 + a_2 + a_8 + 2a_1 a_2 a_8) < 4a_8(a_1 a_2 + 1)$. On the other hand, Lemma 4.10.1 implies that $a_9 \geq 4a_8 a_7$, and we obtain a contradiction.

Let us now consider the case $a_2 < 4a_1$. By Lemma 4.10.1, we have $a_3 > 4a_1$. Thus, by applying the above arguments to the set $\{a_1, a_3, a_4, \ldots, a_{10}\}$, we conclude that there does not exist a Diophantine 10-tuple.

To improve this result and prove that there is no Diophantine 9-tuple in this case too, we will apply Lemma 4.9.2. Let (c_k) and $(\bar{c}_k)$ be the sequences defined in Lemma 4.9.2 with $a = a_1$ and $b = a_2$.

Assume that $a_3 = c_1$. Then $a_3 > 4a_1$ and $a_3 < 4a_2$. Hence, we have

$$a_8 \geq 4^{13} a_3^8 a_2^5 \geq 4^{13} a_3^8 a_3^5 4^{-5} = 4^8 a_3^{13}.$$

If $a_3 \geq 15$, then $4^8 a_3^{13} > 10^{20}$. However, the only Diophantine triples $\{a_1, a_2, a_3\}$ such that $a_1 < a_2 < a_3 \leq 14$ are $\{1, 3, 8\}$ and $\{2, 4, 12\}$. In [24] and [351], it was proved that these triples cannot be extended to quintuples (these are special cases of the result from [93]). Therefore, we may apply Theorem 4.10.11 to the triple $\{a_1, a_3, a_8\}$. As before, we obtain that

$$a_9 \leq 4a_8(a_1 a_3 + 1),$$

which is in contradiction with $a_9 \geq 4a_8 a_7$ and $a_7 \geq 4^7 a_3^5 a_2^3$.

Hence, we may assume that $a_3 > c_1$. It means that $a_3 \geq 4a_1 a_2$. Lemma 4.9.2 implies that one of the sets

$$\{a_3, a_4, \ldots, a_9\} \cap \{c_k : k \geq 2\} \quad \text{and} \quad \{a_3, a_4, \ldots, a_9\} \cap \{\bar{c}_k : k \geq 2\}$$

has at least four elements. Let us denote that set by C' and its elements by $c'_{k_1} < c'_{k_2} < \cdots < c'_{k_j}$, where $c'_{k_i} = c_{k_i}$ or $\bar{c}_{k_i}$. Lemma 4.10.2 implies that $k_2 = k_1 + 1$ or $k_2 \geq 3k_1$, and also $k_3 = k_2 + 1$ or $k_3 \geq 3k_2$. However, $k_3 = k_2 + 1$ is impossible since it implies $c'_{k_3} \leq (4a_1 a_2 + 2)c'_{k_2}$, contradicting the estimate $c'_{k_3} \geq 4c'_{k_2} c'_{k_1}$. Hence, we have $k_3 \geq 3k_2$ and $k_4 \geq 3k_3 \geq 9k_2$.

If $j \geq 5$, then also $k_5 \geq 3k_4 \geq 27k_2$. It is easy to check that $c'_{3k} \geq (c'_k)^3$. It follows that $c'_{k_5} \geq (c'_{k_2})^{27}$. Since $c'_{k_1} \geq 4a_1 a_2$, we have $c'_{k_2} \geq 16a_1^2 a_2^2 \geq a_2^4$ and $c'_{k_5} \geq 10^{50}$. Therefore, we may apply Theorem 4.10.11 to the triple $\{a_1, c'_{k_1}, c'_{k_5}\}$.

As before, we conclude that there does not exist a positive integer $d > c'_{k_5}$ such that

$$\{a_1, a_2, c'_{k_1}, c'_{k_2}, c'_{k_3}, c'_{k_4}, c'_{k_5}, d\}$$

is a Diophantine 8-tuple. It implies that $j \leq 5$. In particular, this finishes the proof in the case $a_2 = a_1 + 2$, since in this case by Lemma 4.9.2 we have $j = 7$.

If $a_2 \neq a_1 + 2$, then $a_2 \geq 8$ and $c'_{k_4} \geq (c'_{k_2})^9 \geq 8^{36} > 10^{30}$. Therefore we may apply Theorem 4.10.11 to the triple $\{a_1, c'_{k_1}, c'_{k_4}\}$. After doing that, we conclude that $j = 4$ and $a_9 = c'_{k_4}$.

Let $C'' = \{a_3, a_4, \ldots, a_9\} \setminus C' = \{c''_{l_1}, c''_{l_2}, c''_{l_3}\}$, where $c''_{l_1} < c''_{l_2} < c''_{l_3}$ and $c''_{l_i} = c_{l_i}$ or $\bar{c}_{l_i}$. As in the proof of $k_3 \geq 3k_2$, we see that we cannot have $k_2 = k_1 + 1$ and $l_2 = l_1 + 1$ simultaneously. It follows that $k_3 \geq 9k_1$ or $l_3 \geq 9l_1$. On the other hand, if both inequalities $k_3 \geq 9k_1$ and $l_3 \geq 9l_1$ are satisfied, then

$$a_8 = \max(c'_{k_3}, c''_{l_3}) \geq \max((c'_{k_1})^9, (c''_{l_1})^9) \geq a_4^9 \geq \bar{c}_3^9 \geq (16a_1^2 a_2^2)^9 \geq a_2^{36} > 10^{30}$$

(we cannot have $a_3 = \bar{c}_2$, $a_4 = c_2$, since $c_2 < 16a_1 a_2^2$ and $4\bar{c}_2 a_2 \geq 16a_1 a_2^2$). Thus, applying Theorem 4.10.11 to the triple $\{a_1, a_4, a_8\}$ contradicts the existence of a_9. Hence, we have $a_3 = c'_{k_1}$, $a_4 = c'_{k_1+1}$ or $a_3 = c''_{l_1}$, $a_4 = c''_{l_1+1}$, which implies

$$a_8 \geq a_5^9 > 10^{30},$$

and an application of Theorem 4.10.11 leads to a contradiction as before.   □

Theorem 4.10.14 *There are only finitely many Diophantine 8-tuples.*

Proof Let $\{a_1, a_2, \ldots, a_8\}$ be a Diophantine 8-tuple such that $a_1 < a_2 < \cdots < a_8$. We have $a_3 > 4a_1$, and also from Lemma 4.10.1,

$$a_4 > a_3, \quad a_5 > 4a_3^2, \quad a_6 > 4^2 a_3^3, \quad a_7 > 4^3 a_3^5.$$

If $a_3 > 10^{343}$, then we may apply Theorem 4.10.12 to the triple $\{a_1, a_3, a_7\}$. This gives that

$$a_8 = a_1 + a_3 + a_7 + 2a_1 a_3 a_7 + 2\sqrt{(a_1 a_3 + 1)(a_1 a_7 + 1)(a_3 a_7 + 1)} \leq 4a_7(a_1 a_3 + 1),$$

which contradicts $a_8 \geq 4a_7 a_6$.

Hence, we proved that there are only finitely many triples $\{a_1, a_2, a_3\}$ which can be extended to a Diophantine 8-tuple, namely it is possible only for triples with $\max\{a_1, a_2, a_3\} \leq 10^{343}$. Since there are only finitely many integer points on the elliptic curve

$$y^2 = (a_1 x + 1)(a_2 x + 1)(a_3 x + 1),$$

the proof is finished.   □

Remark 4.10.15 From the proof of Theorem 4.10.14, it follows that if $\{a_1, \ldots, a_8\}$ is a Diophantine 8-tuple and $a_1 < a_2 < \cdots < a_8$, then $a_4 < 10^{514}$. Since $a_4 > 4a_2$, we may apply (4.77) with $a = a_1$, $b = a_2$, $c = a_4$. We obtain $m < 2 \cdot 10^{23}$, and it implies that $a_8 < 10^{10^{26}}$.

4.11   Diophantine Quintuple Conjecture

4.11.1   There Are No Diophantine Sextuples and Only Finitely Many Diophantine Quintuples

In the previous section, we presented the proof of the first absolute bound ($m \leq 8$) for the size of Diophantine m-tuples, as given in 2001 by Dujella [104]. In 2004, this result was significantly improved by the same author in [107]. The main results of [107] are the following two theorems.

Theorem 4.11.1 *There are only finitely many Diophantine quintuples.*

Theorem 4.11.2 *There does not exist a Diophantine sextuple.*

Theorems 4.11.1 and 4.11.2 improve results from [104], where it was proved that there does not exist a Diophantine 9-tuple and that there are only finitely many Diophantine 8-tuples.

As in [104], the main idea is to prove Conjecture 1.4.3 for a wide class of Diophantine triples, namely, for triples satisfying some gap conditions. However, in [107], these gap conditions are much weaker than in [104]. Thus, the class of Diophantine triples for which Conjecture 1.4.3 can be proved is now so wide that in an arbitrary Diophantine quadruple (with sufficiently large elements), we may find a sub-triple belonging to that class. And this is just what is needed in order to prove Theorem 4.11.1.

In the proof of Conjecture 1.4.3 for a triple $\{a, b, c\}$, the problem is first transformed into solving systems of simultaneous Pellian equations. This reduces to finding intersection of binary recurrence sequences. The next step is the determination of the initial terms of these sequences, under the assumption that they have a non-empty intersection which induces a solution of the original problem. This part is a considerable improvement of the corresponding part of [104]. The improvement comes from the new gap principles.

By applying some congruence relations modulo c^2, lower bounds for solutions are obtained. In obtaining these bounds, it is necessary to assume that the triple satisfies some gap conditions, e.g. $b > 4a$ and $c > b^{2.5}$. Let us note that these conditions are much weaker than conditions used in [104], and this is due to a more precise determination of the initial terms. The comparison of these lower bounds with upper bounds obtained from the Baker's theory on linear forms in logarithms of algebraic numbers (the theorem of Baker and Wüstholz, or more recent theorem of Matveev) yields Theorem 4.11.1, and the comparison with upper bounds obtained from Bennett's theorem on simultaneous approximations of algebraic numbers yields Theorem 4.11.2.

Let us note that to apply Bennett's theorem, it is necessary to assume a large gap between b and c (at least something like $c > b^5$). This is why this approach does not give explicit information about the number of Diophantine quintuples. However, when these gap conditions are satisfied, upper bounds for solutions obtained by Bennett's theorem (i.e. by the hypergeometric method) are much better than those obtained using linear forms in logarithms, and it makes possible to do all necessary computations and to prove that there is no Diophantine sextuple.

Let us give some details concerning the above brief overview of methods used in proving Theorems 4.11.1 and 4.11.2. One crucial ingredient is an improved gap principle given in the following proposition. It is obtained by considering elements with small indices in the sequences (v_m) and (w_n), defined by (4.73) and (4.74). It is not hard to show that if one of the indices m, n is equal to 0, 1 or 2, then d defined by $d = (v_m^2 - 1)/c$ satisfies $d < c$ or $d = d_+$. With additional efforts, it can be shown that $v_3 \neq w_3$.

Proposition 4.11.3 *If $\{a, b, c, d\}$ is a Diophantine quadruple and $a < b < c < d$, then $d = d_+$ or $d > 2.695c^{3.5}a^{2.5}$.*

Proof From Lemma 4.8.1, we get the following estimates for v_1:

$$v_1 = sz_0 + cx_0 \geq cx_0 - s|z_0| = \frac{c^2 - ac - z_0^2}{cx_0 + s|z_0|} > \frac{c^2 - \frac{c^2}{4} - \frac{c\sqrt{c}}{2\sqrt{a}}}{2cx_0} > \frac{c}{4x_0} > \frac{c}{3.364\sqrt[4]{ac}},$$

$$v_1 < 2cx_0 < 1.682c\sqrt[4]{ac}.$$

From this and recurrence (4.73), we obtain

$$\frac{c}{3.364\sqrt[4]{ac}}(2s - 1)^{m-1} < v_m < 1.682c\sqrt[4]{ac}(2s)^{m-1} \quad \text{for } m \geq 1. \qquad (4.131)$$

The above discussion on the equation $v_m = w_n$ for small indices m, n, shows that if $d \neq d_+$, then we have $m \geq 4$ or $n \geq 4$. By (4.131), we have

$$v_4 \geq \frac{c}{3.364\sqrt[4]{ac}}(2s - 1)^3 \geq \frac{125}{8\sqrt{8} \cdot 3.36}\sqrt[4]{a^5 c^5} \cdot c.$$

Therefore, if $m \geq 4$, then

$$d \geq \frac{2.696a^{2.5}c^{4.5} - 1}{c} > 2.695c^{3.5}a^{2.5},$$

and the same conclusion can be obtained if $n \geq 4$. $\square$

Using the gap principle from Proposition 4.11.3, we can improve Lemma 4.10.3 and obtain more precise information on the initial terms of the sequences (v_m) and (w_n).

Lemma 4.11.4

(1) If the equation $v_{2m} = w_{2n}$ has a solution, then $z_0 = z_1$. Furthermore, $|z_0| = 1$ or $|z_0| = cr - st$ or $|z_0| < 0.869\, a^{-5/14} c^{9/14}$.

(2) If the equation $v_{2m+1} = w_{2n}$ has a solution, then $|z_0| = t$, $|z_1| = cr - st$ and $z_0 z_1 < 0$.

(3) If the equation $v_{2m} = w_{2n+1}$ has a solution, then $|z_0| = cr - st$, $|z_1| = s$ and $z_0 z_1 < 0$.

(4) If the equation $v_{2m+1} = w_{2n+1}$ has a solution, then $|z_0| = t$, $|z_1| = s$ and $z_0 z_1 > 0$.

Proof We will prove only the statement (1). The proofs of the other statements of the lemma are analogous.

From Lemma 4.10.3, we have $z_0 = z_1$. Define $d_0 = (z_0^2 - 1)/c$. Then d_0 is an integer and

$$cd_0 + 1 = z_0^2, \quad ad_0 + 1 = \frac{1}{c}(az_0^2 - a + c) = x_0^2, \quad bd_0 + 1 = \frac{1}{c}(bz_1^2 - b + c) = y_1^2.$$

Hence, we have three possibilities: $d_0 = 0$ or $\{a, b, c, d_0\}$ is a regular Diophantine quadruple or $\{a, b, c, d_0\}$ is an irregular Diophantine quadruple. If $d_0 = 0$, then $|z_0| = 1$. If the quadruple $\{a, b, c, d_0\}$ is regular, then (since $d_0 < c$) we have $d_0 = d_-$ and $|z_0| = cr - st$. Otherwise, we may apply Proposition 4.11.3 to obtain $c \geq 2.695 d_0^{3.5} a^{2.5}$. Since $|z_0| \neq 1$, we have $z_0^2 \geq c + 1$. We may assume that $c > 10^6$ (since all quadruples with elements less than 10^6 are regular, see Exercise 12 in Sect. 1.6), and therefore

$$d_0 = \frac{z_0^2 - 1}{c} \geq \frac{z_0^2}{c}\left(1 - \frac{1}{c+1}\right) > 0.999 \frac{z_0^2}{c}.$$

Hence, we obtain $c^{4.5} > 2.685 |z_0^7| a^{2.5}$ and $|z_0| < 0.869 a^{-5/14} c^{9/14}$.   □

Definition 4.11.5 Let $\{a, b, c\}$ be a Diophantine triple and $a < b < c$. We say that $\{a, b, c\}$ is a Diophantine triple of

- *the first kind if $c > b^{4.5}$,*
- *the second kind if $b > 4a$ and $c > b^{2.5}$,*
- *the third kind if $b > 12a$ and $b^{5/3} < c < b^2$,*
- *the fourth kind if $b > 4a$ and $b^2 \leq c < 6ab^2$.*

A triple $\{a, b, c\}$ is called *standard* if it is a Diophantine triple of the first, second, third or fourth kind.

Proposition 4.11.6 *Every Diophantine quadruple contains a standard triple.*

Proof Let $\{a, b, c, d\}$ be a Diophantine quadruple and $a < b < c < d$. If it is not regular, then by Proposition 4.11.3 and Lemma 4.10.1, we have $d > c^{3.5}$ and $c > 4a$. Hence, $\{a, c, d\}$ is a triple of the second kind.

Assume that $\{a, b, c, d\}$ is a regular quadruple. Then

$$c(4ab + 1) < d < 4c(ab + 1).$$

If $b > 4a$ and $c \geq b^{1.5}$, then $d > b^{2.5}$, and we see that $\{a, b, d\}$ is a triple of the second kind.

If $b > 4a$ and $c < b^{1.5}$, then (by Remark 3.8.6) we have two possibilities: if $c = a + b + 2r$, then $c < 4b$, $c^2 < d < 4ac^2$ and therefore $\{a, c, d\}$ is a triple of the fourth kind; if $c \geq 4ab + a + b$, then $d < c^2$, $d > bc > c^{5/3}$ and $\{a, c, d\}$ is a triple of the third kind.

We may now assume that $b < 4a$. By Lemma 4.9.2, we have $c = c_k$ $(k \geq 1)$ or $c = \bar{c}_k$ $(k \geq 2)$. If $c > b^{2.5}$, then $d > 4abc > b^{4.5}$, and we see that $\{a, b, d\}$ is a triple of the first kind.

Since $c_2 = 4r(a + r)(b + r) > 4ab^2 > b^3$, the condition $c \leq b^{2.5}$ implies $c = c_1$ or $c = \bar{c}_2$. If $c = c_1$, then

$$c \leq a + b + \frac{4}{\sqrt{3}}\sqrt{ab} = \sqrt{ab}\left(\sqrt{\frac{a}{b}} + \sqrt{\frac{b}{a}} + \frac{4}{\sqrt{3}}\right) \leq \sqrt{ab}(0.5 + 2 + \frac{4}{\sqrt{3}}) < 4.81\sqrt{ab}$$

and $c \geq 4r$. Hence, $d > 4abc > 0.172c^3 > c^2$ and $d < 4cr^2 \leq c^2r < 2ac^2$. Therefore, the triple $\{a, c, d\}$ is of the fourth kind. If $c \leq b^{2.5}$ and $c = \bar{c}_2 \geq 4ab + 2a + 2b$ (we may assume $b > a + 2$, since otherwise $\bar{c}_2 = c_1$), then $d < c^2$ and $d > 4abc > b^2c > c^{1.8} > c^{5/3}$. Therefore, the triple $\{a, c, d\}$ is of the third kind. $\qquad\square$

By the congruence method (Lemma 4.10.6), we can obtain lower bounds for m and n satisfying $v_m = w_n$, but under milder assumptions than in the previous section. The proof of the following lemma is similar to the proofs of Lemmas 4.10.7, 4.10.8, and 4.10.9, so we omit it. Let us note that in the proof, six possible cases for the initial terms of the sequences (v_m) and (w_n) from Lemma 4.11.4 are considered.

Lemma 4.11.7

(1) Let $\{a, b, c\}$ be a Diophantine triple of the first kind and $c > 10^{100}$. If $v_m = w_n$ and $n > 2$, then $n > c^{0.01}$.

(2) Let $\{a, b, c\}$ be a Diophantine triple of the second kind and $c > 10^{100}$. If $v_m = w_n$ and $n > 2$, then $n > c^{0.04}$.

(3) Let $\{a, b, c\}$ be a Diophantine triple of the third kind and $c > 10^{100}$. If $v_m = w_n$ and $n > 2$, then $n > c^{0.15}$.

(4) Let $\{a, b, c\}$ be a Diophantine triple of the fourth kind and $c > 10^{100}$. If $v_m = w_n$ and $n > 2$, then $n > c^{0.2}$.

In Sect. 4.8.1, we showed how linear forms in logarithms (in particular, the theorem of Baker and Wüstholz) can be applied to obtain an upper bound for solutions of the equation $v_m = w_n$. Let us recall inequality (4.77):

$$\frac{m}{\log m} < 2.628 \cdot 10^{15} \log^2 c.$$

In [107], instead of the theorem of Baker and Wüsthol, the theorem of Matveev [268] was used. Together with certain gap assumptions, this resulted in a smaller constant appearing in the inequality.

Proposition 4.11.8 *Assume that* $c > \max(b^{5/3}, 10^{100})$. *If* $v_m = w_n$, *then*

$$\frac{m}{\log(31.3\,(m+1))} < 3.826 \cdot 10^{12} \log^2 c. \tag{4.132}$$

The assumption $c > \max(b^{5/3}, 10^{100})$ is not essential. It only affects the constant on the right-hand side of (4.132). Baker's method can be applied without any gap assumption. However, gap assumptions are necessary for obtaining lower bounds for solutions using the congruence method.

In Lemmas 4.10.4 and 4.10.5, we proved that $v_m = w_n$ implies $m \geq n$ if $b > 4a$, $c > 100b$ and $n \geq 3$, and also that $m \leq \frac{3}{2}n$, provided $\{a, b, c\}$ satisfies some rather strong gap conditions. Similarly, we can get the following unconditional relationship between m and n.

Lemma 4.11.9 *If* $v_m = w_n$, *then* $n - 1 \leq m \leq 2n + 1$.

Proposition 4.11.10 *Let* $\{a, b, c\}$ *be a standard Diophantine triple and* $c \geq 10^{2171}$. *If* $\{a, b, c, d\}$ *is a Diophantine quadruple and* $d > c$, *then* $d = d_+$.

Proof Let $ad + 1 = x^2$, $bd + 1 = y^2$, $cd + 1 = z^2$. Then there exist integers $m, n \geq 0$ such that

$$z = v_m = w_n.$$

Assume that $d \neq d_+$. Then $m \geq 3$ and $n \geq 3$. Hence, we may apply Lemma 4.11.7. We get that in all cases $n > c^{0.01}$. Now, by Lemma 4.11.9, we have $m + 1 \geq n > c^{0.01}$. If we put this in (4.132), we obtain

$$\frac{m}{\log(31.3\,(m+1))\log^2(m+1)} < 3.826 \cdot 10^{16},$$

which implies $m < 5.108 \cdot 10^{21}$, and finally $c < (m+1)^{100} < 10^{2171}$. $\qquad\square$

Proof of Theorem 4.11.2 Let $\{a, b, c, d, e\}$ be a Diophantine quintuple and $a < b < c < d < e$. Consider the quadruple $\{a, b, c, d\}$. By Proposition 4.11.6, it contains a standard triple, say $\{A, B, C\}$, $A < B < C$.

If $d \geq 10^{2171}$, we may apply Proposition 4.11.10 to the triple $\{A, B, C\}$. We conclude that

$$e = A + B + C + 2ABC + 2\sqrt{(AB+1)(AC+1)(BC+1)} < 4d(bc+1) < d^3.$$

On the other hand, by Proposition 4.11.3, we have

$$e > 2.695 d^{3.5} b^{2.5} > d^3,$$

a contradiction. Hence $d < 10^{2171}$.

Consider now the quadruple $\{A, B, C, e\}$. We have $C \leq d < 10^{2411}$. Let (V_m) and (W_n) be the sequences defined in the same manner as (v_m) and (w_n), using A, B, C instead of a, b, c. Let $e \cdot C + 1 = V_m^2$. By Proposition 4.11.8,

$$\frac{m}{\log(31.3\,(m+1))} < 9.561 \cdot 10^{19}$$

and $m < 5.109 \cdot 10^{21}$.

From (4.131), we obtain

$$V_m < 1.7 C \sqrt[4]{AC}(2\sqrt{AC+1})^{m-1} < 2^m \cdot d^{m+0.5}.$$

Therefore

$$\log_{10} V_m < m \log_{10} 2 + (m+0.5) \cdot 2171 < 1.1094 \cdot 10^{25}$$

and

$$\log_{10} e < 2\log_{10} V_m < 2.2188 \cdot 10^{25}.$$

Hence $e < 10^{10^{26}}$. $\qquad\qquad\square$

Corollary 4.11.11 *If $\{a, b, c, d, e\}$ is a Diophantine quintuple and $a < b < c < d < e$, then $d < 10^{2171}$ and $e < 10^{10^{26}}$.*

Corollary 4.11.11 already gives an upper bound for the number of Diophantine quintuples. By taking into account known results on gaps between elements of a quintuple and bounds for the number of fundamental solutions of Pellian equations (which use the function ω, which counts the number of prime factors of a given positive integer), in [112], it was shown that there are at most 10^{1930} Diophantine quintuples. Fujita significantly improved this bound in [186] by showing that there exist at most 10^{276} Diophantine quintuples. Furthermore, Filipin and Fujita [167] obtained the bound 10^{96}, Elsholtz, Filipin, and Fujita [164] the bound $6.8 \cdot 10^{32}$, Cipu [62] the bound 10^{31}, Trudgian [342] the bound $2.3 \cdot 10^{29}$, and Cipu and Trudgian [68] the bound $1.18 \cdot 10^{27}$. These improvements were based on additional gap results (some will be mentioned in the following subsection) and better handling of certain sums involving the function ω.

Since the bounds on elements of a Diophantine quintuple from Corollary 4.11.11 are huge, it is computationally infeasible to check whether any Diophantine quintuple exists in the remaining range. However, using the theorem of Bennett (Theorem 4.1.20) on simultaneous approximations of algebraic numbers, instead of the theorems by Matveev or Baker-Wüstholz on linear forms in logarithms, we can prove that there is no Diophantine sextuple. We already gave some details on the application of Bennett's theorem in the proofs of Theorem 4.10.13 and the auxiliary results, so here we will briefly sketch the proof of Theorem 4.11.2.

In order to apply Bennett's theorem, we have to assume a bigger gap between b and c, but as a result, we obtain better lower bounds for solutions. Typical such results are the following auxiliary lemmas proved in [107].

Lemma 4.11.12 *Let $\{a, b, c, d\}$, $a < b < c < d$, be a Diophantine quadruple.*

(1) If $c > 344.9\, b^{9.5} a^{3.5}$, then $d < c^{27.62}$.
(2) If $c > 296.4\, b^{11.6} a^{1.4}$, then $d < c^{21.94}$.

Lemma 4.11.13 *Let $\{a, b, c\}$ be a Diophantine triple. Assume that $v_m = w_n$ and $n \geq 3$.*

(1) If $c > \max(b^{11.6}, 2.97 \cdot 10^{16})$, then $n > c^{0.0815}$.
(2) If $b > 4a$ and $c > \max(b^{9.5}, 1.5 \cdot 10^{13})$, then $n > c^{0.11}$.

As a consequence of these two lemmas, we obtain the following result.

Proposition 4.11.14 *If $\{a, b, c, d\}$ is a Diophantine quadruple such that*

$$c > \max(296.4\, b^{11.6} a^{1.4}, 2.97 \cdot 10^{16}) \ \ or$$

$$(b > 4a \ \ and \ \ c > \max(334.9\, b^{9.5} a^{3.5}, 1.5 \cdot 10^{13})),$$

and $d > c$, then $d = d_+$.

Proof Assume that $d \neq d_+$. Then $n \geq 3$, and we may apply Lemma 4.11.13.

If $c > \max(296.4\, b^{11.6} a^{1.4}, 2.97 \cdot 10^{16})$, then $n > c^{0.0815}$. On the other hand, we have

$$z = w_n > \frac{c}{3.132 \sqrt[4]{bc}} (1.999 \sqrt{bc})^{n-1} > c^{\frac{n}{2} + \frac{1}{4}}.$$

From Lemma 4.11.12, which gives $z < c^{11.48}$ in this case, we conclude that $n \leq 22$, and it implies that $c < 2.97 \cdot 10^{16}$, a contradiction.

If $b > 4a$ and $c > \max(334.9\, b^{9.5} a^{3.5}, 1.5 \cdot 10^{13})$, then $n > c^{0.11}$. From Lemma 4.11.12, which gives $z < c^{14.31}$ in this case, it follows that $n \leq 28$ and $c < 1.5 \cdot 10^{13}$, a contradiction. $\square$

Proof of Theorem 4.11.2 To finish the proof of Theorem 4.11.2, assume that $\{a, b, c, d, e, f\}$, where $a < b < c < d < e < f$, is a Diophantine sextuple. By Proposition 4.11.3, we have

$$e > 2.695\, d^{3.5} b^{2.5} > 2.695\, (4abc)^{3.5} b^{2.5} > 344.9\, b^{9.5} a^{3.5}.$$

If $b < 4a$, then $c \geq a + b + 2r > 2.25\, b$, and we obtain

$$e > 2.695 \cdot 2.25^{3.5} b^{9.5} (4a)^{1.4} b^{2.1} > 296.4\, b^{11.6} a^{1.4}.$$

Assume now that

$$e > 2.97 \cdot 10^{16} \quad \text{or} \quad (b < 4a \text{ and } e > 1.5 \cdot 10^{13}). \tag{4.133}$$

Then we may apply Proposition 4.11.14. We conclude that the quadruple $\{a, b, e, f\}$ is regular. It implies that $f \leq 4e(ab + 1) < e^3$. On the other hand, from Proposition 4.11.3, we have $f > e^{3.5}$, a contradiction.

It remains to consider the case when conditions (4.133) are not satisfied. If $e \leq 2.97 \cdot 10^{16}$, then $d < 4 \cdot 10^4$, and it is easy to find all quadruples satisfying $2.695\, d^{3.5} b^{2.5} \leq 2.97 \cdot 10^{16}$ or ($b < 4a$ and $2.695\, d^{3.5} b^{2.5} \leq 1.5 \cdot 10^{13}$). There are exactly ten such quadruples, and using the Baker-Davenport reduction, it is easy to show that they cannot be extended to quintuples (and then, of course, they cannot be extended to sextuples). $\qquad\square$

4.11.2   *On the Proof of Non-existence of Diophantine Quintuples*

An important step toward the proof of the Diophantine quintuple conjecture is a result by Fujita [185], who proved that any Diophantine quintuple contains a regular Diophantine quadruple, i.e. if $\{a, b, c, d, e\}$ is a Diophantine quintuple and $a < b < c < d < e$, then $d = d_+$. In other words, an irregular Diophantine quadruple $\{a, b, c, d\}$ cannot be extended to a Diophantine quintuple $\{a, b, c, d, e\}$ with $e > \max\{a, b, c, d\}$. This gives an alternative proof of the non-existence of Diophantine sextuples. One of the main ingredients in the proof is a better gap principle for irregular Diophantine quadruples $\{a, b, c, d\}$. More precisely, it is proved that if $\{a, b, c, d\}$ is a Diophantine quadruple with $a < b < c < d_+ < d$, then either $d > 60b^8$ or $d > \max(400a^5 b^6, 100b^7)$. Moreover, in the case $b < 2a$, it holds $d > 1400b^{12.5}$ unless the triple $\{a, b, c\}$ is regular. Treating separately the cases $b < 2a$ and $b \geq 2a$, yields a better gap principle in the case $b < 2a$, and a better lower bound for solutions in the case of $b \geq 2a$ (in the case 1 of Lemma 4.11.4 with $|z_0| \neq 1, cr - st$, under the assumption that $c > b^7$, for $b \geq 2a$, the congruence method gives roughly $n > c^{5/28}$, compared with $n > c^{3/28}$ in the general case).

The third ingredient is a slight modification of Bennett's theorem (Theorem 4.1.20), which allows its application if $c > b^7$. Then applying this theorem to the quadruple $\{a, b, d, e\}$ if either $b \geq 2a$ or $c > a + b + 2\sqrt{ab + 1}$ and to the quadruple $\{b, c, d, e\}$ if $b < 2a$ and $c = a + b + 2\sqrt{ab + 1}$, completes the proof.

The next significant contribution is also due to Fujita. In [186], he proved that for a fixed Diophantine triple $\{a, b, c\}$ with $a < b < c$, the number of Diophantine quintuples $\{a, b, c, d, e\}$ with $c < d < e$ is at most four. In a very useful lemma, it was also shown that in the equation $v_n = w_m$, we may assume that m and n are even and the corresponding z_0, z_1 satisfy $|z_0| = |z_1| = 1$ (cf. Lemmas 4.8.5, 4.10.3, and 4.11.4).

In 2016, based on several previous results [62, 64, 65, 164, 342] published from 2014 to 2016, Cipu and Trudgian [68] summarized the status of the Diophantine quintuple conjecture at that time as follows. Any Diophantine quintuple $\{a, b, c, d, e\}$ with $a < b < c < d < e$ must be of one of the following four types:

(A) $4a < b$ and $4ab + a + b < c < b^{3/2}$,
(B) $4a < b$ and $c = a + b + 2\sqrt{ab + 1}$,
(C) $4a < b$ and $c > b^{3/2}$,
(D) $b < 4a$ and $c = a + b + 2\sqrt{ab + 1}$.

Finally, in 2019, He, Togbé, and Ziegler [211] proved the Diophantine quintuple conjecture.

Theorem 4.11.15 *There does not exist a Diophantine quintuple.*

Here, we will give just a rough overview of the proof. An interested reader may consult [211] for details of the proof and also the paper by Bliznac Trebješanin and Filipin [41], where they modified the proof of Theorem 4.11.15 and showed that there does not exist a $D(4)$-quintuple.

The three new key ingredients that lead to the proof of Theorem 4.11.15 are:

- the use of sharp lower bounds for linear forms in three logarithms obtained by applying a result due to Mignotte [276];
- the use of new congruences in the case of the quadruples of Euler's form $\{a, b, a + b + 2r, 4r(r + a)(r + b)\}$;
- the definition of an operator on Diophantine triples and their classification according to the degree defined using the operator.

Mignotte's result on linear forms in logarithms [276] usually produces significantly better upper bounds for solutions compared with the results of Baker and Wüstholz [25] and Matveev [268]. It was recently additionally improved by Mignotte and Voutier [277]. However, in order to apply it, certain non-trivial conditions should be verified. Among the other important applications, it was used in the paper by Bugeaud, Mignotte, and Siksek [51], where they proved that the only Fibonacci numbers which are powers of a positive integer are $F_1 = F_2 = 1^k$, $F_6 = 8 = 2^3$ and $F_{12} = 144 = 12^2$.

The first application of Mignotte's result to Diophantine m-tuples appeared in [50]. In that paper, Bugeaud, Dujella, and Mignotte showed that all quadruples containing the triple $\{k - 1, k + 1, 16k^3 - 4k\}$ for $k \geq 2$ are regular. That was the last remaining unsolved case for triples of the form $\{k - 1, k + 1, c\}$. Since $k + 1 < 4(k - 1)$, by Lemma 4.9.2, we know that $c = c_i$, where $c_1 = 4k$, $c_2 = 16k^3 - 4k$, $c_3 = 64k^5 - 48k^3 + 8k$, ... If $i \geq 3$, then the gap between c and b is large enough to apply the hypergeometric method (a version of theorems of Rickert [307] and Bennett [30]), and that was done by Fujita [184]. On the other hand, if $i = 1$, then we deal with very good simultaneous rational approximations to the numbers $\sqrt{\frac{4a}{c}} = \sqrt{1 - \frac{1}{k}}$ and $\sqrt{\frac{4b}{c}} = \sqrt{1 + \frac{1}{k}}$, which are very close to 1, and Rickert's theorem (Theorem 4.1.19) can be directly applied, as it was done in [93]. Since the hypergeometric method cannot be applied in the remaining case $i = 2$, we rely on linear forms in logarithms. The combination of an improved lower bound for the solutions, coming from a more delicate application of the congruence method, with Matveev's result, gives the bound on k, $k < 3.8 \cdot 10^{10}$, which is too large to be handled by computations (the applications of the Baker-Davenport reduction for values of k in that range). However, the application of Mignotte's result gives the feasible bound $k < 5.4 \cdot 10^8$.

Similarly, in [211], Matveev's result gives the bound $1.55 \cdot 10^{17}$ for the corresponding index, compared to $5.136 \cdot 10^{13}$ obtained by Mignotte's result (however, the first bound is needed in order to check that the conditions of [276] are satisfied). Nevertheless, from a computational point of view, this upper bound is still too large to apply the Baker-Davenport reduction method directly.

Let $\{a, b, c, d\}$, $a < b < c < d$, be a regular Diophantine quadruple and suppose that $e > d$ extends it to a Diophantine quintuple. Then we have $ab + 1 = r^2$, $ac + 1 = s^2$, $bc + 1 = t^2$, $ad + 1 = x^2$, $bd + 1 = y^2$, $cd + 1 = z^2$, $ae + 1 = X^2$, $be + 1 = Y^2$, $ce + 1 = Z^2$, $de + 1 = W^2$, for integers $r, s, t, x, y, z, X, Y, Z, W$. By the result of Fujita [185], we know that $d = d_+$ and we have $x = at + rs$, $y = bs + rt$, $z = cr + st$. By eliminating e from the above equations, we obtain the following system of generalized Pellian equations

$$aY^2 - bX^2 = a - b, \quad aZ^2 - cX^2 = a - c, \quad bZ^2 - cY^2 = b - c,$$

$$aW^2 - dX^2 = a - d, \quad bW^2 - dY^2 = b - d, \quad cW^2 - dZ^2 = c - d.$$

To show the non-extendibility of quadruples of Euler's form

$$\{a, b, a + b + 2r, 4r(r + a)(r + b)\}, \tag{4.134}$$

the authors of [211] introduced a new variant of the congruence method, specific for quadruples of the form (4.134). From the above system, we have that W belongs to the intersection of three binary recurrence sequences. By the above-mentioned result of Fujita [186], we know that the indices of the elements of these sequences that correspond to W are even, say 2ℓ, $2m$ and $2n$, respectively. By the mentioned new congruence method, it was shown that at least one of the following congruences is satisfied:

- $\ell \equiv n \equiv 0 \pmod{s}$,
- $m \equiv n \equiv 0 \pmod{t}$,
- $n \equiv \pm r \pmod{st}$.

Suppose one of first two congruences is satisfied. In that case, we may write the usual linear form in three logarithms as a linear form in two logarithms (let us mention that the idea of replacing three by two logarithms in the context of Diophantine m-tuples was first used by He and Togbé in [210]). Then an application of Laurent's result [246] on lower bounds for linear forms in two logarithms yields $s < 22023$ or $t < 22023$, respectively. The results obtained using Mignotte's result on linear forms in three logarithms yield $r < 900154$ provided that the third congruence is satisfied. These upper bounds are small enough to use a variant of the Baker-Davenport reduction method. This leads to the conclusion that a quadruple of Euler's form cannot be extended to a Diophantine quintuple. In fact, by Fujita [185], this means that a regular triple $\{a, b, a + b + 2r\}$ cannot be extended to a Diophantine quintuple.

It remains to consider the (non)extendability of a regular Diophantine quadruple $\{a, b, c, d\}$, $a < b < c < d$, such that the triple $\{a, b, c\}$ is not regular. For that purpose, in [211], a classification of Diophantine triples was introduced by defining the ∂-*operator* between Diophantine triples. We know that to any Diophantine triple $\{a, b, c\}$, we may add d_+ to obtain a regular Diophantine quadruple $\{a, b, c, d_+\}$. In particular, we obtain three new Diophantine triples $\{a, b, d_+\}$, $\{a, c, d_+\}$ and $\{b, c, d_+\}$ related to $\{a, b, c\}$. We may consider these three new triples to be farther away from being a regular triple than the original triple $\{a, b, c\}$. The idea is to reverse this observation. Thus, for a given non-regular triple $\{a, b, c\}$, we want to get a new Diophantine triple $\{a', b', c'\}$ that is closer to the property of being a regular triple. If d_+ and d_- are obtained starting with a triple $\{a, b, c\}$, we denote them by $d_+(a, b, c)$ and $d_-(a, b, c)$.

Definition 4.11.16 Let ∂ be an operator which sends a non-regular Diophantine triple $\{a, b, c\}$ to a Diophantine triple $\{a', b', c'\}$ such that

$$\partial(\{a, b, c\}) = \{a, b, c, d_-(a, b, c)\} \setminus \{\max\{a, b, c\}\}.$$

For a non-negative integer D, we define the operator ∂_{-D} on Diophantine triples by $\partial_0(\{a, b, c\}) = \{a, b, c\}$ and if $\partial_{-(D-1)}(\{a, b, c\})$ is not a regular triple, then

$$\partial_{-D}(\{a, b, c\}) = \partial(\partial_{-(D-1)}(\{a, b, c\})), \quad \text{for } D \geq 1.$$

Furthermore, we put $d_{-D}(a, b, c) = d_-(\partial_{-D+1}(\{a, b, c\}))$. In particular, we have $\partial = \partial_{-1}$ and $\partial_{-2}(\{a, b, c\}) = \partial(\partial_{-1}(\{a, b, c\}))$.

Since, by (1.13), we know that $0 \leq d_- < c$ and $d_- = 0$ if and only if the triple $\{a, b, c\}$ is regular, we see that the ∂-operator is well defined, i.e. a Diophantine triple $\{a, b, c\}$ is indeed mapped to another Diophantine triple, unless $\{a, b, c\}$ is a regular triple.

Definition 4.11.17 We say that a Diophantine triple $\{a, b, c\}$ is of *degree D* and is *generated by a regular triple* $\{a', b', c'\}$, if $d_{-(D+1)}(a, b, c) = 0$ and $\partial_{-D}(\{a, b, c\}) = \{a', b', c'\}$. If the triple $\{a, b, c\}$ is of degree D, we write $\deg(a, b, c) = D$.

For example, the triples $\{1, 3, 120\}$, $\{1, 8, 120\}$, $\{3, 8, 120\}$ have degree 1 and are generated by the regular triple $\{1, 3, 8\}$.

The proof of the non-extendibility of non-regular triples to quintuples now distinguishes two cases: $\deg(a, b, c) = 1$ and $\deg(a, b, c) \geq 2$. If $\deg(a, b, c) = 1$, then $d_-(a, b, c) = a + b \pm 2r$ and $c = 4r(r \pm a)(b \pm r)$. Then the estimates obtained by applying Mignotte's result give reasonable bounds on a and r, namely, $a \leq 93595$, $r \leq 2315167$, and the remaining cases can be eliminated by the Baker-Davenport reduction.

The case $\deg(a, b, c) \geq 2$ is more involved, so we very briefly sketch it. Since $\{a, b, c\}$ is not regular, by Remark 3.8.6, we have $c > 4ab$. On the other hand, a result of Fujita and Miyazaki [187] implies that $ac < 180.45b^3$. Hence,

$$4ab < c < 180.45 \, \frac{b^3}{a},$$

and the proof continues by splitting the interval $(4ab, 180.45\, b^3/a)$ into five subintervals. In each case, inequalities obtained for a, b, c (and also for the quantities d_{-1} and d_{-2}, whose existence follows from the assumption that $\deg(a, b, c) \geq 2$), leave a feasible number of triples to by eliminated by the Baker-Davenport reduction.

4.12 Exercises

1. Let $a + b\sqrt{d}$ be a solution of Pell's equation $x^2 - dy^2 = 1$ and $a > \frac{1}{2}b^2 - 1$. Prove that $a + b\sqrt{d}$ is the fundamental solution.
2. Find the smallest solutions in positive integers of the equations:

   (a) $x^2 - 13y^2 = \pm 1$,
   (b) $x^2 - 14y^2 = \pm 1$,
   (c) $x^2 - 31y^2 = \pm 1$.

3. Find the fundamental solution of Pell's equation $x^2 - 991y^2 = 1$.
4. Prove that for an arbitrary $k \in \mathbb{Z}$, the equation $x^2 - (k^2 - 1)y^2 = -1$ does not have solutions in integers x and y.
5. Let m be an arbitrary positive integer. Prove that there are infinitely many solutions of Pell's equation $x^2 - dy^2 = 1$, which satisfy the additional condition that $y \equiv 0 \pmod{m}$.

6. Let $k > 1$ be an odd integer. Prove that the equation $x^2 - (k^2 - 4)y^2 = 4k$ has no solutions in coprime integers (x, y) (see [262]).

7. Find all positive integers $k \leq 100$ such that the Pellian equation $x^2 - (k^2 + 1)y^2 = k^2$ has exactly five fundamental solutions.

8. Find positive integers q, p_1, p_2 such that $q > 10^9$ and

$$\left| \sqrt{2} - \frac{p_1}{q} \right| < q^{-3/2}, \quad \left| \sqrt{3} - \frac{p_2}{q} \right| < q^{-3/2}.$$

9. Prove that the equation $y^2 = x^3 + 7$ has no integer solutions.

10. Let $a = 119, b = 472, c = 26288040$. Check that $\{a, b, c\}$ is a Diophantine triple and that the elliptic curve $y^2 = (x + ab)(x + ac)(x + bc)$ has more than 100 integer points.

11. Find all solutions of the system of equations

$$y^2 - 2x^2 = 1, \quad z^2 - 3x^2 = 1$$

in integers x, y, z.

12. Find a positive integer $d > 2$ such that the system of Pell's equations $x^2 - 2y^2 = 1, z^2 - dy^2 = 1$ (with common unknown y) has at least one solution in positive integers, and then find all integer solutions of the obtained system.

13. Prove that for any positive integer n, the set

$$\{F_{2n}, \ F_{2n+4}, \ 5F_{2n+2}, \ 4L_{2n+1}F_{2n+2}L_{2n+3}\}$$

is a Diophantine quadruple.

14. Let k be a positive integer. Prove that

$$F_{6k+5} < 4F_{2k+1}F_{2k+2}F_{2k+3} < F_{6k+6}.$$

15. Prove that the set

$$\{P_{2n}, \ P_{2n+8}, \ 36P_{2n+4}, \ P_{2n+2}P_{2n+4}P_{2n+6}\}$$

has the property $D(144)$.

16. For $2 \leq g \leq 200$, find all triples $\{a, b, c\}$ with the property that $ab + 1, ac + 1$ and $bc + 1$ are repdigits in base g (see [33]). (An integer N is a repdigit in base g if all base g digits of N are the same, i.e. if $N = d \left(\frac{g^k - 1}{g - 1} \right)$ for some $d \in \{1, 2, \ldots, g - 1\}$.)

17. Prove that there are infinitely many Diophantine triples $\{a, b, c\}$ such that $b = (a + c)/2$ (see [80]).

18. Let us denote by $D_m(N)$ the number of Diophantine m-tuples with elements $\leq N$. Prove that $D_2(N) = \frac{6}{\pi^2}N \log N + O(N)$ and $D_3(N) = \frac{3}{\pi^2}N \log N + O(N)$ (see [112]).

19. Prove that for any positive integer k, there is a Diophantine triple $\{a, b, c\}$ such that $a \equiv b \equiv c \equiv 0 \pmod{k}$ (see [158]).
20. Find a pair of positive integers (k, ℓ) with the property that there does not exist a Diophantine triple $\{a, b, c\}$ such that $a \equiv b \equiv c \equiv \ell \pmod{k}$.
21. Compute the degree of the Diophantine triple $\{120, 11781, 16988440\}$.
22. Prove that for any fixed Diophantine triple $\{a, b, c\}$, there exists a unique non-negative integer $D < \frac{\log(abc)}{\log 12}$ such that $d_{-(D+1)}(a, b, c) = 0$ (see [211]).
23. Prove that for an arbitrary but fixed regular triple $\{a', b', c'\}$, there are at most 3^D Diophantine triples $\{a, b, c\}$ generated by $\{a', b', c'\}$ with $\deg(a, b, c) = D$.

Chapter 5
Sets with the Property $D(n)$

5.1 Existence of $D(n)$-quadruples

Let n be a (non-zero) integer. A set of m distinct non-zero integers $\{a_1, a_2, \ldots, a_m\}$ such that $a_i a_j + n$ is a perfect square for all $1 \leq i < j \leq m$ is called a *Diophantine m-tuple with the property $D(n)$ or a $D(n)$-m-tuple*.

It is usual to exclude the case $n = 0$ from the definition of $D(n)$-m-tuples since it is trivial to see that there are infinite sets with the property $D(0)$ (we can take an arbitrarily large set consisting of squares, or of squares multiplied by the same factor), but when this condition is combined with other non-trivial conditions, then interesting problems may arise (as we will see in Sects. 5.4.3 and 5.4.4). In that context, it makes sense to allow $n = 0$ in the definition.

Sometimes it is convenient to suppose that elements of a $D(n)$-m-tuple are positive integers. In the case $n = 1$, this is no loss of generality, since a $D(1)$-m-tuple cannot have elements of mixed signs, while by multiplying all elements of a $D(1)$-m-tuple with negative elements by -1, we obtain a $D(1)$-m-tuple with positive elements. Also, m-tuples with mixed signs are not possible for negative n. However, already with $n = 4$, we have a $D(4)$-triple $\{-1, 3, 4\}$ with mixed signs, and we observed in Sect. 3.8 that the elliptic curve induced by this triple has an interesting property. Diophantine tuples with mixed signs appeared also in other places within Chap. 3.

It is easy to see that for every integer n, there are infinitely many $D(n)$-triples. Namely, if $\{a, b\}$ is any $D(n)$-pair and if $ab + n = x^2$, then $\{a, b, a + b + 2x\}$ is a $D(n)$-triple, which we call a *regular $D(n)$-triple*. If we take, for example, $a = 1$, then we get an explicit infinite family of $D(n)$-triples $\{1, x^2 - n, (x + 1)^2 - n\}$.

However, already with $D(n)$-quadruples, the situation is significantly more interesting, and soon we arrive to open problems.

In Theorem 1.5.1, we proved that if $n \equiv 2 \pmod 4$, then there is no $D(n)$ quadruple. Now we will prove a kind of converse of that statement. The idea is to "cover" the set of integers that are not of the form $4k + 2$ with several arithmetic

A. Dujella, *Diophantine m-tuples and Elliptic Curves*,
Developments in Mathematics 79, https://doi.org/10.1007/978-3-031-56724-7_5

progressions, i.e. linear polynomials of the form $\alpha k + \beta$, for which there are $D(\alpha k + \beta)$-quadruples whose elements are polynomials of small degree in the variable k. Such quadruples can be obtained by specializing one of the parameters in two-parameter families of quadruples containing two regular triples. We met one such family in Sect. 3.6.1. We showed that the quadruple

$$\{a,\ ak^2 - 2k - 2,\ a(k+1)^2 - 2k,\ a(2k+1)^2 - 8k - 4\} \tag{5.1}$$

has the property $D(2a(2k+1)+1)$. Let us recall how we got to those quadruples. We started from the pair $\{a, b\}$ which satisfies $ab + n = x^2$, and observed the set $\{a, b, a+b+2x, a+4b+4x\}$, which contains two regular $D(n)$-triples: $\{a, b, a+b+2x\}$ and $\{b, a+b+2x, a+4b+4x\}$. It remains to satisfy the condition $a(a+4b+4x)+n = y^2$, and we wrote it in the form

$$3n = (a + 2x - y)(a + 2x + y). \tag{5.2}$$

After that, we equated the factors on the right-hand side of (5.2) with 3 and n and obtained the equation $a(b+2) = (x-1)(x-3)$, for which we assumed that $x = ak + 1$ to finally obtain quadruple (5.1).

If we do not want to introduce additional assumptions on possible factorizations of n, condition (5.2) can also be satisfied so that the factors on the right-hand side are equated with 1 and $3n$. Thus we get the system of equations

$$a + 2x - y = 1,$$
$$a + 2x + y = 3n.$$

By adding, we get $3n = 2a + 4x - 1$, so replacing $3n$ with $3(x^2 - ab)$ gives us

$$a(3b + 2) = (3x - 1)(x - 1).$$

Let us put $x = am + 1$. Now from $3b + 2 = m(3x - 1)$, we conclude that m is of the form $m = 3k + 1$. By inserting, we get $b = a(3k+1)^2 + 2k,\ n = 2a(2k+1)+1$ and the quadruple

$$\{a,\ a(3k+1)^2 + 2k,\ a(3k+2)^2 + 2k + 2,\ 9a(2k+1)^2 + 8k + 4\} \tag{5.3}$$

with the property $D(2a(2k+1)+1)$.

So, we got two formulas (5.1) and (5.3) for $D(2a(2k+1)+1)$-quadruples. From them, by specializing the parameters a or k, we get formulas for $D(n)$-quadruples where n is a linear polynomial, and the elements of the quadruples are polynomials of small degree (in the case when we specialize a, one polynomial is a constant, and the other three are quadratic polynomials in the variable k, and when we specialize k, all four polynomials are linear in the variable a).

As we announced, we want the obtained linear polynomials for n to cover all integers that are not of the form $4k + 2$. It is easy to see that an integer that is not of the form $4k + 2$ has one of the following forms:

$$4k + 3, \quad 8k + 1, \quad 8k + 5, \quad 8k, \quad 16k + 4, \quad 16k + 12.$$

Let us show that by specialization of the parameter a and admissible transformations on the elements of the quadruple and the corresponding n, from the formulas for $D(2a(2k + 1) + 1)$-quadruples, we can obtain $D(n)$-quadruples for each of these six forms.

Putting $a = 1$ in (5.1) and (5.3), we get quadruples with the property $D(4k + 3)$. For $a = 2$, we get quadruples with the property $D(8k+5)$. For $a = 4$, by substitution $k' = 2k + 1$ (which preserves the integrality of elements), we get quadruples with the property $D(8k' + 1)$. By multiplying all elements of (5.1) and (5.3) by 2, putting $a = \frac{1}{2}$ and substituting $k' = k + 1$, we get quadruples with the property $D(8k')$. Furthermore, by putting $a = 2$ and by substitution $k' = 2k + 1$, after multiplying all elements by 2, we get quadruples with the property $D(16k' + 4)$. Finally, quadruples with the property $D(16k + 12)$ are obtained by multiplying by 2 all the elements of previously obtained quadruples with the property $D(4k + 3)$.

So, we have two families of quadruples for each of the six cases. More precisely,

- $\{1, k^2 - 2k - 2, k^2 + 1, 4k^2 - 4k - 3\}$ and $\{1, 9k^2 + 8k + 1, 9k^2 + 14k + 6, 36k^2 + 44k + 13\}$ are $D(4k + 3)$-quadruples,
- $\{4, k^2 - 3k, k^2 + k + 2, 4k^2 - 4k\}$ and $\{4, 9k^2 - 5k, 9k^2 + 7k + 2, 36k^2 + 4k\}$ are $D(8k + 1)$-quadruples,
- $\{2, 2k^2 - 2k - 2, 2k^2 + 2k + 2, 8k^2 - 2\}$ and $\{2, 18k^2 + 14k + 2, 18k^2 + 26k + 10, 72k^2 + 80k + 22\}$ are $D(8k + 5)$-quadruples,
- $\{1, k^2 - 6k + 1, k^2 - 4k + 4, 4k^2 - 20k + 9\}$ and $\{1, 9k^2 - 8k, 9k^2 - 2k + 1, 36k^2 - 20k + 1\}$ are $D(8k)$-quadruples,
- $\{4, k^2 - 4k - 1, k^2 + 3, 4k^2 - 8k\}$ and $\{4, 9k^2 - 4k - 1, 9k^2 + 8k + 3, 36k^2 + 8k\}$ are $D(16k + 4)$-quadruples,
- $\{2, 2k^2 - 4k - 4, 2k^2 + 2, 8k^2 - 8k - 6\}$ and $\{2, 18k^2 + 16k + 2, 18k^2 + 28k + 12, 72k^2 + 88k + 26\}$ are $D(16k + 12)$-quadruples.

These formulas show that for all but finitely many integers n, such that $n \not\equiv 2$ (mod 4), there are at least two $D(n)$-quadruples. Finitely many exceptions occur because for some (small) values of the parameter k, some of these quadruples may have two equal elements (so they are not "proper" quadruples). Another case that, by definition, should be excluded is when one of the elements in the quadruple is equal to 0. However, in that case, n must be a perfect square, and we already know that then there are infinitely many $D(n)$-quadruples. Exceptions are easy to find by calculating the roots of the corresponding quadratic or linear polynomials. After that, some exceptions can be eliminated by searching for $D(n)$-quadruples with relatively small elements (for example, $\{1, 12, 477, 23052\}$ is a $D(52)$-quadruple). In that way, the following results are obtained (see [88] for details).

Theorem 5.1.1 *Let n be an integer such that $n \not\equiv 2 \pmod 4$ and $n \notin S = \{-4, -3, -1, 3, 5, 8, 12, 20$. Then there exists at least one $D(n)$-quadruple.*

Theorem 5.1.2 *Let n be an integer such that $n \not\equiv 2 \pmod 4$ and $n \notin S \cup T$, where $T = \{-15, -12, -7, 7, 13, 15, 21, 24, 28, 32, 48, 60, 84\}$. Then there exist at least two distinct $D(n)$-quadruples.*

A conjecture is that there is no $D(n)$-quadruple for the elements of the set S. At the moment, it is known that the conjecture is valid for $n = -1$ and $n = -4$. Namely, Bonciocat, Cipu, and Mignotte [44] proved in 2022 that there is no $D(-1)$-quadruple. And the following proposition shows that the nonexistence of $D(-4)$-quadruples follows from their result since the elements of a $D(-4)$-quadruple would all have to be even, so by dividing elements of that quadruple with 2, we would get a $D(-1)$-quadruple.

Proposition 5.1.3 *Let n be an integer of the form $16k + 12$ and let $\{b_1, b_2, b_3, b_4\}$ be a $D(n)$-quadruple. Then all numbers b_i are even.*

Proof Assume that, e.g., b_1 is odd. Squares, when divided by 16, give remainders $0, 1, 4$ or 9. Therefore, we have $b_i b_j \equiv 4, 5, 8$ or $13 \pmod{16}$. This implies that if any of the numbers b_2, b_3, b_4 is even, then it is divisible by 4, and no two of these numbers can be divisible by 4. We conclude that among the numbers b_2, b_3, b_4, there is at most one even, i.e. at least two odd. So, we may assume that b_1, b_2, b_3 are odd. From the condition $b_i b_j \equiv 5$ or $13 \pmod{16}$, we have $b_i b_j \equiv 5 \pmod 8$, i.e.

$$b_1 b_2 \equiv 5 \,(\mathrm{mod}\ 8), \quad b_1 b_3 \equiv 5 \,(\mathrm{mod}\ 8), \quad b_2 b_3 \equiv 5 \,(\mathrm{mod}\ 8).$$

By multiplying these three congruences, we get $(b_1 b_2 b_3)^2 \equiv 5 \pmod 8$, which is a contradiction, because squares, when divided by 8, give remainders $0, 1$ or 4. $\square$

In addition to formulas (5.1) and (5.3), in [91], some other formulas for two-parameter quadruples were derived. E.g. it was shown that the quadruple

$$\{a, \ ak^2 + 2k - 2, \ a(k + 1)^2 + 2k + 4, \ a(2k + 1)^2 + 8k + 4\} \tag{5.4}$$

has the property $D(2a(2k + 1) + 9)$, the quadruple

$$\{9a + 4(3k - 1), \ (3k - 2)^2 a + 2(k - 1)(6k^2 - 4k + 1),$$

$$(3k + 1)^2 a + 2k(6k^2 + 2k - 1), \ (6k - 1)^2 a + 4k(2k - 1)(6k - 1)\} \tag{5.5}$$

has the property $D(2a(6k - 1) + (4k - 1)^2)$, the quadruple

$$\{a, \ ak^2 + 2(k^2 + k + 1), \ a(k - 1)^2 + 2k(k - 1), \ a(k + 1)^2 + 2(k + 1)(k + 2)\} \tag{5.6}$$

has the property $D(2a(k^2 - 1) + (2k + 1)^2)$, and the quadruple

$$\{a,\ a(3k+1)^2+2(3k^2+3k+1),\ a(3k+2)^2+2(k+1)(3k+2),\ 9ak^2+2k(3k+1)\} \tag{5.7}$$

has the property $D(2ak(3k + 2) + (2k + 1)^2)$.

Using formulas (5.4)–(5.7), the statements of Theorems 5.1.1 and 5.1.2 can be improved, if not in full generality, then at least for integers n of special forms. It is known (see [94]) that if $|n|$ is large enough and $n \equiv 1$ (mod 8) or $n \equiv 4$ (mod 32) or $n \equiv 0$ (mod 16), then there are at least six, and if $n \equiv 8$ (mod 16) or $n \equiv 13, 21$ (mod 24) or $n \equiv 3, 7$ (mod 12), then there are at least four distinct Diophantine quadruples with the property $D(n)$.

However, if we denote by U the set of all integers n, $n \not\equiv 2$ (mod 4), for which there are at most two distinct Diophantine quadruples with the property $D(n)$, it is an open question whether U is finite or infinite. It is known that (sufficiently large) elements of U must have one of the forms: $4k + 3$, $16k + 12$, $8k + 5$, $32k + 20$. Furthermore, for each of these forms, additional limitations are known. For example, if n is of the form $4k + 3$, then the numbers $\frac{|n-1|}{2}, \frac{|n-9|}{2}$ and $\frac{|9n-1|}{2}$ must be primes (because each of their non-trivial factorizations gives an additional quadruple by the methods sketched above), and furthermore, either the number $|n|$ is prime or $n = pq$, where p and $q = p + 2$ are twin primes and $p \equiv 3$ (mod 4) (because (almost every) non-trivial factorization of n gives an additional quadruple). This result was proved in [96], but the case $p \equiv 1$ (mod 4) was later eliminated in [134], via the following formula for a $D((4k + 1)(4k + 3))$-quadruple (for which it is not yet known whether it has an analogue for $p \equiv 3$ (mod 4)):

$$\{1,\ 144k^4 + 216k^3 + 113k^2 + 20k + 1,\ 144k^4 + 360k^3 + 329k^2 + 134k + 22,$$

$$576k^4 + 1152k^3 + 848k^2 + 272k + 33\}.$$

5.2 Bounds for the Size of $D(n)$-tuples

Let n be a non-zero integer. We may ask how large a set with the property $D(n)$ can be. Let us define

$$M_n = \sup\{|S| : S \text{ has the property } D(n)\},$$

where $|S|$ denotes the number of elements in the set S. By the result of He, Togbé, and Ziegler [211] discussed in Sect. 4.11.2, we know that $M_1 = 4$, while Bliznac-Trebješanin and Filipin [41] proved that $M_4 = 4$. By Theorem 1.5.1, we know that $M_n = 3$ for $n \equiv 2$ (mod 4), while Bonciocat, Cipu, and Mignotte [44] proved that $M_{-1} = 3$ and $M_{-4} = 3$. These are the only values of M_n for which the exact values are known.

Since the number of integer points on an elliptic curve

$$y^2 = (a_1 x + n)(a_2 x + n)(a_3 x + n) \tag{5.8}$$

is finite, we conclude that there does not exist an infinite set with the property $D(n)$ (there are only finitely many extensions of a triple to a quadruple). However, known bounds for the size and the number of solutions of (5.8) depend not only on n but also on a_1, a_2, a_3.

On the other hand, we may consider a hyperelliptic curve

$$y^2 = (a_1 x + n)(a_2 x + n)(a_3 x + n)(a_4 x + n)(a_5 x + n) \tag{5.9}$$

of genus $g = 2$. Caporaso, Harris, and Mazur [53] proved that the Lang conjecture on varieties of general type implies that for $g \geq 2$, the number $B(g, \mathbb{K}) = \max_C |C(\mathbb{K})|$ is finite. Here C runs over all curves of genus g over a number field $\mathbb{K}$, and $C(\mathbb{K})$ denotes the set of all $\mathbb{K}$-rational points on C. However, even the question of whether $B(2, \mathbb{Q}) < \infty$ is still open. An example of Stoll [337] (building on the work of Elkies) shows that $B(2, \mathbb{Q}) \geq 642$. Since obviously $M_n \leq 5 + B(2, \mathbb{Q})$ (by [214], we have also $M_n \leq 4 + B(4, \mathbb{Q})$), we see that the Lang conjecture implies that

$$M = \sup\{M_n : n \in \mathbb{Z} \setminus \{0\}\}$$

is finite. The existence of rational Diophantine sextuples shows that $M \geq 6$.

At present, we only know that M_n is finite for all $n \in \mathbb{Z} \setminus \{0\}$. However, all known upper bounds for M_n are unbounded as $|n|$ tends to infinity. Here we will derive upper bounds for M_n, based on papers [106, 108]. In the proof, we estimate the number of "large" (greater than $|n|^3$), "small" (between n^2 and $|n|^3$) and "very small" (less than n^2) elements of a set with the property $D(n)$. Let us introduce the following notation:

$$A_n = \sup\{|S \cap [|n|^3, +\infty)| : S \text{ has the property } D(n)\},$$

$$B_n = \sup\{|S \cap (n^2, |n|^3)| : S \text{ has the property } D(n)\},$$

$$C_n = \sup\{|S \cap [1, n^2]| : S \text{ has the property } D(n)\}.$$

In the above-mentioned papers, it was supposed that all elements of a $D(n)$-tuple are positive (and this is certainly the case if $n < 0$). So we have $M_n \leq A_n + B_n + C_n$ in that case. However, a minor modification of the arguments provides also upper bounds for M_n in the more general case when elements of a $D(n)$-tuple are non-zero integers. Indeed, if a $D(n)$-tuple has elements with mixed signs, then all its elements are in absolute value less than or equal to n. So, we have $M_n \leq 2C_n$ in that case. Thus, in the general case, we have

$$M_n \leq \max(A_n + B_n + C_n, 2C_n).$$

In estimating the number of "large" elements, we use the result of Bennett [30] (Theorem 4.1.20) on simultaneous approximations of algebraic numbers and a gap principle. In fact, we follow a similar strategy as in Sect. 4.10.

Proposition 5.2.1 *We have $A_n \leq 21$ for all non-zero integers n.*

For the estimate of the number of "small" elements, we use a variant of the gap principle.

Proposition 5.2.2 *We have $B_n < 0.65 \log |n| + 2.24$ for all non-zero integers n.*

Finally, in the estimate of the number of "very small" elements we use the large sieve method due to Gallagher [189] and an estimate for the size of $D(n)$-tuples in finite fields $\mathbb{F}_p$ obtained from an inequality for a sum of Legendre symbols due to Vinogradov [352, Problem V.8.c)].

Let us mention here that the problems related to $D(1)$ and $D(n)$-m-tuples in finite fields are of independent interest. Dujella and Kazalicki [136] studied the number $N^{(m)}(p)$ of Diophantine m-tuples in $\mathbb{F}_p$ for $m = 2$, 3 and 4. The cases $m = 2$ and $m = 3$ can be handled by considering sums of Legendre symbols (see Exercise 12 in Sect. 5.5), while Fourier coefficients of certain modular forms appear in the case $m = 4$. An asymptotic formula for $N^{(m)}(p)$ given in [136] has been improved in [265, 325]. A variant of the problem in which products of two elements are replaced by products of k elements is considered in [207].

Proposition 5.2.3 *For $|n| > 400$, we have $C_n < 11 \log |n|$.*

It is easy to verify with a computer that $C_n \leq 5$ holds for $|n| \leq 400$ (even if we allow elements with mixed signs). More precisely, the equality $C_n = 5$ in this range holds if and only if $n \in \{-299, -255, 256, 400\}$ (one of $D(400)$-quintuples has elements with mixed signs). Furthermore, Proposition 5.2.2 implies $B_n \leq 6$ for $|n| \leq 400$. So, we may combine Propositions 5.2.1, 5.2.2 and 5.2.3 to obtain

Theorem 5.2.4

$$M_n \leq 32 \quad for \ |n| \leq 400,$$

$$M_n < 22 \log n \quad for \ n > 400,$$

$$M_n < 11.65 \log |n| + 23.24 < 15.53 \log |n| \quad for \ n < -400.$$

In the following subsections, we will prove Propositions 5.2.1, 5.2.2 and 5.2.3.

5.2.1 Large Elements

Assume that $\{a, b, c, d\}$ is a $D(n)$-quadruple. Let $ab + n = r^2$, $ac + n = s^2$, $bc + n = t^2$, where r, s, t are nonnegative integers. Eliminating d from the system

$$ad + n = x^2, \quad bd + n = y^2, \quad cd + n = z^2,$$

we obtain the following system of generalized Pellian equations

$$az^2 - cx^2 = n(a - c), \tag{5.10}$$

$$bz^2 - cy^2 = n(b - c). \tag{5.11}$$

We will apply Theorem 4.1.20 to the numbers

$$\theta_1 = \frac{s}{a}\sqrt{\frac{a}{c}} = \sqrt{\frac{ac + n}{ac}} = \sqrt{1 + \frac{n}{ac}} = \sqrt{1 + \frac{nb}{abc}},$$

$$\theta_2 = \frac{t}{b}\sqrt{\frac{b}{c}} = \sqrt{\frac{bc + n}{bc}} = \sqrt{1 + \frac{n}{bc}} = \sqrt{1 + \frac{na}{abc}}.$$

Lemma 5.2.5 *Assume that $0 < a < b < c$ and $3ac > n$. Then all positive integer solutions x, y, z of the system (5.10) and (5.11) satisfy*

$$\max\left(\left|\theta_1 - \frac{sbx}{abz}\right|, \left|\theta_2 - \frac{zay}{abz}\right|\right) < \frac{c \cdot |n|}{a} z^{-2}.$$

Proof We have

$$\left|\frac{s}{a}\sqrt{\frac{a}{c}} - \frac{sbx}{abz}\right| = \frac{s}{az\sqrt{c}}|z\sqrt{a} - x\sqrt{c}| = \frac{s}{az\sqrt{c}} \cdot \frac{|n(c - a)|}{z\sqrt{a} + x\sqrt{c}}.$$

If $n < 0$, then $s = \sqrt{ac - |n|} < \sqrt{ac}$ and we obtain

$$\left|\theta_1 - \frac{sbx}{abz}\right| < \frac{\sqrt{ac} \cdot |n| \cdot c}{a\sqrt{ac}z^2} = \frac{c|n|}{a} z^{-2}.$$

If $n > 0$, then $x\sqrt{c} > z\sqrt{a}$ and we obtain

$$\left|\theta_1 - \frac{sbx}{abz}\right| < \frac{\sqrt{ac + n} \cdot n \cdot c}{2a\sqrt{ac}z^2} = \sqrt{1 + \frac{n}{ac}} \cdot \frac{cn}{2a} z^{-2} < \frac{cn}{a} z^{-2}.$$

In the same manner, we obtain $\left|\theta_2 - \frac{tay}{abz}\right| < \frac{c|n|}{b} z^{-2} < \frac{c|n|}{a} z^{-2}$. $\square$

Lemma 5.2.6 *Let $\{a, b, c, d\}$ be a $D(n)$-quadruple such that $0 < a < b < c < d$. If $c > b^{11}|n|^{11}$, then $d \le c^{131}$.*

Proof Let r, s, t, x, y, z be defined as in the beginning of this subsection. We apply Theorem 4.1.20 with $(a_0, a_1, a_2) = (0, na, nb)$, $N = abc$, $M = |nb|$, $q = abz$, $p_1 = sbx$, $p_2 = tay$. Since $abc > |n|^9 b^9$, the condition $N > M^9$ is satisfied. For the quantity γ from Theorem 4.1.20, we have $\gamma = \frac{b^2(b-a)^2}{2b-a}|n|^3$ if $b \geq 2a$ and $\gamma = \frac{a^2 b^2}{a+b}|n|^3$ if $a < b \leq 2a$. In both cases, we have

$$\frac{b^3}{6}|n|^3 \leq \gamma < \frac{b^3}{2}|n|^3.$$

For the quantity λ from Theorem 4.1.20, we have

$$\lambda = 1 + \frac{\log(32.04abc\gamma)}{\log(1.68c^2(b-a)^{-2}n^{-6})} = 2 - \lambda_1,$$

where

$$\lambda_1 = \frac{\log\left(\frac{1.68c}{32.04ab(b-a)^2 n^6 \gamma}\right)}{\log(1.68c^2(b-a)^{-2}n^{-6})}.$$

Theorem 4.1.20 and Lemma 5.2.5 imply

$$\frac{c|n|}{az^2} > (130abc\gamma)^{-1}(abz)^{\lambda_1-2} > (130abc\gamma)^{-1}a^{-2}b^{-2}z^{\lambda_1-2}.$$

This implies

$$z^{\lambda_1} < 130a^2 b^3 c^2 |n|\gamma$$

and

$$\log z < \frac{\log\left(130a^2 b^3 c^2 |n|\gamma\right)\log\left(1.68c^2(b-a)^{-2}n^{-6}\right)}{\log\left(\frac{1.68c}{32.04ab(b-a)^2 n^6\gamma}\right)}. \tag{5.12}$$

Let us estimate the right-hand side of (5.12). We have

$$130a^2 b^3 c^2 |n|\gamma < 65a^2 b^6 c^2 n^4 < c^3 \cdot \frac{65a^2}{b^5 |n|^7} < c^3,$$

unless $n = -1$, $a = 1$, $b = 2$. However, we can exclude the case $n = -1$, since, by [44] we know that a $D(-1)$-quadruple does not exist.

Furthermore, we have

$$1.68c^2(b-a)^{-2}n^{-6} < c^2$$

(note that for $n = 1$, $b - a \geq 2$).

Finally,

$$\frac{1.68c}{32.04ab(b-a)^2 n^6 \gamma} > 0.106 a^{-1} b^{-6} c n^{-9} > c^{\frac{1}{11}} \cdot \frac{b^4 |n|}{9.71a} > c^{\frac{1}{11}}.$$

The last estimate shows that $\lambda_1 > 0$, which we implicitly used in (5.12).

Putting these three estimates in (5.12), we obtain

$$\log z < \frac{3 \log c \cdot 2 \log c}{\frac{1}{11} \log c} = 66 \log c.$$

Hence, $z < c^{66}$ and

$$d = \frac{z^2 - n}{c} \leq \frac{z^2 + |n|}{c} < \frac{c^{132} + c^{\frac{1}{11}}}{c} < c^{131} + 1.$$

$\square$

Now, we will develop a useful gap principle for the elements of a $D(n)$-m-tuple. The principle is based on Lemma 3.8.1, which was already used in Sect. 3.8 for obtaining certain gaps between elements in $D(1)$ and $D(4)$-m-tuples.

Lemma 5.2.7 *If $\{a, b, c, d\}$ is a $D(n)$-quadruple and $|n|^3 \leq a < b < c < d$, then*

$$d > \frac{3.847\, bc}{n^2}.$$

Proof Let e be the integer obtained by applying Lemma 3.8.1 to the triple $\{a, c, d\}$. Since $ce + n^2$ is a perfect square, we have that $ce + n^2 \geq 0$. On the other hand, the assumption is that $c > |n|^3$. Hence, if $e \leq -1$, then $ce + n^2 < -|n|^3 + n^2 \leq 0$, a contradiction. Since e is an integer, we have $e \geq 0$. If $e = 0$, then $d = a + c + 2s$. If $e \geq 1$, then

$$d > a + c + \frac{2ac}{n^2} + \frac{2s\sqrt{ac}}{n^2} > \frac{2ac}{n^2}. \tag{5.13}$$

(Note that, in the notation from the proof of Lemma 3.8.1, if $n > 0$, then $x > 0$, $y > 0$, and if $n < 0$ and $b > |n|$, then $x < 0$, $y < 0$.)

Analogously, applying Lemma 3.8.1 to the triple $\{b, c, d\}$, we obtain that $d = b + c + 2t$ or $d > b + c + \frac{2bc}{n^2} + \frac{2t\sqrt{bc}}{n^2}$. However, $d = b + c + 2t$ is impossible since $b + c + 2t > a + c + 2s$ and

$$b + c + 2t \leq b + c + 2\sqrt{c(c-1) + n} < 4c \leq \frac{2ac}{n^2},$$

unless $a < 2n^2$. But if $|n|^3 \leq a < 2n^2$, then $|n| = 1$, $a = 1$, and in that case we have

$$a + c + \frac{2ac}{n^2} + \frac{2s\sqrt{ac}}{n^2} > 3c + 2\sqrt{c(c-1)} > 4c.$$

Hence, we proved that

$$d > b + c + \frac{2bc}{n^2} + \frac{2t\sqrt{bc}}{n^2}. \tag{5.14}$$

From [283], we know that the triples $\{1, 2, 3\}$ and $\{1, 2, 4\}$ cannot be extended to Diophantine quadruples (see Exercise 1 in Sect. 5.5). Thus $bc \geq 10$, and this implies

$$t^2 = bc + n \geq bc - |n| > bc - \sqrt[6]{bc} > 0.853\, bc.$$

If we put this in (5.14), we obtain $d > \frac{3.847\, bc}{n^2}$. $\qquad\square$

Proof of Proposition 5.2.1 Assume that $\{a_1, a_2, \ldots, a_{22}\}$ has the property $D(n)$ and $|n|^3 \leq a_1 < a_2 < \cdots < a_{22}$. By Lemma 5.2.7, we find that

$$a_4 > \frac{a_2^2}{n^2}, \qquad a_5 > \frac{a_2^3}{n^4}, \qquad a_6 > \frac{a_2^5}{n^8}, \qquad a_7 > \frac{a_2^8}{n^{14}},$$

$$a_8 > \frac{a_2^{13}}{n^{24}}, \qquad a_9 > \frac{a_2^{21}}{n^{40}}, \qquad a_{10} > \frac{a_2^{34}}{n^{66}}, \qquad a_{11} > \frac{a_2^{55}}{n^{108}}.$$

Since $a_2 > |n|^3$, we have $\frac{a_2^{55}}{n^{108}} > a_2^{11}|n|^{11}$, and we may apply Lemma 5.2.6 with $a = a_1$, $b = a_2$, $c = a_{11}$. We conclude that $a_{22} \leq a_{11}^{131}$. However, Lemma 5.2.7 implies

$$a_{12} > |n|a_{11}, \qquad a_{13} > \frac{a_{11}^2}{|n|}, \qquad a_{14} > \frac{a_{11}^3}{n^2}, \qquad a_{15} > \frac{a_{11}^5}{|n|^5},$$

$$a_{16} > \frac{a_{11}^8}{|n|^9}, \qquad a_{17} > \frac{a_{11}^{13}}{n^{16}}, \qquad a_{18} > \frac{a_{11}^{21}}{|n|^{27}}, \qquad a_{19} > \frac{a_{11}^{34}}{|n|^{45}},$$

$$a_{20} > \frac{a_{11}^{55}}{n^{74}}, \qquad a_{21} > \frac{a_{11}^{89}}{|n|^{121}}, \qquad a_{22} > \frac{a_{11}^{144}}{|n|^{197}}.$$

Since $a_{11} > a_2^{11}|n|^{11} > n^{44}$, we obtain

$$a_{22} > \frac{a_{11}^{144}}{|n|^{197}} \geq a_{11}^{144 - \frac{197}{44}} > a_{11}^{139} > a_{11}^{131},$$

a contradiction. $\qquad\square$

5.2.2   Small Elements

Lemma 5.2.8 *If $\{a, b, c, d\}$ is a $D(n)$-quadruple and $n^2 \le a < b < c < d$, then $c > 3.88\, a$ and $d > 4.89\, c$.*

Proof We apply Lemma 3.8.1 again. Since $b > n^2$, we have $e \ge 0$. Thus Lemma 3.8.1 implies that

$$c \ge a + b + 2r.$$

Assume that $n \neq 1$. Then we have $ab \ge 20$ and $r^2 \ge ab - \sqrt[4]{ab} > 0.89ab > 0.89a^2$. Hence, $c > 3.88\, a$. However, it is easy to check that the same inequality also holds for all regular $D(1)$-triples with $ab < 20$ (there are five such triples).

Since $d \ge b + c + 2t > a + c + 2s$, from (5.13), we conclude that

$$d > a + c + \frac{2ac}{n^2} + \frac{2s\sqrt{ac}}{n^2}.$$

Assume that $n \neq 1$. Then we have $ac \ge 24$ and $s^2 \ge ac - \sqrt[4]{ac} > 0.9\, ac$. Therefore,

$$d > a + c + \frac{3.89\, ac}{n^2} > 4.89\, c.$$

Again, we can check that the same inequality also holds for both $D(1)$-triples with $ac < 24$ ($\{1, 3, 8\}$ and $\{1, 8, 15\}$). $\square$

Proof of Proposition 5.2.2 We may assume that $|n| \ge 2$ since $B_1 = B_{-1} = 0$. Let $\{a_1, a_2, \ldots, a_m\}$ be a $D(n)$-m-tuple and $n^2 < a_1 < a_2 < \cdots < |n|^3$. By Lemma 5.2.8, we have

$$a_3 > 3.88a_1, \quad a_4 > 3.88 \cdot 4.89a_1, \quad \ldots, \quad a_m > 3.88 \cdot 4.89^{m-3}a_1.$$

Therefore

$$3.88 \cdot 4.89^{m-3} \cdot n^2 < |n|^3,$$

and from $m - 3 < \dfrac{\log \frac{|n|}{3.88}}{\log 4.89}$, we obtain $m < 0.65 \log |n| + 2.24$. $\square$

5.2.3   Very Small Elements

We are left with the task of estimating the number of "very small" elements in a Diophantine m-tuple.

We will apply the so-called larger sieve method based on the following theorem of Gallagher [189].

Theorem 5.2.9 *If all but $g(p)$ residue classes* $\bmod\,p$ *are removed for each prime p in a finite set S, then the number of integers which remain in any interval of length N is at most*

$$\left(\sum_{p\in S}\log p-\log N\right)\Big/\left(\sum_{p\in S}\frac{\log p}{g(p)}-\log N\right),\qquad(5.15)$$

provided the denominator is positive.

The sieve is called "larger" because it is very suitable in cases when a very large part of residue classes is removed. In contrast to the well-known sieve of Eratosthenes in which we remove only one residue class, or the case of squares modulo p when half of the classes is removed, we will see that in our situation, only approximately $\sqrt{p}$ classes will not be removed.

To estimate the size of $D(n)$-tuples modulo p, i.e. $D(n)$-tuples in finite fields $\mathbb{F}_p$, we will use a result of Vinogradov concerning a double sum of Legendre symbols.

Let us recall that for coprime integers a and m, if the congruence $x^2\equiv a$ (mod m) has a solution, we say that a is a *quadratic residue* modulo m, while otherwise, we say that a is a *quadratic non-residue* modulo m. Let p be an odd prime number. The *Legendre symbol* $(\frac{a}{p})$ is equal to 1 if a is a quadratic residue modulo p, it is equal to -1 if a is a quadratic non-residue modulo p, and it is equal to 0 if $p\mid a$. It is easy to see that there are exactly $(p-1)/2$ quadratic residues and $(p-1)/2$ quadratic non-residues modulo p. In other words, $\sum_{x=0}^{p-1}(\frac{x}{p})=0$.

The proof of the following lemma can be found in [352, Problem V.8.c)] and [116, Example 4.5] (for a more general result, in which the Legendre symbol is replaced by any non-trivial Dirichlet character modulo p, see [29, 204]).

Lemma 5.2.10 *Let p be an odd prime and* $\gcd(n,p)=1$. *If $A,B\subseteq\{0,1,\ldots,p-1\}$ and*

$$T=\sum_{x\in A}\sum_{y\in B}\left(\frac{xy+n}{p}\right),$$

then $|T|<\sqrt{p|A|\cdot|B|}$.

In the proof of Proposition 5.2.3, we will also need some explicit estimates for certain functions related to the distribution of primes. For $x>0$, we denote by $\pi(x)$ the number of primes p such that $p\le x$. Furthermore, let $\vartheta(x)=\sum_{p\le x}\log p$ (Chebyshev ϑ function), and let p_n denote the n-th prime number. By the prime number theorem, we have $\pi(x)\sim\frac{x}{\log(x)}$, or equivalently, $\vartheta(x)\sim x$. However, for our purpose, instead of asymptotic behaviours, we will need explicit estimates valid in certain concrete ranges. A standard reference for such kind of results is the paper by Rosser and Schoenfeld [310]. We collect the needed estimates in the following lemma.

Lemma 5.2.11 *We have*

(*i*) $\frac{x}{\log(x)} < \pi(x)$ *for* $x \geq 17$ ([310, formula (3.5)]),

(*ii*) $n \log(n) < p_n$ *for* $n \geq 1$ ([310, formula (3.12)]),

(*iii*) $x \left(1 - \frac{1}{\log(x)}\right) < \vartheta(x)$ *for* $n \geq 41$ ([310, formula (3.16)]),

(*iv*) $\vartheta(x) < 1.01624x$ *for* $x > 0$ ([310, formula (3.32)]).

Proof of Proposition 5.2.3 Let $N \geq n^2$ be a positive integer. Since $|n| > 400$, we have $N > 1.6 \cdot 10^5$. Let $D = \{a_1, a_2, \ldots, a_m\} \subseteq \{1, 2, \ldots, N\}$ be a $D(n)$-m-tuple. We would like to find an upper bound for m depending on N. We will use Gallagher's sieve (Theorem 5.2.9). Let

$$S = \{p \ : \ p \text{ is prime, } \gcd(n, p) = 1 \text{ and } p \leq Q\},$$

where Q is sufficiently large. For a prime $p \in S$, let C denote the set of integers b such that $b \in \{0, 1, 2, \ldots, p-1\}$ and there is at least one $a \in D$ such that $b \equiv a$ (mod p). Then $\left(\frac{xy+n}{p}\right) \in \{0, 1\}$ for all distinct $x, y \in C$. If $0 \in C$, then $\left(\frac{n}{p}\right) = 1$. For given $x \in C \setminus \{0\}$, we have $\left(\frac{xy_0+n}{p}\right) = 0$ for at most one $y_0 \in C$. If $y \in C$, $y \neq x, y_0$, then $\left(\frac{xy+n}{p}\right) = 1$. Therefore,

$$T = \sum_{x,y \in C} \left(\frac{xy+n}{p}\right) = \sum_{x \in C}\left(\sum_{y \in C}\left(\frac{xy+n}{p}\right)\right)$$

$$\geq \sum_{x \in C}(|C| - 3) = |C|(|C| - 3).$$

On the other hand, Lemma 5.2.10 implies

$$T < |C| \cdot \sqrt{p}.$$

Thus, $|C| < \sqrt{p} + 3$, and we may apply Theorem 5.2.9 with

$$g(p) = \min(\lfloor\sqrt{p}\rfloor + 3, p).$$

Note that $g(p) = \lfloor\sqrt{p}\rfloor + 3$ for $p \geq 7$.

Let us denote the numerator and denominator from (5.15) by E and F, respectively. By Lemma 5.2.11(iv), we have

$$E = \sum_{p \in S} \log p - \log N < \vartheta(Q) < 1.01624\, Q. \qquad (5.16)$$

It is easy to check that the function $f(x) = \frac{\log x}{\min(\sqrt{x+3},x)}$ is strictly decreasing for $x > 25$. Moreover, if $Q \geq 118$, then $f(p) \geq f(Q)$ for all $p \leq Q$.

For $p \in S$, we have $\gcd(n, p) = 1$. This condition comes from the assumptions of Lemma 5.2.10. However, we will show later that n can be divisible only by a small proportion of the primes $\leq Q$. Assume that n is divisible by at most 5% of primes $\leq Q$. Then, for $Q \geq 118$, we have

$$F \geq \sum_{p \in S} f(p) - \log N \geq \frac{\log Q}{\sqrt{Q+3}} \cdot |S| - \log N$$

$$\geq \frac{\log Q}{\sqrt{Q+3}} \cdot \frac{19}{20} \pi(Q) - \log N > \frac{0.95\, Q}{\sqrt{Q+3}} - \log N. \tag{5.17}$$

Since F has to be positive in the applications of Theorem 5.2.9, we choose Q of the form $c_1 \cdot \log^2 N$, say

$$Q = 6 \cdot \log^2 N. \tag{5.18}$$

We have to check whether our assumption on the proportion of primes that divide n is correct. Suppose that n is divisible by at least 5% of the primes $\leq Q$. Then $|n| \geq p_1 p_2 \cdots p_{\lceil \pi(Q)/20 \rceil}$, where p_i denotes the i-th prime. By parts (i) and (ii) of Lemma 5.2.11, we have $p_{\lceil \pi(Q)/20 \rceil} > R$, where

$$R = \frac{1}{20} \frac{Q}{\log Q} \log \left(\frac{1}{20} \frac{Q}{\log Q} \right).$$

By our choice of Q, i.e. (5.18), we have $Q > 860$ and $R > 11.77$. Using Lemma 5.2.11(iii), it is easy to check that

$$\log |n| > \sum_{p \leq R} \log p > R \left(1 - \frac{1.136}{\log R} \right).$$

Furthermore, $\frac{1}{20} \frac{Q}{\log Q} > Q^{0.273}$ and $R > 0.0136\, Q$. We have $\log R > 2.465$ and

$$\log N \geq 2 \log |n| > 0.01466\, Q \geq 0.08796 \log^2 N,$$

contradicting the assumption that $N > 1.6 \cdot 10^5$. Therefore, we conclude that n is divisible by at most 5% of the primes $\leq Q$, and hence, we have justified estimate (5.17).

Inequality (5.17) implies

$$F > 0.861 \sqrt{Q} - \log N > 1.109 \log N,$$

and combining this with $E < 6.097 \log^2 N$, which follows from (5.16), we obtain

$$\frac{E}{F} < 5.5 \log N. \tag{5.19}$$

Putting $N = n^2$ in (5.19), we obtain the statement of Proposition 5.2.3. $\square$

Remark 5.2.12 It is not surprising that in Theorem 5.2.4, the main contribution comes from C_n. Namely, if we define $C = \sup\{C_n : n \in \mathbb{Z} \setminus \{0\}\}$, then we have $M = C$. Indeed, if $\{a_1, a_2, \ldots, a_m\}$ is a Diophantine m-tuple with the property $D(n)$, then $\{a_1 c, a_2 c, \ldots, a_m c\}$ has the property $D(nc^2)$ and for sufficiently large c we have $a_i c \leq (nc^2)^2$, $i = 1, 2, \ldots, m$. It means that in order to prove $M < \infty$, it suffices to prove $C < \infty$.

We may define also $A = \sup\{A_n : n \in \mathbb{Z} \setminus \{0\}\}$ and $B = \sup\{B_n : n \in \mathbb{Z} \setminus \{0\}\}$. Gibbs' example of $D(2985984)$-sextuple $\{99, 315, 9920, 32768, 44460, 19534284\}$ shows that $C \geq 6$ and $M \geq 6$. For any integer n, we have $B_n \geq 3$ since $\{n^2 + n + 1, n^2 + 3n + 4, 4n^2 + 8n + 9\}$ is a $D(n)$-triple. Hence $B \geq 3$. Finally, since $\{k, k + 2, 4k + 4, 16k^3 + 48k^2 + 44k + 12\}$ is a $D(1)$-quadruple for any positive integer k, we have $A \geq A_1 = 4$.

Remark 5.2.13 The proof of Theorem 5.2.1 suggests that for large $|n|$ the constants in the estimates for B_n and C_n (and thus also for M_n) can be improved. That was sketched in [108, Remark 1] and worked out by Becker and Murty in [29]. They proved that

$$M_n \leq 2.6071 \log |n| + O\left(\frac{\log |n|}{(\log \log |n|)^2}\right),$$

where the implied constant in O is effectively computable.

Remark 5.2.14 Several authors considered the higher power variant of the problems studied in this section. For $k \geq 2$ and a non-zero integer n, we call a set of m distinct positive integers a *k-th power Diophantine m-tuple* or a *Diophantine m-tuple with the property $D_k(n)$* if the product of any two distinct elements increased by n is a k-th power, and we define

$$M_k(n) = \sup\{|A| : A \text{ satisfies the property } D_k(n)\}.$$

By the result of He, Togbé, and Ziegler [211] on the nonexistence of Diophantine quintuples, we know that $M_k(1) \leq 4$ for any even $k \geq 2$. In [49], Bugeaud and Dujella showed that $M_3(1) \leq 7$, $M_5(1) \leq 5$, $M_k(1) \leq 4$ for $6 \leq k \leq 176$, and $M_k \leq 3$ for $k \geq 177$ (as pointed out in [31], there was a minor inaccuracy in the proof, but it only affected the upper bound on $M_5(1)$). For general n, Bérczes, Dujella, Hajdu, and Luca [31] proved that $M_k(n) \leq 2|n|^5 + 3$ for $k \geq 5$. The dependence on n has been significantly improved by Dixit, Kim, and Murty [84], Bhattacharjee, Dixit, and Saikia [36], Kim, Yip, and Yoo [237], and Yip [360] and the bounds of the form $M_k(n) \ll_k \log |n|$ (for fixed $k \geq 2$ and $|n| \to \infty$) were obtained. Thus, these results can be understood as generalizations of Theorem 5.2.4.

5.2.4 Diophantine m-Tuples for Primes

It should be emphasized that there is a huge gap between the best-known lower bound for M, namely $M \geq 6$, and the known upper bounds for M_n. Although we expect that M is finite, i.e. that there is an absolute upper bound for M_n, and even the guess that $M = 6$ seems reasonable, we are very far from proving such a result.

However, in 2005, Dujella and Luca [142] proved absolute upper bounds for sets with properties $D(p)$ and $D(-p)$, where p is a prime. More precisely, they proved that $M_p < 3 \cdot 2^{168}$ and $M_{-p} < 3 \cdot 2^{168}$ holds for all primes p.

Furthermore, they showed that for square-free positive integers n, the numbers M_n and M_{-n} are bounded in terms of the number of prime factors $\omega(n)$ of n. They also showed that for every $\varepsilon > 0$, the set of positive integers n with the property that there exists a Diophantine m-tuple with the property $D(n)$ or $D(-n)$ and with $m > (1 + \varepsilon) \log \log n$, is of asymptotic density zero.

In this section, we will briefly sketch the proof of mentioned upper bounds on M_p and M_{-p}. We will concentrate on M_p and mention some minor differences for M_{-p}. Let $\mathcal{A} = \{a_1, \ldots, a_m\}$ be a $D(p)$-m-tuple, and let $a_i a_j + p = x_{ij}^2$ with $x_{ij} \geq 0$. Since our goal is to obtain an absolute upper bound for M_p (and we expect to obtain very large bound), we may ignore certain small subsets of $\mathcal{A}$. Also, we may assume that all elements of $\mathcal{A}$ are positive (mixed signs are not possible in the $D(-p)$ case, while in the $D(p)$ case, allowing mixed signs means, at worst, multiplying the final bound by 2; however, we will see below that in this case the contribution of mixed signs can be neglected). Note that there exists at most one element of $\mathcal{A}$ which is a multiple of p. Indeed, if two such elements, say a_i and a_j, exist, then reducing the equation $a_i a_j + p = x_{ij}^2$ modulo p^2, we get $p \equiv x_{ij}^2 \pmod{p^2}$, which is a contradiction. Eliminating such an element from $\mathcal{A}$, we may assume that p does not divide a for any $a \in \mathcal{A}$. Assume that $0 < a_1 < a_2 < a_3 < \cdots < a_m$. We will show that $a_3 > p^{1/4}$. Indeed, note that in $a_1 a_3 + p = x_{13}^2$, $a_2 a_3 + p = x_{23}^2$, we have $x_{23} > x_{13} > p^{1/2}$. In particular,

$$a_3^2 > a_3(a_2 - a_1) = x_{23}^2 - x_{13}^2 = (x_{23} - x_{13})(x_{23} + x_{13}) > 2p^{1/2}.$$

Hence, $a_3 > p^{1/4}$. When p is replaced by $-p$, we have $a_2^2 > a_1 a_2 = p + x_{12}^2 > p$, therefore $a_2 > p^{1/2}$, which is an even better inequality than in the case of p. Thus, eliminating the two smallest elements, if needed, we may assume that $a > p^{1/4}$ holds for all $a \in \mathcal{A}$. Furthermore, from Proposition 5.2.1, we know that there are at most 21 elements of $a \in \mathcal{A}$ such that $a > p^3$. Eliminating those elements, we may assume that $p^{1/4} < a < p^3$ holds for all $a \in \mathcal{A}$.

To finish the proof, it suffices to obtain a gap principle of the form

$$a_{i+\ell} > p^{\gamma} a_i. \tag{5.20}$$

Indeed, from (5.20), by induction on i, we obtain the inequality $a_i > p^{1/4+\lfloor i/\ell \rfloor \gamma}$ for $i = 1, 2, \ldots, m$, and since $a_m < p^3$, we conclude that $\lfloor m/\ell \rfloor \gamma + 1/4 < 3$ and

$$m < \frac{(11 + 4\gamma)\ell}{4\gamma}.$$

The starting point in deriving the gap principle (5.20) is Lemma 3.8.1, which was already very useful in several places in this book. Let $a < b < c$ be three elements of $\mathcal{A}$ and $ab + p = r^2$, $ac + p = s^2$, $bc + p = t^2$, where r, s, t are positive integers. We define the quantities e (as in the proof of Lemma 3.8.1) and $\bar{e}$ by

$$e = p(a + b + c) + 2abc - 2rst,$$

$$\bar{e} = p(a + b + c) + 2abc + 2rst.$$

A straightforward computation shows that

$$e\bar{e} = p^2(c - a - b + 2r)(c - a - b - 2r). \tag{5.21}$$

Notice that (5.21) is a direct analogue of formula (1.13) for the product $d_+ d_-$ in the case of $D(1)$-triples. It is clear that (5.21) holds if p is replaced by an arbitrary integer n, and even in more general situations when $D(n)$-triples are considered in commutative rings. In fact, formula (5.21) played an important role in studying polynomial $D(n)$-m-tuples for linear or quadratic polynomials n (see [127, 128, 134, 225]).

Both sides of (5.21) are equal to zero if and only if $c = a + b \pm 2r$ (i.e. the triple $\{a, b, c\}$ is regular). Thus, there is at most one value of $c > b$ (namely, $c = a + b + 2r$) for which both sides of the equation vanish, and thus we proceed with the assumption that both sides are non-zero. Furthermore, note that e and $\bar{e}$ cannot be simultaneously divisible by p, since otherwise, from $2abc - 2rst \equiv 2abc + 2rst \equiv 0$ (mod p), we would obtain $4abc \equiv 0$ (mod p), which contradicts our assumption that the elements of $\mathcal{A}$ are not divisible by p (here we used that $p > 2$, which we can assume since we know that $M_2 = M_{-2} = 3$).

Thus, we conclude that either $p^2 \mid e$ or $p^2 \mid \bar{e}$. Let us first apply this conclusion to handle the case of $D(p)$-m-tuples with mixed signs (a $D(-p)$-m-tuple obviously cannot have elements with mixed signs). Let $\mathcal{A} = \{a_1, a_2, \ldots, a_m\}$ and $\mathcal{B} = \{b_1, b_2, \ldots, b_{m'}\}$, with $m, m' \geq 1$, be sets of positive integers with the property $D(p)$ such that $\mathcal{A} \cup -\mathcal{B} = \{a_1, a_2, \ldots, a_m, -b_1, -b_2, \ldots, -b_{m'}\}$ also has the property $D(p)$. Assume that $a_m = \max(\mathcal{A} \cup \mathcal{B})$. Then, from $-a_m b_j + p \geq 0$, it follows that $b_1, \ldots, b_{m'} \leq \sqrt{p}$. However, this implies that for sufficiently large p,

p^2 cannot divide e and $\bar{e}$. Indeed, we have $e \le \bar{e} \le 11p\sqrt{p} < p^2$ for $p > 121$ (and by Theorem 5.2.4, we have $M_p \le 32$ for $p < 121$).

It remains to consider the cases $p^2 \mid e$ or $p^2 \mid \bar{e}$ for m-tuples with positive elements. In what follows, it will be convenient to assume that p is sufficiently large. We justify this by applying Theorem 5.2.4, which implies, e.g. that for $p \le 2^{2^{76}}$, we have $M_p \le 2^{80}$. Thus, we may assume that $p > 2^{2^{76}}$.

If $p^2 \mid e$, then $|e| \ge p^2$, and Eq. (5.21) implies

$$4abc < 2rst + 2abc + p(a+b+c) \le |(c-a-b+2r)(c-a-b-2r)|$$

$$= |a^2 + b^2 + c^2 - 2ab - 2ac - 2bc - 4p| < 6c^2 + 4p. \tag{5.22}$$

Suppose that $4p > 6c^2$. Then (5.22), combined with $p(a+b+c) > 3p^{5/4}$, gives $8p > 3p^{5/4}$ and $p^{1/4} < 8/3$, contradicting our assumption on p. Thus $6c^2 > 4p$ and (5.22) implies $4abc < 12c^2$, which leads to

$$c > ab/3 > p^{1/4}b/3 > p^{1/8}b. \tag{5.23}$$

With minor modification of arguments, the same inequality can be obtained for the case of $D(-p)$-m-tuples. Inequality (5.23) is exactly of the desired form (5.20). However, we should also take into account the possibility that both sides of (5.21) are equal to zero, i.e. that certain subtriples of an m-tuple are regular. By taking all possible cases into account, it can be shown that, under the assumption that $p^2 \mid e$ holds for at least one of the corresponding triples, we have

$$a_{i+16} > p^{1/8}a_i,$$

which is the desired gap inequality with $\gamma = 1/8$ and $\ell = 16$ (see [142] for the omitted details).

It remains to consider the more demanding case when $p^2 \mid \bar{e}$. We will give here only a rough idea of how to attack this case, and we advise an interested reader to consult the paper [142]. An important step is a lemma on congruence properties of $D(p)$-quadruples [142, Lemma 3.1]. Using it, to any $D(p)$-quadruple, an integer Λ can be assigned. In that way, to any $D(p)$-quintuple, we can assign a quintuple of such integers $(\Lambda_1, \Lambda_2, \Lambda_3, \Lambda_4, \Lambda_5)$. By elimination of variables (see, e.g. [314, Chapter 9]), one gets a polynomial of bounded height which must satisfy a congruence modulo p for rather small values of the variables if the gap principle we want is not fulfilled. The smallness of the variables yields the key fact that the congruence in question must be in fact an equation. The final step consists of deducing from this the existence of a polynomial of bounded degree M in five variables, which vanishes on a set which is a cartesian product $\mathcal{M}^5$, with $\mathcal{M}$ a set of cardinality larger than M, which is impossible. This establishes the desired gap principle (with $\gamma = 2^{-80}$ and $\ell = 2^{88}$).

5.3 Existence of Rational $D(q)$-quintuples

In this section, we will consider rational $D(n)$-m-tuples. In the rational case, it is usual to talk of $D(q)$-m-tuples, so we will do the same.

We have already seen in Sect. 1.1 that for every rational number q, there exist infinitely many rational $D(q)$-quadruples. We will now give another proof of this result.

Theorem 5.3.1 *For every rational number q, there are infinitely many rational $D(q)$-quadruples.*

Proof Consider the set

$$\{k,\ 16k+8,\ 25k+14,\ 36k+20\}. \tag{5.24}$$

It is a (rational) $D(16k+9)$-quadruple for all but finitely many integer (rational) values of k. We can obtain it from the two-parameter formula (5.7) so that we first exchange a and k and then insert $a = -2$.

Now let $s \neq 0$ be an arbitrary rational number. We define the rational number k by $16k+9 = qs^2$, i.e. $k = \frac{qs^2-9}{16}$. From (5.24), we get a rational $D(q)$-quadruple

$$\left\{ \frac{qs^2-9}{16s},\ \frac{qs^2-1}{s},\ \frac{25qs^2-1}{16s},\ \frac{9qs^2-1}{4s} \right\}$$

(for all but finitely many values of s). We claim that we obtained infinitely many distinct rational $D(q)$-quadruples in this way. Indeed, only finitely many distinct rationals s can induce the same quadruple. For $\alpha \in \mathbb{Q}$, the condition $\frac{qs^2-9}{16s} = \alpha$ gives a quadratic equation in s, which has at most two solutions. The same applies to the other elements of the quadruple. $\square$

The question arises naturally for which rational numbers q there are infinitely many rational $D(q)$-quintuples. That question is still open, although recent results by Dražić [86] support the conjecture that the answer is: for all rational numbers q.

It is clear that in this question, we may assume that q is a square-free integer since by multiplying the elements of a $D(q)$-m-tuple by u, we get a $D(qu^2)$-m-tuple.

In the case $q = 1$, we saw that already Euler showed that there are infinitely many rational $D(1)$-quintuples. In [100], it was shown that there are infinitely many rational $D(-3)$-quintuples (the fact that the elliptic curve $y^2 = x^3 + 42x^2 + 432x + 1296$ has positive rank was used in the proof), while in [105] it was shown that there are infinitely many rational $D(-1)$-quintuples (here the fact that an elliptic curve associated to the curve $y^2 = -(x^2 - x - 3)(x^2 + 2x - 12)$ has positive rank was used in the proof). For example, $\{\frac{5}{4}, \frac{12}{5}, \frac{73}{20}, \frac{217}{20}, \frac{133}{5}\}$ is a rational $D(-3)$-quintuple, while $\{10, \frac{25}{8}, \frac{37}{10}, \frac{13}{40}, \frac{533}{40}\}$ is a rational $D(-1)$-quintuple.

In 2012, Dujella and Fuchs [125] significantly expanded the set of rationals q for which it is known that there are infinitely many $D(q)$-quintuples, and this result was recently significantly improved by Dražić [86].

In the proof of Theorem 5.3.1, we used a polynomial Diophantine quadruple with the property $D(n)$, where n was a linear polynomial, and all the elements of the quadruple were also linear polynomials. Many examples of such polynomial quadruples are known. Therefore, the question arises whether we can find a polynomial quintuple, all elements of which are linear polynomials. The answer is negative, which was proved by Dujella, Fuchs, and Walsh [128] in 2006 (improving the results from [127]).

We will sketch the proof of this result. Let $\{ax + b, cx + d, ex + f\}$ be a polynomial $D(ux + v)$-triple, where all coefficients are integers. Then $\{a^2x + ab, acx + ad, aex + af\}$ is a polynomial $D(a^2ux + a^2v)$-triple. By the substitution $ax + b = z$ we get a polynomial $D(auz + v')$-triple $\{az, cz + d', ez + f'\}$. We can assume that $\gcd(a, c, e) = 1$ (otherwise, we introduce the substitution $z' = z \gcd(a, c, e)$), so we get that a, c, e are perfect squares: $a = A^2$, $c = C^2$, $e = E^2$. By inserting $z = 0$, we see that v' is also a perfect square: $v' = V^2$. But, $v' = a^2v - abu = A^4v - A^2bu$, so $V = AW$, with $W^2 = A^2v - bu$. Now from

$$A^2z(C^2z + d') + (A^2uz + A^2W^2) = (ACz \pm AW)^2,$$

by comparing the coefficients of z, we get that $d' = \pm 2CW - u$. Analogously, we find that $f' = \pm 2EW - u$. So, we have a $D(A^2uz + A^2W^2)$-triple

$$\{A^2z, \ C^2z \pm 2CW - u, \ E^2z \pm 2EW - u\}.$$

We also have the condition that

$$(C^2z \pm 2CW - u)(E^2z \pm 2EW - u) + (A^2uz + A^2W^2)$$

is the square of a linear polynomial (in the variable z), which implies that the discriminant of this quadratic polynomial is equal to 0. By factoring the discriminant, we get the condition

$$(C-E-A)(C-E+A)(\pm 2CEW-Cu-Eu+Au)(\pm 2CEW-Cu-Eu-Au) = 0.$$

The conditions $\pm 2CEW - Cu - Eu \pm Au = 0$ can be written as

$$(\pm 2CW - u)(\pm 2EW - u) = u^2 \pm 2AWu.$$

So, if a polynomial quintuple with linear polynomials exists, then there exists a $D(A^2uz + A^2W^2)$-quintuple with one element equal to A^2z and the others of the form

$$m_i^2z + 2m_i W - u, \quad i = 1, 2, 3, 4.$$

Here for $i \neq j$ we have $|m_i - m_j| = A$ (this corresponds to the general construction of triples of the form $\{a, b, a + b + 2r\}$) or $(\pm 2m_i W - u)(\pm 2m_j W - u) = u^2 \pm 2AWu$.

In [128], it was shown that this leads to a contradiction, which means that there is no polynomial quintuple with the required property.

However, in [125], it was also shown that these conditions could be met if we omit one condition (for example, $(i, j) = (3, 4)$). Moreover, such "almost quintuple" is essentially unique (all others can be obtained from it by "admissible" transformations). That set is

$$\{x, \, 9x + 8, \, 25x + 20, \, 4x + 2, \, 16x + 14\}$$

which contains two polynomial $D(10x+9)$-quadruples: $\{x, 9x+8, 25x+20, 4x+2\}$ and $\{x, 9x + 8, 25x + 20, 16x + 14\}$.

So, for a rational number x, the set

$$\{x, \, 4x + 2, \, 9x + 8, \, 16x + 14, \, 25x + 20\}$$

will be a rational $D(10x + 9)$-quintuple if

$$(4x + 2)(16x + 14) + 10x + 9 = y^2 \tag{5.25}$$

for $y \in \mathbb{Q}$. This defines a curve of genus 0. If we put $y = 8x + t$ in (5.25), we get a parametric solution

$$x = \frac{t^2 - 37}{2(49 - 8t)}.$$

After simplifying (removing the denominator), we get the following result.

Theorem 5.3.2 *The set*

$$\{t^2 - 37, \, 4t^2 - 32t + 48, \, 9t^2 - 128t + 451, \, 16t^2 - 224t + 780, \, 25t^2 - 320t + 1035\} \tag{5.26}$$

is a $D(4(8 - t)(5t - 32)(8t - 49))$-quintuple.

Now let q be a non-zero rational. We are interested in whether Theorem 5.3.2 can be used to obtain a rational $D(q)$-quintuple. If there are rational numbers $s \neq 0$ and t such that

$$4(8 - t)(5t - 32)(8t - 49) = qs^2, \tag{5.27}$$

then by dividing all the elements of (5.26) by s, we get exactly a $D(q)$-quintuple.

Equation (5.27) defines an elliptic curve over $\mathbb{Q}$. So, we are interested in whether that curve has a point with the s-coordinate different from zero, and this leads us to consider curves of positive rank in the family of elliptic curves (depending on the parameter q). It is the family of (quadratic) twists of the elliptic curve

$$s^2 = 4(8 - t)(5t - 32)(8t - 49).$$

With the substitutions $t = -x/40 + 49/8$, $s = y/20$, we get the equation of the curve in the Weierstrass form

$$E: \ y^2 = x(x + 11)(x + 75) = x^3 + 86x^2 + 825x. \tag{5.28}$$

The curve E has the discriminant $D = 2^{16}3^2 5^4 11^2$ and the conductor $C = 330 = 2 \cdot 3 \cdot 5 \cdot 11$. Its torsion group is isomorphic to $\mathbb{Z}/2\mathbb{Z} \times \mathbb{Z}/2\mathbb{Z}$ (the only non-trivial torsion points are those with y-coordinate equal to 0), and the rank is equal to 1 with a generator $(x, y) = (-15, 60)$.

We now ask which q-twists of the curve E have positive rank. Here we can also assume that q is a square-free integer. We observe the family of elliptic curves E_q given by

$$E_q: \quad qy^2 = x^3 + 86x^2 + 825x. \tag{5.29}$$

In general, on the curve of the form $f(t)y^2 = f(x)$, there is a rational point $(x, y) = (t, 1)$ of infinite order (see [129]). If we write $t = u/v$, $\gcd(u, v) = 1$, we get that if q is of the form

$$q = uv(u^2 + 86uv + 825v^2), \tag{5.30}$$

for integers $u, v \neq 0$, then $(u/v, 1/v^2)$ is a point of infinite order on E_q. This gives us infinitely many values of q for which the rank is positive, and therefore there are infinitely many rational $D(q)$-quintuples for them. It can be shown (see [197]) that for $\varepsilon > 0$ and sufficiently large N, there are at least $N^{1/2-\varepsilon}$ square-free numbers q, $|q| \leq N$, of the form (5.30).

If we assume that the parity conjecture is valid for twists of E, then we can precisely describe those integers q for which the rank of the q-twist is odd (and therefore positive). The parity conjecture (for E) says that the rank of the Mordell-Weil group of E is of the same parity as the vanishing order of the associated L-function $L(E, s)$ in $s = 1$. This conjecture is a weaker form of the well-known Birch and Swinerton-Dyer conjecture. The parity conjecture implies that for an odd square-free integer q which is relatively prime to the conductor C of E, the ranks of E and E_q are of the same parity if and only if $\chi_q(-C) = 1$, where χ_q is the quadratic Dirichlet character associated to the quadratic field $\mathbb{Q}(\sqrt{q})$ (see [197]).

Let us also mention Goldfeld's conjecture, which predicts that the average rank of all twists of E is equal to $1/2$. Together with the parity conjecture, this implies that the number of square-free integers q, $|q| \leq N$, such that E_q has rank 1 (resp. 0), is asymptotically equal to $6N/\pi^2$ when $N \to \infty$. For our purpose, it is enough that the rank is positive, and the parity conjecture implies that the number of such square-free integers q with $|q| \leq N$ is $\geq 6N/\pi^2$.

Theorem 5.3.3 *For infinitely many square-free integers q there exist infinitely many rational $D(q)$-quintuples. If we assume that the parity conjecture holds, then for all square-free positive integers q in at least 497 residue classes modulo 1320 and all square-free negative integers q in at least 493 residue classes modulo 1320, there are infinitely many rational $D(q)$-quintuples.*

Proof We have actually already proved the first statement from the theorem. It remains to verify that at most finitely many different points on the elliptic curve E_q can induce the same quintuple. For $\alpha \in \mathbb{Q}$, the condition $\frac{t^2-37}{s} = \alpha$ for a point (t, s) on (5.27) gives a polynomial of the fourth degree in t, so at most four points can satisfy that condition. The same applies to the other elements of the quintuple.

Let us assume now that the parity conjecture holds. If $\gcd(q, 330) = 1$, then the parity conjecture implies that the ranks of E and its q-twist have the same parity if and only if $\chi_q(-330) = 1$, where χ_q is the quadratic Dirichlet character associated to the quadratic field $\mathbb{Q}(\sqrt{q})$. In particular, if $q \equiv 1 \pmod{4}$, then

$$\chi_q(-330) = \left(\frac{-330}{|q|} \right),$$

where $(\div)$ denotes the Jacobi symbol, defined by $(\frac{a}{q}) = \prod_{j=1}^{k}(\frac{a}{q_j})$, where $q = q_1 \cdot \ldots \cdot q_k$, q_j $(j = 1, \ldots, k)$ are odd primes and $(\frac{a}{q_j})$ are the Legendre symbols. We obtain that for all q satisfying

$$\gcd(q, 330) = 1, \quad q \equiv 1 \pmod{4} \quad \text{and} \quad \left(\frac{-330}{|q|} \right) = 1,$$

the q-twist has odd rank. We find that 80 residue classes modulo $330 \cdot 4 = 1320$ satisfy these conditions (for both positive and negative integers q). For $q \equiv 3 \pmod{4}$, we understand the q-twist as the $(-q)$-twist of the (-1)-twist of E. We calculate that (-1)-twist has the conductor $2^4 \cdot 3 \cdot 5 \cdot 11$ and the root number 1 (so the rank is conditionally even; it is easy to check that the rank is indeed equal to 0). Therefore, we get that for all integers q that satisfy

$$\gcd(q, 330) = 1, \quad q \equiv 3 \pmod{4} \quad \text{and} \quad \left(\frac{-165}{|q|} \right) = -1,$$

the rank of the q-twist is odd. This condition is satisfied for 40 residue classes modulo $165 \cdot 4$, i.e. 80 residue classes modulo 1320. If $\gcd(q, 330) = g > 1$, we proceed similarly. We put $q = gh$, and then consider q-twist as h-twist of g-twist of E (or $(-h)$-twist of $(-g)$-twist). For $g \in \{\pm2, \pm3, \pm5, \pm11, \pm6, \pm10, \pm15, \pm22, \pm30, \pm33, \pm55, \pm66, \pm110, \pm165, \pm330\}$, we calculate the conductor and root number of the g-twist. In each of the cases, we get that the conductor is of the form $2^k \cdot |g| \cdot 330$ for $k \in \{0, 3, 4\}$. This implies that the conditions for q can, in all cases, be written in terms of residue classes modulo 1320. Finally, for positive integers q, we get exactly 497 residue classes modulo 1320 ($q \equiv i \pmod{1320}$, $i = 1, 7, 9, 10, 11, 18, 21, 22, 23, 30, \ldots$) which satisfy the condition that the rank of the q-twist is odd (we observe only those classes whose elements are not divisible by 4, because we are only interested in square-free numbers), while for negative integers q, we get 493 residue classes modulo 1320 ($-q \equiv i \pmod{1320}$, $i = 2, 3, 9, 14, 17, 18, 22, 25, 26, 27, 30, \ldots$) with the same property. $\square$

In the observed family, there are also curves of rank greater than 1, e.g., E_{-21} has rank 2, E_{-551} has rank 3, E_{5217} has rank 4, $E_{19712449}$ has rank 5, $E_{19712449}$ has rank 5, while $E_{18427939089}$ has rank 6.

Example 5.3.4 The smallest positive integer $q > 1$ for which the above construction gives a $D(q)$-quintuple is $q = 7$. We have the twist $7y^2 = x(x + 11)(x + 75)$ of rank 1 and a generator $P = (x, y) = (-25, 50)$. It induces the point $(t, s) = (27/4, 5/2)$ on (5.27). Now from Theorem 5.3.2, we get the rational $D(7)$-quintuple

$$\left\{ \frac{137}{40}, \frac{57}{10}, -\frac{47}{40}, -\frac{6}{5}, \frac{45}{8} \right\}.$$

By multiplying the elements of this quintuple by 40, we get an integer $D(11200)$-quintuple. Now, we look at the point $2P = (4/7, 414/49)$. It induces the point $(t, s) = (1711/280, 207/490)$ on (5.27). From Theorem 5.3.2, we get a rational $D(7)$-quintuple with positive elements

$$\left\{ \frac{2969}{3680}, \frac{35681}{8280}, \frac{383849}{33120}, \frac{42401}{2070}, \frac{205285}{6624} \right\}.$$

The largest negative number with the same property is $q = -2$. The (-2)-twist has rank 1 and a generator $(x, y) = (-297/2, 3465/4)$. It induces the point $(t, s) = (787/80, 693/16)$ on (5.27). From Theorem 5.3.2, we obtain the following rational $D(-2)$-quintuple

$$\left\{ \frac{11593}{8400}, \frac{5833}{2100}, \frac{4059}{2800}, \frac{1513}{525}, \frac{2377}{336} \right\}.$$

Let us mention that the smallest positive integer n for which the construction from Theorem 5.3.2 gives an integer $D(n)$-quintuple is $n = 1309$, for which we get the $D(1309)$-quintuple $\{2, 30, 106, 186, 270\}$, while the largest negative integer with the same property is $n = -299$ via the $D(-299)$-quintuple $\{14, 22, 30, 42, 90\}$ (this quintuple appeared also in [282]). These examples motivate the question of what are, in general, the smallest positive integer n_1 and the largest negative integer n_2 for which there are Diophantine quintuples with properties $D(n_i)$, $i = 1, 2$. It is known that $n_1 \leq 256$ and $n_2 \geq -255$, since the sets $\{1, 33, 105, 320, 18240\}$ and $\{5, 21, 64, 285, 6720\}$ have the property $D(256)$, while the set $\{8, 32, 77, 203, 528\}$ has the property $D(-255)$ (see [92]).

In Theorem 5.3.3, we saw that under the assumption that the parity conjecture holds, there are infinitely many rational $D(q)$-quintuples for approximately 50% square-free numbers (because 990 residue classes modulo 1320 contain square-free numbers). This result was significantly improved in 2022 by Dražić [86]. Namely, in that paper, it was shown that in addition to elliptic curve (5.28), seven more elliptic curves could be used analogously by considering their twists for which the parity conjecture implies that they have an odd rank to obtain the conclusion on the existence of $D(q)$-quintuples. Let us list those curves in their short Weierstrass forms:

$$y^2 = x^3 - 33210675x + 6964980750,$$

$$y^2 = x^3 - 24651x + 1453194,$$

$$y^2 = x^3 - 97227x + 10789254,$$

$$y^2 = x^3 - 7155x + 187650,$$

$$y^2 = x^3 + 274725x + 126596250,$$

$$y^2 = x^3 - 24003x + 1296702,$$

$$y^2 = x^3 - 1196883x + 46619118.$$

Furthermore, the conductors of all eight curves are "in agreement" in the sense that in their factorizations, some common factors appear. This enables the following conclusion (assuming that the parity conjecture is valid):

- for all square-free positive integers q in at least 295026 residue classes modulo 394680, there are infinitely many rational $D(q)$-quintuples;
- for all square-free negative integers q in at least 295435 residue classes modulo 394680, there are infinitely many rational $D(q)$-quintuples.

So, assuming that the parity conjecture holds, the density of square-free integers for which there are infinitely many rational $D(q)$-quintuples is at least $295026/296010 \approx 99.7\% \approx 1 - \frac{1}{2^8}$. However, the question of whether for every rational number q there are infinitely many $D(q)$-quintuples (and also the question of whether there is at least one $D(q)$-quintuple) remains open. Let us mention that $q = 1579$ is the smallest positive integer for which no rational $D(q)$-quintuple is known.

5.4 Sets with the Property $D(n)$ for Several Values of n

5.4.1 $D(n)$-Triples for Several Values of n

When studying $D(n)$-m-tuples, the integer n is usually fixed in advance. However, we can ask whether it is possible for the same set to simultaneously have the property $D(n)$ for several distinct values of n. This question was asked for the first time in 2001 by A. Kihel and O. Kihel [235]. For example, $\{8, 21, 55\}$ is a $D(1)$-triple and a $D(4321)$-triple, while $\{1, 8, 120\}$ is a $D(1)$-triple and a $D(721)$-triple [362].

In this subsection, we will sketch the result by Adžaga, Dujella, Kreso, and Tadić [4] in which it was shown that there are infinitely many triples $\{a, b, c\}$, which are simultaneously $D(1)$-, $D(n_2)$-, and $D(n_3)$-triples for $1 < n_2 < n_3$.

The construction uses integer points on elliptic curves associated with Diophantine triples. Let $\{a, b, c\}$ be a Diophantine triple and let $ab + 1 = r^2, ac + 1 = s^2$, $bc + 1 = t^2$. We are interested in the integer solutions x of the system of equations

$$x + ab = \square, \quad x + ac = \square, \quad x + bc = \square. \tag{5.31}$$

Let us consider the elliptic curve obtained by multiplying the three conditions from (5.31):

$$E: \quad y^2 = (x + ab)(x + ac)(x + bc).$$

Since E has only finitely many integer points, we conclude that there are only finitely many integers n such that $\{a, b, c\}$ is a $D(n)$-triple. We know that E has several obvious rational points:

$$A = (-bc, 0), \quad B = (-ca, 0), \quad C = (-ab, 0), \quad P = (0, abc), \quad S = (1, rst).$$

Furthermore, we know from Theorem 2.4.9 that for $T \in E(\mathbb{Q})$, the number $x = x(T)$ is a rational solution of the system (5.31) if and only if $T \in 2E(\mathbb{Q})$. So, we are interested in points in the intersection $2E(\mathbb{Q}) \cap \mathbb{Z}^2$. One such point is the point S, which corresponds to $x = 1$. Indeed, we saw in Sect. 3.1 that $S = 2R$, where

$$R = (rs + rt + st + 1, (r + s)(r + t)(s + t)) \in E(\mathbb{Q}).$$

Points A, B, C are of order 2, so they are not interesting to us here. In our search for Diophantine triples that are $D(n)$-triples for several distinct integers n, we should observe the triples $\{a, b, c\}$ for which $2kP + \ell S \in \mathbb{Z}^2$ for some "small" integers k and ℓ. In general, we can expect that the points P and S will be independent points of infinite order. However, we showed in Sect. 3.1 that if $c = a + b \pm 2r$, where

$ab + 1 = r^2$ (i.e. if the triple $\{a, b, c\}$ is regular), then $2P = \pm S$. We have

$$x(2P) = \frac{1}{4}(a + b + c)^2 - ab - ac - bc,$$

which leads us to the following result.

Proposition 5.4.1 *Let a, b, c be non-zero integers such that the number $a + b + c$ is even. Then $\{a, b, c\}$ is a $D(n)$-triple for*

$$n = \frac{1}{4}(a + b + c)^2 - ab - ac - bc.$$

Furthermore, $n = 0$ is equivalent to $c = a + b \pm 2\sqrt{ab}$ (and therefore impossible if $\{a, b, c\}$ is a $D(1)$-triple), while $n = 1$ is equivalent to $c = a + b \pm 2\sqrt{ab + 1}$.

Every Diophantine triple $\{a, b, c\}$ such that $a + b + c$ is even and $c \neq a + b \pm 2\sqrt{ab + 1}$ is also a $D(n)$-triple for some $n \neq 1$.

A computer search shows that for Diophantine triples $\{a, b, c\}$, where $a, b \leq 1000$, $c \leq 1000000$, the corresponding points $S - 2P$ and $4P$ never have integer coordinates, while the point $S + 2P = 2(R + P)$ has integer coordinates for 14 triples in the observed range, e.g. for $\{4, 12, 420\}$, $\{4, 420, 14280\}$, $\{12, 24, 2380\}$, $\{24, 40, 7812\}$, $\{40, 60, 19404\}$. We will show that there are infinitely many such examples.

Let us first notice that all the above examples satisfy the additional condition that $x(S + 2P) = a + b + c$. Direct calculation shows that the condition $x(S + 2P) = a + b + c$ is equivalent to $q_1 q_2 q_3 = 0$, where

$$q_1 = a^2 + b^2 + c^2 - 2ab - 2ac - 2bc - 4,$$

$$q_2 = a^2 + b^2 + c^2 - 8abc - 2ab - 2ac - 2bc - 4a - 4b - 4c,$$

$$q_3 = a^4 + b^4 + c^4 - 2a^2b^2 - 2a^2c^2 - 2b^2c^2 + 2a^3 + 2b^3 + 2c^3$$
$$\qquad - 2a^2b - 2a^2c - 2ab^2 - 2ac^2 - 2b^2c - 2bc^2 - 4abc$$
$$\qquad + a^2 + b^2 + c^2 - 2ab - 2ac - 2bc - 4a - 4b - 4c.$$

The condition $q_1 = 0$ is equivalent to $c = a + b \pm 2\sqrt{ab + 1}$, but in that case $x(2P) = 1$, so in this way, we do not get a Diophantine triple which is a $D(n)$-triple for two distinct integers $n \neq 1$. It can be shown (similarly as in [58], where an analogous problem for $D(-1)$-triples in $\mathbb{Z}[i]$ was considered) that the equation $q_3 = 0$ is not satisfied for any Diophantine triple $\{a, b, c\}$.

So, the only interesting condition is $q_2 = 0$. It is equivalent to

$$c = 2 + a + b + 4ab \pm 2\sqrt{(2a + 1)(2b + 1)(ab + 1)}, \tag{5.32}$$

and this is precisely the condition that $\{2, a, b, c\}$ is a regular Diophantine quadruple. Notice that, by Proposition 1.4.4, if $(2a+1)(2b+1)(ab+1)$ is a perfect square, then $\{2, a, b\}$ is a Diophantine triple.

It can be verified that for such triples, $n_2 = x(S + 2P)$ and $n_3 = x(2P)$ satisfy $n_2 \neq n_3$, $n_2 \neq 1$, $n_3 \neq 1$. Note that for c given by (5.32), the number $a + b + c$ is even.

Thus we proved the following theorem.

Theorem 5.4.2 *Let $\{2, a, b, c\}$ be a regular Diophantine quadruple. Then the Diophantine triple $\{a, b, c\}$ is also a $D(n)$-triple for two distinct integers n such that $n \neq 1$.*

From Theorem 5.4.2, we can obtain explicit families of Diophantine triples with the desired property. One such example is the infinite family of triples

$$a = 2(i + 1)i, \quad b = 2(i + 2)(i + 1), \quad c = 4(2i^2 + 4i + 1)(2i + 3)(2i + 1),$$

with

$$n_2 = 32i^4 + 128i^3 + 172i^2 + 88i + 16,$$

$$n_3 = 256i^8 + 2048i^7 + 6720i^6 + 11648i^5 + 11456i^4$$

$$+ 6400i^3 + 1932i^2 + 280i + 16,$$

for a positive integer i. We get this family by putting $2a + 1 = (2i + 1)^2$, then extending the pair $\{2, a\}$ to the regular triple $\{2, a, b\}$, and then this triple to the regular quadruple $\{2, a, b, c\}$.

A computer search finds seven examples of triples $\{a, b, c\}$ which are $D(n)$-triples for $n_1 = 1 < n_2 < n_3 < n_4$, given in Table 5.1. However, the question of whether there are infinitely many such triples remains open.

If we omit the condition that $1 \in N$, then the set N, for which there is a triple $\{a, b, c\}$ of non-zero integers, which is a $D(n)$-triple for all $n \in N$, can be arbitrarily large. Indeed, let us take a triple $\{a, b, c\}$ such that the induced elliptic curve E, given by the equation $y^2 = (x + ab)(x + ac)(x + bc)$, has positive rank. Then we

Table 5.1 Examples of $D(n_2)$-, $D(n_3)$-, and $D(n_4)$-triples

$\{a, b, c\}$	n_2, n_3, n_4
$\{4, 12, 420\}$	436, 3796, 40756
$\{10, 44, 21252\}$	825841, 6921721, 112338361
$\{4, 420, 14280\}$	14704, 950896, 47995504
$\{40, 60, 19404\}$	19504, 3680161, 93158704
$\{78, 308, 7304220\}$	242805865, 4770226465, 13336497750865
$\{4, 485112, 16479540\}$	16964656, 2007609136, 63955397832496
$\{15, 528, 32760\}$	66609, 5369841, 15984081

have infinitely many rational points on E. For an arbitrarily large positive integer m, we can choose m distinct rational points $R_1, \ldots, R_m \in 2E(\mathbb{Q})$, such that

$$x(R_i) + ab = \square, \quad x(R_i) + ac = \square, \quad x(R_i) + bc = \square.$$

We can achieve this by taking points of the form $2m_1 P_1 + 2m_2 P_2 + \cdots + 2m_r P_r$, where $P_1, \ldots, P_r$ are generators of $E(\mathbb{Q})$. Let us take $z \in \mathbb{Z} \setminus \{0\}$ so that $z^2 x(R_i) \in \mathbb{Z}$ holds for all $i = 1, 2, \ldots, m$. Then $\{az, bz, cz\}$ is a $D(n)$-triple for $n = x(R_i)z^2$, $i = 1, 2, \ldots, m$.

Example 5.4.3 Let us consider the Diophantine triple $\{1, 8, 120\}$, whose induced elliptic curve E has rank equal to 3. Following the procedure described above, we find the points $R_1, \ldots, R_5 \in 2E(\mathbb{Q})$ such that

$$x(R_1) = 1, \quad x(R_2) = 721, \quad x(R_3) = 12289/4,$$

$$x(R_4) = 769/9, \quad x(R_5) = 1921/36.$$

We may take $z = 6$. We get the triple $\{az, bz, cz\} = \{6, 48, 720\}$, which is a $D(n)$-triple for $n = 36, 1921, 3076, 25956, 110601$.

5.4.2   Diophantine Quadruples with Properties $D(n_1)$ and $D(n_2)$

After studying triples with the property $D(n)$ for several values of n, we will now consider an analogous problem for quadruples. This problem was investigated for the first time in 2020 by Dujella and Petričević [156], where a positive answer was given to the question of whether there are sets of four non-zero integers, which are $D(n_i)$-quadruples for two distinct integers n_1 and n_2.

If $\{a, b, c, d\}$ is a $D(n_1)$- and a $D(n_2)$-quadruple and u is a non-zero rational number such that $au, bu, cu, du, n_1 u^2$ and $n_2 u^2$ are integers, then $\{au, bu, cu, du\}$ is a $D(n_1 u^2)$- and a $D(n_2 u^2)$-quadruple. We say that these two quadruples are equivalent. The following theorem is the main result of [156].

Theorem 5.4.4 *There are infinitely many non-equivalent quadruples $\{a, b, c, d\}$ with the property that there exist two distinct non-zero integers n_1 and n_2 such that $\{a, b, c, d\}$ is a $D(n_1)$-quadruple and $D(n_2)$-quadruple.*

The idea of the proof of this theorem came from analyzing the results of experiments in which individual examples were sought with the property from the theorem. First, $D(n_1)$-quadruples were searched, where $-500{,}000 \leq n_1 \leq 500{,}000$. For a fixed integer n_1, divisors of numbers of the form $m^2 - n_1$ were observed in the range $m \leq 333{,}333$. Then, for a $D(n_1)$-quadruple $\{a, b, c, d\}$, the search for integer points on the hyperelliptic curve $y^2 = (ab + x)(ac +$

Table 5.2 $D(n_1)$- and $D(n_2)$-quadruples with $a/b = -1/7$

$\{a, b, c, d\}$	$\{n_1, n_2\}$	$\{a, b, c, d\}$	$\{n_1, n_2\}$
$-27, 189, -32, 133$	$6192, 8352$	$-3, 21, 2597, 3132$	$11512, 80152$
$-27, 189, 28, 493$	$13752, 61272$	$-1, 7, 64, 119$	$128, 848$
$-27, 189, 4189, 6364$	$194328, 1325304$	$-1, 7, 4484, 4879$	$6248, 43688$
$-3, 21, 1152, 1517$	$5392, 37312$	$-1, 7, 22532, 23407$	$30632, 214376$

$x)(ad + x)(bc + x)(bd + x)(cd + x)$ with $|x| \leq 10^8$, $x \neq n_1$, was performed using the program `ratpoints` by Michael Stoll [338] (in PARI, the algorithm is implemented in the function `hyperellratpoints`), and it was then checked whether x satisfies the condition for n_2.

In that way, 26 examples of quadruples that are both $D(n_1)$- and $D(n_2)$-quadruples for $n_1 \neq n_2$ were found. Looking for some regularity among these examples, it can be observed that eight quadruples $\{a, b, c, d\}$ satisfy the condition $a/b = -1/7$. These are the quadruples given in Table 5.2. Looking for other regularities among these selected quadruples, we can notice that they contain regular triples. As we have already used several times, if $AB + n = R^2$, then $\{A, B, A + B + 2R\}$ and $\{A, B, A + B - 2R\}$ are $D(n)$-triples, because $A(A + B \pm 2R) + n = (A \pm R)^2$, $B(A + B \pm 2R) + n = (B \pm R)^2$. Let $cd + n_1 = r^2$ and $cd + n_2 = s^2$. If $c + d - 2r = 7$ and $c + d - 2s = -1$, then $\{7, c, d\}$ is a $D(n_1)$-triple, while $\{-1, c, d\}$ is a $D(n_2)$-triple. The remaining six conditions from the definition of $D(n_i)$-quadruples remain to be satisfied. In [156], it was shown that these conditions can be satisfied parametrically. Thus, we get rational quadruples of the form $\{-1, 7, c, d\}$ that depend on one rational parameter u. If we put $u = v/w$ and eliminate the denominators, we get the following result.

Proposition 5.4.5 *Let v and w be coprime integers such that*

$$v/w \notin \{0, 1, -1, 2, -2, 3, 4, -5, 7, -7, 7/2, -7/2, 7/3, 7/4, -7/5, \infty\}.$$

Then the set

$$\{-(-v^2 + 7w^2)^2, \; 7(-v^2 + 7w^2)^2, \; -(-2v^2 + vw + 7w^2)(2v^2 - 3vw + 7w^2),$$
$$(v^2 - 3vw + 14w^2)(-v^2 - vw + 14w^2)\} \tag{5.33}$$

is a $D(n_1)$-quadruple and a $D(n_2)$-quadruple for

$$n_1 = 4(-v^2 + 7w^2)^2(2v^4 - v^3w - 20v^2w^2 - 7vw^3 + 98w^4),$$
$$n_2 = 4(-v^2 + 7w^2)^2(2v^2 - 7vw + 14w^2)(v^2 + 7w^2).$$

It is clear that if we choose v and w to be solutions of Pell's equation $v^2 - 7w^2 = 1$, then formula (5.4.5) gives integer quadruples of the form $\{-1, 7, c, d\}$.

Table 5.3 $D(n_1)$- and $D(n_2)$-quadruples of the form $\{-1, 7, c, d\}$

$\{a, b, c, d\}$	$\{n_1, n_2\}$
$-1, 7, 119, 64$	$128, 848$
$-1, 7, 4879, 4484$	$6248, 43688$
$-1, 7, 23407, 22532$	$30632, 214376$
$-1, 7, 1191959, 1185664$	$1585088, 11095568$
$-1, 7, 5840864, 5826919$	$7778528, 54449648$
$-1, 7, 302003332, 301903007$	$402604232, 2818229576$
$-1, 7, 1481896324, 1481674079$	$1975713608, 13829995208$
$-1, 7, 76695715424, 76694116519$	$102259887968, 715819215728$
$-1, 7, 376369378007, 376365836032$	$501823476032, 3512764332176$
$-1, 7, 19480219444399, 19480193962244$	$25973608937768, 181815262564328$
$-1, 7, 95595918622159, 95595862172804$	$127461187196648, 892228310376488$

Furthermore, if we take v and w to be solutions of the Pellian equation $v^2 - 7w^2 = 2$ (which has infinitely many solutions because $(v, w) = (3, 1)$ is a solution), then all elements of (5.4.5) are divisible by 4, while the corresponding n_1 and n_2 are divisible by 16, and then dividing all the elements of the quadruple by the common factor 4, we get again an integer quadruple of the form $\{-1, 7, c, d\}$. In Table 5.3, we give some examples of quadruples obtained in the described manner:

5.4.3   Doubly Regular Diophantine Quadruples

Let us notice that in the quadruples $\{a, b, c, d\}$ from the previous subsection, the elements a and b always had the opposite signs. Therefore, we can ask whether there are quadruples with positive elements that are simultaneously $D(n_1)$- and $D(n_2)$-quadruples for $n_1 \neq n_2$, and whether there are infinitely many such quadruples. Furthermore, we can ask whether there are quadruples that are $D(n)$-quadruples for three distinct integers n. An affirmative answer to these questions was given in 2020 by Dujella and Petričević [157]. However, we emphasize that in order to get an affirmative answer to the second question, it should be allowed for one of the values of n to be equal to 0. More precisely, the following result was obtained.

Theorem 5.4.6 *There are infinitely many non-equivalent quadruples of positive integers $\{a, b, c, d\}$ which are $D(n_1)$- and $D(n_2)$-quadruples for distinct non-zero squares n_1 and n_2. Moreover, we can take that all the elements of these quadruples are squares, so they are also $D(0)$-quadruples.*

In [157], explicit formulas for quadruples which satisfy the conditions from Theorem 5.4.6 were derived. We give here one of them.

Let t be an integer such that $t \neq 0, \pm 1, \pm 2$, and let

$$a = (t-1)^2(t-2)^2(t+2)^2(3t^6 - 2t^5 - 13t^4 + 8t^3 + 16t^2 - 16)^2$$
$$\times (5t^6 - 6t^5 - 27t^4 + 40t^3 + 32t^2 - 64t + 16)^2,$$
$$b = 64t^2(t-1)^2(t-2)^2(t+2)^2(t^3 - t^2 - 3t + 4)^2(t^2 - 2)^2$$
$$\times (t^3 - t^2 - 2t + 4)^2(2t^4 - t^3 - 7t^2 + 4t + 4)^2,$$
$$c = t^2(t-1)^2(t^2 - 3)^2(t^6 - 6t^5 - 3t^4 + 28t^3 - 8t^2 - 32t + 16)^2$$
$$\times (4t^7 - 5t^6 - 26t^5 + 39t^4 + 48t^3 - 88t^2 - 16t + 48)^2,$$
$$d = (t+1)^2(t^3 - t^2 - 3t + 4)^2(t^6 + 2t^5 - 7t^4 + 8t^2 - 16t + 16)^2$$
$$\times (4t^7 - 7t^6 - 22t^5 + 49t^4 + 20t^3 - 88t^2 + 32t + 16)^2.$$

Then $\{a, b, c, d\}$ is a $D(n_1)$-, $D(n_2)$-, and $D(n_3)$-quadruple, where

$$n_1 = 16t^2(t+1)^2(t-2)^4(t+2)^4(t-1)^6(t^2 - 3)^2$$
$$\times (t^3 - t^2 - 2t + 4)^2(t^3 - t^2 - 3t + 4)^2(2t^4 - t^3 - 7t^2 + 4t + 4)^2$$
$$\times (3t^6 - 2t^5 - 13t^4 + 8t^3 + 16t^2 - 16)^2$$
$$\times (5t^6 - 6t^5 - 27t^4 + 40t^3 + 32t^2 - 64t + 16)^2,$$
$$n_2 = 4t^2(t^2 - 2)^2(t^3 - t^2 - 3t + 4)^2(t^6 + 2t^5 - 7t^4 + 8t^2 - 16t + 16)^2$$
$$\times (t^6 - 6t^5 - 3t^4 + 28t^3 - 8t^2 - 32t + 16)^2$$
$$\times (4t^7 - 5t^6 - 26t^5 + 39t^4 + 48t^3 - 88t^2 - 16t + 48)^2$$
$$\times (4t^7 - 7t^6 - 22t^5 + 49t^4 + 20t^3 - 88t^2 + 32t + 16)^2,$$
$$n_3 = 0.$$

For example, if we take $t = 3$, after canceling common factors, we get the set

$$\{46190^2, \ 120120^2, \ 126684^2, \ 297388^2\}$$

which is a $D(19022889600^2)$-, $D(10988337906^2)$-, and $D(0)$-quadruple.

The main idea of the proof consists of the construction of the so-called *double regular quadruples*. A rational $D(1)$-quadruple $\{a, b, c, d\}$ is said to be *regular* if it satisfies the equation

$$(a + b - c - d)^2 = 4(ab + 1)(cd + 1). \tag{5.34}$$

A quadruple $\{a,b,c,d\}$ is doubly regular if it is a rational $D(1)$ and $D(x^2)$-quadruple for $x^2 \neq 1$, such that $\{a,b,c,d\}$ and $\{a/x, b/x, c/x, d/x\}$ are both regular rational $D(1)$-quadruples.

The regularity condition for the quadruple $\{a/x, b/x, c/x, d/x\}$ implies the following equation for x:

$$4x^4 + (-a^2 + 2ab + 2ad - b^2 + 2bc + 2ac - c^2 + 2cd - d^2 + 2bd)x^2 + 4abcd = 0.$$
$$(5.35)$$

By inserting the regularity condition for $\{a,b,c,d\}$, i.e. (5.34), in the coefficient of x^2 in (5.35), we get

$$4(x^2 - 1)(x^2 - abcd) = 0.$$

Since we are looking for solutions where $x^2 \neq 1$, we conclude that $x^2 = abcd$. Now $ab + x^2 = ab(1 + cd) = \square$ implies that ab is a square (and analogously ac, ad, bc, bd and cd are squares, which means that $\{a,b,c,d\}$ is a $D(0)$-quadruple). We use the previously mentioned parametrization of rational $D(1)$-triples from [228],

$$a = \frac{2t_1(1 + t_1 t_2(1 + t_2 t_3))}{(-1 + t_1 t_2 t_3)(1 + t_1 t_2 t_3)},$$

$$b = \frac{2t_2(1 + t_2 t_3(1 + t_3 t_1))}{(-1 + t_1 t_2 t_3)(1 + t_1 t_2 t_3)},$$

$$c = \frac{2t_3(1 + t_3 t_1(1 + t_1 t_2))}{(-1 + t_1 t_2 t_3)(1 + t_1 t_2 t_3)},$$

modified by the following substitutions:

$$t_1 = \frac{k}{t_2 t_3},$$

$$t_2 = m - \frac{1}{t_3}.$$

After calculating d from the regularity condition, the remaining condition that $abcd$ is a perfect square can be expressed in terms of an elliptic curve over $\mathbb{Q}(t)$ of positive rank. One of the points of infinite order on that curve gives the above-mentioned parametric family of quadruples with the required property.

Somewhat simpler concrete examples can be found by a "brute force" search over the parameters k, m, t_3 with small numerators and denominators that satisfy the condition that $abcd$ is a perfect square. The simplest example obtained in this way is $\{1458, 66248, 5000, 14112\}$ which is a $D(16769025)$-, $D(406425600)$-, and $D(0)$-quadruple. By multiplying the elements of this set by 2, we get the quadruple

$$\{54^2, 364^2, 100^2, 168^2\}$$

whose elements are squares.

5.4.4 $D(n)$-Quintuples with Square Elements

Having considered triples and quadruples, we can now ask if there are quintuples that are simultaneously $D(n_1)$- and $D(n_2)$-quintuples for $n_1 \neq n_2$. To give an affirmative answer to this question, we will have to allow one of the numbers n_1 and n_2 to be equal to 0. So it remains an open question whether there are such examples of quintuples in which both n_1 and n_2 are different from zero.

Let us also note here that if $\{a, b, c, d, e\}$ is a $D(n_1)$-quintuple and u is a rational number different from zero, then $\{ua, ub, uc, ud, ue\}$ is a $D(n_1 u^2)$-quintuple and we say that two such quintuples are equivalent.

The following theorem is the main result of the paper [141] by Dujella, Kazalicki, and Petričević from 2021.

Theorem 5.4.7 *There are infinitely many non-equivalent quintuples that have the property $D(n_1)$ for some $n_1 \in \mathbb{N}$, such that all elements of the quintuple are squares. So there are infinitely many non-equivalent quintuples that are simultaneously $D(n_1)$-quintuples and $D(0)$-quintuples.*

To prove Theorem 5.4.7, it suffices to prove that there are infinitely many rational Diophantine quintuples with the property that the product of any two of their elements is a perfect square. Namely, it is clear that every rational Diophantine quintuple is equivalent to some $D(u^2)$-quintuple, where u is a multiple of the common denominator of the elements of the rational quintuple.

The construction is motivated by an experimentally found example of a rational Diophantine quintuple with the specified property:

$$\left\{ \frac{3375}{32032}, \frac{143}{840}, \frac{2457}{1760}, \frac{5632}{1365}, \frac{4459}{330} \right\}. \tag{5.36}$$

By multiplying all elements of quintuple (5.36) by 480480, we get the $D(480480^2)$-quintuple

$$\{225^2, 286^2, 819^2, 1408^2, 2548^2\}$$

with the desired property.

Analyzing quintuple (5.36), we can see that it contains the sub-quadruple $\{\frac{3375}{32032}, \frac{143}{840}, \frac{5632}{1365}, \frac{4459}{330}\}$ which has the property that the product of all its elements is equal to 1, and two regular sub-quadruples $\{\frac{3375}{32032}, \frac{143}{840}, \frac{2457}{1760}, \frac{5632}{1365}\}$ and $\{\frac{3375}{32032}, \frac{2457}{1760}, \frac{5632}{1365}, \frac{4459}{330}\}$.

Motivated by this example, we will say that a rational Diophantine quintuple $\{a, b, c, d, e\}$ is *exotic* if $abcd = 1$, the quadruples $\{a, b, d, e\}$ and $\{a, c, d, e\}$ are regular, and if the product of any two elements of the quintuple is a perfect square. We will show that there are infinitely many exotic quintuples.

Our starting point is the following characterization of the rational Diophantine quadruples $\{a, b, c, d\}$ such that $abcd = 1$.

Proposition 5.4.8 *Let $\{a, b, c, d\}$ be a rational Diophantine quadruple such that $abcd = 1$. Then there exist $r, s, t \in \mathbb{Q}$ such that*

$$a = xyz, \quad b = \frac{x}{yz}, \quad c = \frac{y}{xz}, \quad d = \frac{z}{xy},$$

where $x = \frac{t^2-1}{2t}, y = \frac{s^2-1}{2s}$ and $z = \frac{r^2-1}{2r}$. In particular, the product of any two elements of the quadruple is a perfect square.

Proof From $ab + 1 = ab + abcd = ab(1 + cd)$, it follows that ab is a perfect square, and similarly for other products. Let us put $ab = x^2, ac = y^2$ and $ad = z^2$, for $x, y, z \in \mathbb{Q}$. It follows that $a^2 = \frac{ab \cdot ac}{bc} = \frac{x^2 y^2}{1/z^2}$, so $a = xyz$. Analogously, we get $b = \frac{x}{yz}, c = \frac{y}{xz}$ and $d = \frac{z}{xy}$ (with the appropriate choice of signs). Since $x^2 + 1$ is a perfect square, there exists $t \in \mathbb{Q}$ such that $x = \frac{t^2-1}{2t}$, and analogously for y and z. $\qquad\square$

To extend the quadruple $\{a, b, c, d\}$, given in terms of $r, s, t \in \mathbb{Q}$ as in Proposition 5.4.8, to an exotic quintuple, it is sufficient that the triples $\{a, b, d\}$ and $\{a, c, d\}$ have a common regular extension e and that ae is a perfect square.

It can be shown that the regularity condition leads to the condition

$$s = \frac{-1 + r^2 + t + r^2 t}{-1 - r^2 - t + r^2 t}.$$

It remains to satisfy the condition that ae is a perfect square. This condition leads to several curves of genus 0. One of them is the curve given by the equation

$$r^2 t^2 + 3r^2 - t^2 + 2t - 1 = 0,$$

whose parametric solution is

$$(r, t) = \left(-\frac{2u + 1}{u^2 + u + 1}, \frac{u^2 + 4u + 1}{(u - 1)(u + 1)}\right).$$

Then the condition that ae is a square gives the following quartic

$$v^2 = -48(u^2 - 3u - 1)(u^2 + 5u + 3).$$

It has a rational point $(u, v) = (0, 12)$, so it is equivalent to the elliptic curve

$$y^2 + xy + y = x^3 - x^2 - 41x + 96,$$

of rank 1 (a generator is $P = (2, 3)$) and torsion group isomorphic to $\mathbb{Z}/2\mathbb{Z} \times \mathbb{Z}/2\mathbb{Z}$. The point P (and the same applies to the points $P + T$, where T is a torsion point) does not solve our problem because the corresponding values of u are roots of the denominators that appear in the expressions for a, b, c, d. The point $2P = (9, 15)$ on the elliptic curve gives the point $(u, v) = (3, 36)$ on the quartic which corresponds to the parameters $(r, s, t) = (-\frac{7}{13}, -\frac{7}{8}, \frac{11}{4})$ and gives (permuted) our initial $D(1)$-quintuple

$$\left\{ \frac{3375}{32032}, \frac{4459}{330}, \frac{143}{840}, \frac{5632}{1365}, \frac{2457}{1760} \right\}.$$

The points $2P + T$, where T is a torsion point, give equivalent solutions (with permuted elements or elements multiplied by -1). The point $3P = (-\frac{262}{49}, \frac{4782}{343})$ on the elliptic curve gives the point $(u, v) = (-\frac{28}{37}, -\frac{5916}{1369})$ on the quartic corresponding to the parameters $(r, s, t) = (\frac{703}{1117}, -\frac{703}{585}, \frac{1991}{585})$ and gives the $D(1)$-quintuple

$$\left\{ \frac{126066472448}{914609323485}, \frac{388078111459}{22127635530}, \frac{131212873}{529696440}, \frac{398601435375}{238691175968}, \right.$$

$$\left. \frac{19085447/0299}{39338018720} \right\}.$$

It is clear that this construction gives infinitely many quintuples with the desired property by taking the multiples $mP = (x, y), m \geq 2$, calculating the corresponding values for u from $u = \frac{-3x+18}{-y+3x-15}$ and inserting them in the formulas for a, b, c, d from Proposition 5.4.8 and the formula for e from the regularity condition:

$$a = \frac{(u + 2)^3 (u^2 - 2u - 2)u^3}{2(2u + 1)(u - 1)(u + 1)(u^2 + u + 1)(u^2 + 4u + 1)},$$

$$b = \frac{8(u^2 + u + 1)(2u + 1)^3}{u(u - 1)(u + 2)(u + 1)(u^2 - 2u - 2)(u^2 + 4u + 1)},$$

$$c = \frac{(u^2 + u + 1)(u^2 - 2u - 2)(u^2 + 4u + 1)}{2u(u - 1)(u + 2)(u + 1)(2u + 1)},$$

$$d = \frac{(u+1)^3(u^2+4u+1)(u-1)^3}{2u(u+2)(u^2-2u-2)(2u+1)(u^2+u+1)},$$

$$e = -\frac{3(2u^7+7u^6-17u^5-60u^4-85u^3-64u^2-23u-3)}{2u(u^7+4u^6-6u^5-32u^4-17u^3+24u^2+22u+4)}.$$

Although we found an infinite number of rational quintuples with the property $D(0)$, i.e. such that the product of any two of their elements is a perfect square, it remains an open question whether there is a rational Diophantine quintuple whose elements are all squares. On the other hand, it is known that there are infinitely many rational Diophantine quadruples whose all elements are squares. For example, the following two-parameter family of quadruples has this property:

$$a = \frac{3^2(s-1)^2(s+1)^2v^2}{2^2(2s^3-2s+v^2)^2},$$

$$b = \frac{v^2(-4s^3+4s+v^2)^2}{2^2(s+1)^2(s-1)^2(-s^3+s+v^2)^2},$$

$$c = \frac{(2s^3-2s+v^2)^2}{3^2v^2s^2},$$

$$d = \frac{4^2(-s^3+s+v^2)^2s^2}{v^2(-4s^3+4s+v^2)^2}.$$

This family is obtained by taking that $t = 1/(r-1)$ in the notation from Proposition 5.4.8 so that it satisfies $abcd = 1$. There are also examples of rational Diophantine quadruples $\{a, b, c, d\}$ whose elements are squares, and for which $abcd \neq 1$, e.g.

$$\left\{\left(\frac{18}{77}\right)^2, \left(\frac{55}{96}\right)^2, \left(\frac{56}{15}\right)^2, \left(\frac{340}{77}\right)^2\right\}.$$

Let us recall here that in [139], Dujella, Kazalicki, and Petričević showed that there are infinitely many rational Diophantine sextuples such that the denominators of all elements (in the lowest terms) are perfect squares (see Sect. 3.5).

5.5 Exercises

1. Prove that there is no integer n and $D(n)$-quadruple $\{a, b, c, d\}$ such that $a \equiv 1$ (mod 4), $b \equiv 2$ (mod 4), $c \equiv 3$ (mod 4) or $a \equiv 0$ (mod 4), $b \equiv 1$ (mod 4), $c \equiv 2$ (mod 4) (see [283]).

2. Prove that for any positive integer n, the set

$$\{F_{2n},\ F_{2n+6},\ 4F_{2n+2},\ 4F_{2n+1}F_{2n+3}F_{2n+4}\}$$

   is a $D(4)$-quadruple.
3. Prove that for any positive integer k, the sets

$$\{k,\ 9k+8,\ 16k+14,\ 25k+20\} \quad \text{and} \quad \{k,\ 4k+2,\ 9k+8,\ 25k+20\}$$

   are $D(10k+9)$-quadruples.
4. Find all $D(n)$-quintuples for $|n| \le 400$ with all elements less than n^2.
5. Find a non-zero integer n and a $D(n)$-quintuple $\{a, b, c, d, e\}$ (mixed signs allowed) such that $a \equiv b \equiv c \equiv d \equiv e \equiv 1 \pmod 4$.
6. Find an example of positive integers a, b, c, n such that $\{a, b, c\}$ is a $D(n)$-triple and $c < \frac{2ab}{n^2}$.
7. Show that there exist infinite sequences (a_k), (b_k), (c_k), (n_k) of positive integers such that $\{a_k, b_k, c_k\}$ is a $D(n_k)$-triple and $\lim_{k \to \infty} \frac{cn_k^2}{ab} = \frac{4}{3}$. (Hint: It may be taken that a_k, b_k, c_k are polynomials of degree 4 and n_k a polynomial of degree 2 in k.)
8. Let $n \ge 2$ be an integer. Prove that there is a $D((n^2+1)(n^2+9))$-quadruple of the form $\{a, b, -a, -b\}$, with positive integers a and b.
9. Let k be a positive integer. Let

$$b = 256k^4 - 128k^3 - 48k^2 + 16k,$$

$$c = 4096k^6 - 4096k^5 - 512k^4 + 1152k^3 + 16k^2 - 88k - 7,$$

$$d = 4096k^6 - 2048k^5 - 1792k^4 + 640k^3 + 288k^2 - 48k - 15.$$

   Show that there exists an integer n such that $|n| < 64k^2$ and $\{1, b, c, d\}$ is a $D(n)$-quadruple (see [155]).
10. Find a positive integer $n > 10^5$ and a $D(n)$-quadruple $\{a, b, c, d\}$ (mixed signs are allowed) such that $\max\{|a|, |b|, |c|, |d|\} < 2|n|^{2/5}$.
11. Prove that the function $f(x) = \frac{\log x}{\sqrt{x+3}}$ is strictly decreasing for $x > 25$.
12. Let p be an odd prime. Determine the number of Diophantine pairs and Diophantine triples in $\mathbb{F}_p$, depending on whether p is congruent to 1 or 3 modulo 4 (see [136]).
13. Prove that the rational $D(16t+9)$-quadruple $\{t, 16t+8, 25t+14, 36t+20\}$ can be extended to a rational $D(16t+9)$-quintuple for infinitely many $t \in \mathbb{Q}$ (see [312]).
14. Prove that for any integer n, there is an infinite set of integers A_n with the property that for all $a, b \in A_n$, the number $ab + n$ is not square-free.
15. Find integers $n_1 < n_2 < n_3 < n_4 < n_5 < n_6$ such that $\{28, 168, 1848\}$ is a $D(n_i)$-triple for $i = 1, 2, 3, 4, 5, 6$.
16. Find positive integers $n_1 \ne n_2$ such that $\{-176, -169, 169, 176\}$ is a $D(n_i)$-quadruple for $i = 1, 2$.

References

1. Adédji, K.N., He, B., Pintér, Á., Togbé, A.: On the Diophantine pair $\{a, 3a\}$. J. Number Theory **227**, 330–351 (2021)
2. Adžaga, N.: On the size of Diophantine m-tuples in imaginary quadratic number rings. Bull. Math. Sci. **11**(1), Paper No. 1950020 (2021)
3. Adžaga, N., Dražić, G., Dujella, A., Pethő, A.: Asymptotics of $D(q)$-pairs and triples via L-functions of Dirichlet characters. Preprint (2023). arXiv: 2304.01775
4. Adžaga, N., Dujella, A., Kreso, D., Tadić, P.: Triples which are $D(n)$-sets for several n's. J. Number Theory **184**, 330–341 (2018)
5. Aguirre, J., Dujella, A., Jukić Bokun, M., Peral, J.C.: High rank elliptic curves with prescribed torsion group over quadratic fields. Period. Math. Hungar. **68**, 222–230 (2014)
6. Aguirre, J., Dujella, A., Peral, J.C.: On the rank of elliptic curves coming from rational Diophantine triples. Rocky Mountain J. Math. **42**, 1759–1776 (2012)
7. Aguirre, J., Lozano-Robledo, Á., Peral, J.C.: Elliptic curves of maximal rank. In: Proceedings of the Segundas Jornadas de Teoria de Numeros, pp. 1–28. Bibl. Rev. Mat. Iberoamericana, Madrid (2008)
8. Alaca, S., Williams, K.S.: Introductory Algebraic Number Theory. Cambridge University Press, Cambridge (2004)
9. Aleksentsev, Y.M.: The Hilbert polynomial and linear forms in the logarithms of algebraic numbers. Izv. Math. **72**, 1063–1110 (2008)
10. Alp, M., Irmak, N., Szalay, L.: Reduced diophantine quadruples with the binary recurrence $G_n = AG_{n-1} - G_{n-2}$. An. Ştiinţ. Univ. "Ovidius" Constanţa Ser. Mat. **23**, 23–31 (2015)
11. Andreescu, T., Andrica, D.: Quadratic Diophantine Equations. Springer, New York (2015)
12. Arkin, J., Bergum, G.E.: More on the problem of Diophantus. In: Philippou, A.N., Horadam, A.F., Bergum, G.E. (eds.) Application of Fibonacci Numbers, vol. 2, pp. 177–181. Kluwer, Dordrecht (1988)
13. Arkin, J., Hoggatt, V.E., Strauss, E.G.: On Euler's solution of a problem of Diophantus. Fibonacci Q. **17**, 333–339 (1979)
14. Artin, M., Rodriguez-Villegas, F., Tate, J.: On the Jacobians of plane cubics. Adv. Math. **198**, 366–382 (2005)
15. Atkin, A.O.L., Morain, F.: Finding suitable curves for the elliptic curve method of factorization. Math. Comput. **60**, 399–405 (1993)
16. Baćić, Lj., Filipin, A.: On the extendibility of $D(4)$-pair $\{k - 2, k + 2\}$. J. Comb. Number Theory **5**, 181–197 (2013)
17. Baker, A.: Rational approximations to $\sqrt[3]{2}$ and other algebraic numbers. Q. J. Math. Oxford Ser. (2) **15**, 375–383 (1964)

18. Baker, A.: Simultaneous rational approximations to certain algebraic numbers. Proc. Camb. Philos. Soc. **63**, 693–702 (1967)
19. Baker, A.: Linear forms in the logarithms of algebraic numbers. IV. Mathematika **15**, 204–216 (1968)
20. Baker, A.: The diophantine equation $y^2 = ax^3 + bx^2 + cx + d$. J. Lond. Math. Soc. **43**, 1–9 (1968)
21. Baker, A.: Transcendental Number Theory. Cambridge University Press, Cambridge (1990)
22. Baker, A.: A Concise Introduction to the Theory of Numbers. Cambridge University Press, Cambridge (1994)
23. Baker, A.: A Comprehensive Course in Number Theory. Cambridge University Press, Cambridge (2012)
24. Baker, A., Davenport, H.: The equations $3x^2 - 2 = y^2$ and $8x^2 - 7 = z^2$. Q. J. Math. Oxford Ser. (2) **20**, 129–137 (1969)
25. Baker, A., Wüstholz, G.: Logarithmic forms and group varieties. J. Reine Angew. Math. **442**, 19–62 (1993)
26. Baker, A., Wüstholz, G.: Logarithmic Forms and Diophantine Geometry. Cambridge University Press, Cambridge (2008)
27. Barbeau, E.J.: Pell's Equation. Springer, New York (2003)
28. Bashmakova, I.G.: Diophantus of Alexandria. Arithmetica and the Book of Polygonal Numbers. Introduction and Comments. Nauka, Moscow (1974), (in Russian)
29. Becker, R., Murty, M.R.: Diophantine m-tuples with the property $D(n)$. Glas. Mat. Ser. III **54**, 65–75 (2019)
30. Bennett, M.A.: On the number of solutions of simultaneous Pell equations. J. Reine Angew. Math. **498**, 173–199 (1998)
31. Bérczes, A., Dujella, A., Hajdu, L., Luca, F.: On the size of sets whose elements have perfect power n-shifted products. Publ. Math. Debrecen **79**, 325–339 (2011)
32. Bérczes, A., Dujella, A., Hajdu, L., Tengely, S.: Finiteness results for F-Diophantine sets. Monatsh. Math. **180**, 469–484 (2016)
33. Bérczes, A., Luca, F., Pink, I., Ziegler, V.: Finiteness results for Diophantine triples with repdigit values. Acta Arith. **172**, 133–148 (2016)
34. Bernstein, D.J., Lange, T.: Faster addition and doubling on elliptic curves. In: Lecture Notes in Comput. Sci., vol. 4833, pp. 29–50. Springer, Berlin (2007)
35. Bhargava, M., Shankar, A,: The average size of the 5-Selmer group of elliptic curves is 6, and the average rank is less than 1. Preprint (2013). arXiv: 1312.7859
36. Bhattacharjee, S., Dixit, A.B., Saikia, D.: An effective bound on generalized Diophantine m-tuples. Bull. Aust. Math. Soc. **109**, 242–253 (2024)
37. Bilu, Yu.F., Hanrot, G.: Solving Thue equations of high degree. J. Number Theory **60**, 373–392 (1996)
38. Birch, B.J.: Elliptic curves and modular functions. In: Symposia Mathematica, vol. IV, pp. 27–32. Academic Press, London (1970)
39. Birch, B.J., Swinnerton-Dyer, H.P.F.: Notes on elliptic curves. I. J. Reine Angew. Math. **212**, 7–25 (1963)
40. Blake, I., Seroussi, G., Smart, N.: Elliptic Curves in Cryptography. Cambridge University Press, Cambridge (1999)
41. Bliznac Trebješanin, M., Filipin, A.: Nonexistence of $D(4)$-quintuples. J. Number Theory **194**, 170–217 (2019)
42. Bober, J.W.: Conditionally bounding analytic ranks of elliptic curves. In: ANTS X – Proceedings of the Tenth Algorithmic Number Theory Symposium, pp. 135–144. Mathematical Sciences Publishers, Berkeley (2013)
43. Bombieri, E., Gubler, W.: Heights in Diophantine Geometry. Cambridge University Press, Cambridge (2006)
44. Bonciocat, N.C., Cipu, M., Mignotte, M.: There is no Diophantine $D(-1)$-quadruple. J. Lond. Math. Soc. **105**, 63–99 (2022)

45. Bosma, W., Cannon, J.J., Fieker, C., Steel, A.: Handbook of Magma Functions, Edition 2.28 (2023)

46. Bosman, J., Bruin, P., Dujella, A., Najman, F.: Ranks of elliptic curves with prescribed torsion over number fields. Int. Math. Res. Not. IMRN **2014**, 2885–2923 (2014)

47. Brown, E.: Sets in which $xy + k$ is always a square. Math. Comput. **45**, 613–620 (1985)

48. Bugeaud, Y.: Linear Forms in Logarithms and Applications. IRMA Lectures in Mathematics and Theoretical Physics, vol. 28. European Mathematical Society, Zürich (2018)

49. Bugeaud, Y., Dujella, A.: On a problem of Diophantus for higher powers. Math. Proc. Camb. Philos. Soc. **135**, 1–10 (2003)

50. Bugeaud, Y., Dujella, A., Mignotte, M.: On the family of Diophantine triples $\{k - 1, k + 1, 16k^3 - 4k\}$. Glasgow Math. J. **49**, 333–344 (2007)

51. Bugeaud, Y., Mignotte, M., Siksek, S.: Classical and modular approaches to exponential Diophantine equations. I. Fibonacci and Lucas perfect powers. Ann. Math. (2) **163**, 969–1018 (2006)

52. Bundschuh, P.: Einführung in die Zahlentheorie. Springer, Berlin (2008)

53. Caporaso, L., Harris, J., Mazur, B.: Uniformity of rational points. J. Amer. Math. Soc. **10**, 1–35 (1997)

54. Carmichael, R.D.: Diophantine Analysis. Dover, New York (1959)

55. Cassels, J.W.S.: Arithmetic on curves of genus 1. III. The Tate-Šafareviè and Selmer groups. Proc. Lond. Math. Soc. (3) **12**, 259–296 (1962)

56. Cassels, J.W.S.: Lectures on Elliptic Curves. Cambridge University Press, Cambridge (1995)

57. Chahal, J.S.: Topics in Number Theory. Plenum Press, New York (1988)

58. Chakraborty, K., Gupta, S., Hoque, A.: Diophantine triples with the property $D(n)$ for distinct n's. Mediterr. J. Math. **20**, Article 31 (2023)

59. Chakraborty, K., Gupta, S., Hoque, A.: On a conjecture of Franušić and Jadrijević: Counterexamples. Results Math. **78**, Article 18 (2023)

60. Chebyshev, P.L., Sur les formes quadratiques. J. Math. Pures Appl. **16**, 257–282 (1851)

61. Christianidis, J., Oaks, J.: The Arithmetica of Diophantus. A Complete Translation and Commentary. Routledge, London (2023)

62. Cipu, M.: Further remarks on Diophantine quintuples. Acta Arith. **168**, 201–219 (2015)

63. Cipu, M., Dujella, A., Fujita, Y.: Extensions of a Diophantine triple by adjoining smaller elements. Mediterr. J. Math. **19**, Article 187 (2022)

64. Cipu, M., Filipin, A., Fujita, Y.: Bounds for Diophantine quintuples II. Publ. Math. Debrecen **88**, 59–78 (2016)

65. Cipu, M., Fujita, Y.: Bounds for Diophantine quintuples. Glas. Mat. Ser. III **50**, 25–34 (2015)

66. Cipu, M., Fujita, Y.: On the length of $D(\pm 1)$-tuples in imaginary quadratic rings. Bull. Lond. Math. Soc. **56**, 274–287 (2024)

67. Cipu, M., Fujita, Y., Miyazaki, T.: On the number of extensions of a Diophantine triple. Int. J. Number Theory **14**, 899–917 (2018)

68. Cipu, M., Trudgian, T.: Searching for Diophantine quintuples. Acta Arith. **173**, 365–382 (2016)

69. Cohen, H.: A Course in Computational Algebraic Number Theory. Springer, Berlin (1993)

70. Cohen, H.: Number Theory. Volume I: Tools and Diophantine Equations. Springer, New York (2007)

71. Cohn, J.H.E.: Lucas and Fibonacci numbers and some Diophantine equations. Proc. Glasgow Math. Assoc. **7**, 24–28 (1965)

72. Cohn, J.H.E.: The length of the period of the simple continued fraction of $d^{1/2}$. Pacific J. Math. **71**, 21–32 (1977)

73. Connell, I.: Elliptic Curve Handbook. McGill University, Montreal (1999)

74. Cremona, J.E.: Algorithms for Modular Elliptic Curves. Cambridge University Press, Cambridge (1997)

75. Cremona, J.E.: Elliptic Curve Data (2020). http://johncremona.github.io/ecdata/

76. Davenport, H., Roth, K.F.: Rational approximations to algebraic numbers. Mathematika **2**, 160–167 (1955)

77. David, S.: Minorations de formes linéaires de logarithmes elliptiques. Mém. Soc. Math. France (N.S.) **62**, 1–143 (1995)
78. de Weger, B.M.M.: Algorithms for Diophantine Equations. Centrum voor Wiskunde en Informatica, Amsterdam (1989)
79. Denton, A.D.: A holiday brain teaser. The Sunday Times, 4th August 1957, 18th August 1957
80. Deshpande, M.N.: One interesting family of Diophantine triplets. Int. J. Math. Educ. Sci. Technol. **33**, 253–256 (2002)
81. Deshpande, M.N.: Wonderland of Families of Diophantine Triples. Nagpur (2022)
82. Deshpande, M.N, Dujella, A.: An interesting property of a recurrence related to the Fibonacci sequence. Fibonacci Q. **40**, 157–160 (2002)
83. Dickson, L.E.: History of the Theory of Numbers, Volume 2: Diophantine Analysis. Chelsea, New York (1966)
84. Dixit, A.B., Kim, S., Murty, M.R.: Generalized Diophantine m-tuples. Proc. Amer. Math. Soc. **150**, 1455–1465 (2022)
85. Dörge, K.: Über die Seltenheit der reduziblen Polynome und der Normalgleichungen. Math. Ann. **95**, 247–256 (1926)
86. Dražić, G.: Rational $D(q)$-quintuples. Rev. R. Acad. Cienc. Exactas Fis. Nat. Ser. A Math. RACSAM **116**, Article 9 (2022)
87. Dražić, G., Kazalicki, M.: Rational $D(q)$-quadruples. Indag. Math. (N.S.) **33**, 440–449 (2022)
88. Dujella, A.: Generalization of a problem of Diophantus. Acta Arith. **65**, 15–27 (1993)
89. Dujella, A.: Diophantine quadruples for squares of Fibonacci and Lucas numbers. Portugaliae Math. **52**, 305–318 (1995)
90. Dujella, A.: Generalized Fibonacci numbers and the problem of Diophantus. Fibonacci Q. **34**, 164–175 (1996)
91. Dujella, A.: Some polynomial formulas for Diophantine quadruples. Grazer Math. Ber. **328**, 25–30 (1996)
92. Dujella, A.: On Diophantine quintuples. Acta Arith. **81**, 69–79 (1997)
93. Dujella, A.: The problem of the extension of a parametric family of Diophantine triples. Publ. Math. Debrecen **51**, 311–322 (1997)
94. Dujella, A.: Some estimates of the number of Diophantine quadruples. Publ. Math. Debrecen **53**, 177–189 (1998)
95. Dujella, A.: A problem of Diophantus and Pell numbers. In: Bergum, G.E., Philippou, A.N., Horadam, A.F. (eds.) Application of Fibonacci Numbers, vol. 7, pp. 61–68. Kluwer, Dordrecht (1998)
96. Dujella, A.: A problem of Diophantus and Dickson's conjecture. In: Győry, K., Pethő, A., Sós, V.T. (eds.) Number Theory, Diophantine, Computational and Algebraic Aspects, pp. 147–156. Walter de Gruyter, Berlin (1998)
97. Dujella, A.: An extension of an old problem of Diophantus and Euler. Fibonacci Q. **37**, 312–314 (1999)
98. Dujella, A.: A proof of the Hoggatt-Bergum conjecture. Proc. Amer. Math. Soc. **127**, 1999–2005 (1999)
99. Dujella, A.: Diophantine triples and construction of high-rank elliptic curves over $\mathbb{Q}$ with three non-trivial 2-torsion points. Rocky Mountain J. Math. **30**, 157–164 (2000)
100. Dujella, A.: A note on Diophantine quintuples. In: Halter-Koch, F., Tichy, R.F. (eds.) Algebraic Number Theory and Diophantine Analysis, pp. 123–127. Walter de Gruyter, Berlin (2000)
101. Dujella, A.: A parametric family of elliptic curves. Acta Arith. **94**, 87–101 (2000)
102. Dujella, A.: Irregular Diophantine m-tuples and elliptic curves of high rank. Proc. Japan Acad. Ser. A Math. Sci. **76**, 66–67 (2000)
103. Dujella, A.: Diophantine m-tuples and elliptic curves. J. Théor. Nombres Bordeaux **13**, 111–124 (2001)
104. Dujella, A.: An absolute bound for the size of Diophantine m-tuples. J. Number Theory **89**, 126–150 (2001)

105. Dujella, A.: An extension of an old problem of Diophantus and Euler. II. Fibonacci Q. **40**, 118–123 (2002)
106. Dujella, A.: On the size of Diophantine m-tuples. Math. Proc. Camb. Philos. Soc. **132**, 23–33 (2002)
107. Dujella, A.: There are only finitely many Diophantine quintuples. J. Reine Angew. Math. **566**, 183–214 (2004)
108. Dujella, A.: Bounds for the size of sets with the property $D(n)$. Glas. Mat. Ser. III **39**, 199–205 (2004)
109. Dujella, A.: Continued fractions and RSA with small secret exponent. Tatra Mt. Math. Publ. **29**, 101–112 (2004)
110. Dujella, A.: On Mordell-Weil groups of elliptic curves induced by Diophantine triples. Glas. Mat. Ser. III **42**, 3–18 (2007)
111. Dujella, A.: Conjectures and results on the size and number of Diophantine tuples. In: Komatsu, T. (ed.) Diophantine Analysis and Related Fields (DARF 2007/2008), AIP Conf. Proc., vol. 976, pp. 58–61. Amer. Inst. Phys., Melville (2008)
112. Dujella, A.: On the number of Diophantine m-tuples. Ramanujan J. **15**, 37–46 (2008)
113. Dujella, A.: Rational Diophantine sextuples with mixed signs. Proc. Japan Acad. Ser. A Math. Sci. **85**, 27–30 (2009)
114. Dujella, A.: What is … a Diophantine m-tuple? Notices Amer. Math. Soc. **63**, 772–774 (2016)
115. Dujella, A.: History of elliptic curves rank records (2020). https://web.math.pmf.unizg.hr/~duje/tors/rankhist.html
116. Dujella, A.: Number Theory. Školska knjiga, Zagreb (2021)
117. Dujella, A.: Diophantine m-tuples and elliptic curves. Course webpage (2022). https://web.math.pmf.unizg.hr/~duje/diophell.html
118. Dujella, A.: High rank elliptic curves with prescribed torsion (2022). https://web.math.pmf.unizg.hr/~duje/tors/tors.html
119. Dujella, A.: High rank elliptic curves with prescribed torsion over quadratic fields (2023). https://web.math.pmf.unizg.hr/~duje/tors/torsquad.html
120. Dujella, A.: Infinite families of elliptic curves with high rank and prescribed torsion (2023). https://web.math.pmf.unizg.hr/~duje/tors/generic.html
121. Dujella, A.: Diophantine m-tuples page (2023). https://web.math.pmf.unizg.hr/~duje/dtuples.html
122. Dujella, A., Filipin, A., Fuchs, C.: Effective solution of the $D(-1)$-quadruple conjecture. Acta Arith. **128**, 319–338 (2007)
123. Dujella, A., Fuchs, C.: Complete solution of the polynomial version of a problem of Diophantus. J. Number Theory **106**, 326–344 (2004)
124. Dujella, A., Fuchs, C.: Complete solution of a problem of Diophantus and Euler. J. Lond. Math. Soc. **71**, 33–52 (2005)
125. Dujella, A., Fuchs, C.: On a problem of Diophantus for rationals. J. Number Theory **132**, 2075–2083 (2012)
126. Dujella, A., Fuchs, C., Luca, F.: A polynomial variant of a problem of Diophantus for pure powers. Int. J. Number Theory **4**, 57–71 (2008)
127. Dujella, A., Fuchs, C., Tichy, R.F.: Diophantine m-tuples for linear polynomials. Period. Math. Hungar. **45**, 21–33 (2002)
128. Dujella, A., Fuchs, C., Walsh, P.G.: Diophantine m-tuples for linear polynomials. II. Equal degrees. J. Number Theory **120**, 213–228 (2006)
129. Dujella, A., Gusić, I., Lasić, L.: On quadratic twists of elliptic curves $y^2 = x(x-1)(x-\lambda)$. Rad Hrvat. Akad. Znan. Umjet. Mat. Znan. **18**, 27–34 (2014)
130. Dujella, A., Gusić, I., Petričević, V., Tadić, P.: Strong Eulerian triples. Glas. Mat. Ser. III **53**, 33–42 (2018)
131. Dujella, A., Jadrijević, B.: A family of quartic Thue inequalities. Acta Arith. **111**, 61–76 (2004)

132. Dujella, A., Jukić Bokun, M., Soldo, I.: On the torsion group of elliptic curves induced by Diophantine triples over quadratic fields. Rev. R. Acad. Cienc. Exactas Fis. Nat. Ser. A Math. RACSAM **111**, 1177–1185 (2017)
133. Dujella, A., Jurasić, A.: On the size of sets in a polynomial variant of a problem of Diophantus. Int. J. Number Theory **6**, 1449–1471 (2010)
134. Dujella, A., Jurasić, A.: Some Diophantine triples and quadruples for quadratic polynomials. J. Comb. Number Theory **3**(2), 123–141 (2011)
135. Dujella, A., Kazalicki, M.: More on Diophantine sextuples. In: Elsholtz, C., Grabner, P. (eds.) Number Theory - Diophantine Problems, Uniform Distribution and Applications, Festschrift in Honour of Robert F. Tichy's 60th Birthday, pp. 227–235. Springer, Cham (2017)
136. Dujella, A., Kazalicki, M.: Diophantine m-tuples in finite fields and modular forms. Res. Number Theory **7**, Article 3 (2021)
137. Dujella, A., Kazalicki, M., Mikić, M., Szikszai, M.: There are infinitely many rational Diophantine sextuples. Int. Math. Res. Not. IMRN **2017**(2), 490–508 (2017)
138. Dujella, A., Kazalicki, M., Peral, J.C.: Elliptic curves with torsion groups $\mathbb{Z}/8\mathbb{Z}$ and $\mathbb{Z}/2\mathbb{Z} \times \mathbb{Z}/6\mathbb{Z}$. Rev. R. Acad. Cienc. Exactas Fis. Nat. Ser. A Math. RACSAM **115**, Article 169 (2021)
139. Dujella, A., Kazalicki, M., Petričević, V.: Rational Diophantine sextuples with square denominators. J. Number Theory **205**, 340–346 (2019)
140. Dujella, A., Kazalicki, M., Petričević, V.: Rational Diophantine sextuples containing two regular quadruples and one regular quintuple. Acta Math. Spalatensia **1**, 19–27 (2021)
141. Dujella, A., Kazalicki, M., Petričević, V.: $D(n)$-quintuples with square elements. Rev. R. Acad. Cienc. Exactas Fis. Nat. Ser. A Math. RACSAM **115**, Article 172 (2021)
142. Dujella, A., Luca, F.: Diophantine m-tuples for primes. Int. Math. Res. Not. **47**, 2913–2940 (2005)
143. Dujella, A., Luca, F.: On a problem of Diophantus with polynomials. Rocky Mountain J. Math. **37**, 131–157 (2007)
144. Dujella, A., Mikić, M.: On the torsion group of elliptic curves induced by $D(4)$-triples. An. Ştiinţ. Univ. "Ovidius" Constanţa Ser. Mat. **22**, 79–90 (2014)
145. Dujella, A., Mikić, M.: Rank zero elliptic curves induced by rational Diophantine triples. Rad Hrvat. Akad. Znan. Umjet. Mat. Znan. **24**, 29–37 (2020)
146. Dujella, A., Paganin, M., Sadek, M.: Strong rational Diophantine $D(q)$-triples. Indag. Math. (N.S.) **31**, 505–511 (2020)
147. Dujella, A., Peral, J.C.: High rank elliptic curves with torsion $\mathbb{Z}/2\mathbb{Z} \times \mathbb{Z}/4\mathbb{Z}$ induced by Diophantine triples. LMS J. Comput. Math. **17**, 282–288 (2014)
148. Dujella, A., Peral, J.C.: Elliptic curves with torsion group $\mathbb{Z}/8\mathbb{Z}$ or $\mathbb{Z}/2\mathbb{Z} \times \mathbb{Z}/6\mathbb{Z}$. In: Trends in Number Theory, Contemp. Math., vol. 649, pp. 47–62 (2015)
149. Dujella, A., Peral, J.C.: Elliptic curves induced by Diophantine triples. Rev. R. Acad. Cienc. Exactas Fis. Nat. Ser. A Math. RACSAM **113**, 791–806 (2019)
150. Dujella, A., Peral, J.C.: High rank elliptic curves induced by rational Diophantine triples. Glas. Mat. Ser. III **55**, 237–252 (2020)
151. Dujella, A., Peral, J.C.: Construction of high rank elliptic curves. J. Geom. Anal. **31**, 6698–6724 (2021)
152. Dujella, A., Pethő, A.: A generalization of a theorem of Baker and Davenport. Q. J. Math. Oxf. Ser. (2) **49**, 291–306 (1998)
153. Dujella, A., Pethő, A.: Integer points on a family of elliptic curves. Publ. Math. Debrecen **56**, 321–335 (2000)
154. Dujella, A., Petričević, V.: Strong Diophantine triples. Exp. Math. **17**, 83–89 (2008)
155. Dujella, A., Petričević, V.: On the largest element in $D(n)$-quadruples. Indag. Math. (N.S.) **30**, 1079–1086 (2019)
156. Dujella, A., Petričević, V.: Diophantine quadruples with the properties $D(n_1)$ and $D(n_2)$. Rev. R. Acad. Cienc. Exactas Fis. Nat. Ser. A Math. RACSAM **114**, Article 21 (2020)
157. Dujella, A., Petričević, V.: Doubly regular Diophantine quadruples. Rev. R. Acad. Cienc. Exactas Fis. Nat. Ser. A Math. RACSAM **114**, Article 189 (2020)

158. Dujella, A., Saradha, N.: Diophantine m-tuples with elements in arithmetic progressions. Indag. Math. (N.S.) **25**, 131–136 (2014)

159. Dujella, A., Soydan, G.: On elliptic curves induced by rational Diophantine quadruples. Proc. Japan Acad. Ser. A Math. Sci. **98**, 1–6 (2022)

160. Edwards, H.M.: A normal form for elliptic curves. Bull. Amer. Math. Soc. (N.S.) **44**, 393–422 (2007)

161. Elkies, N.D.: $\mathbb{Z}^{28}$ in $E(\mathbb{Q})$, etc. Number Theory Listserver (2006). https://listserv.nodak.edu/cgi-bin/wa.exe?A2=NMBRTHRY;99f4e7cd.0605

162. Elkies, N.D.: Three lectures on elliptic surfaces and curves of high rank. Lecture Notes, Oberwolfach (2007). arXiv: 0709.2908

163. Elkies, N.D., Klagsbrun, Z.: New rank records for elliptic curves having rational torsion. In: Proceedings of the Fourteenth Algorithmic Number Theory Symposium, pp. 233–250. Mathematical Sciences Publishers, Berkeley (2020)

164. Elsholtz, C., Filipin, A., Fujita, Y.: On Diophantine quintuples and $D(-1)$-quadruples. Monatsh. Math. **175**, 227–239 (2014)

165. Eroshkin, Y.G.: Personal communication (2008, 2009)

166. Fermigier, S.: Exemples de courbes elliptiques de grand rang sur $\mathbb{Q}(t)$ et sur $\mathbb{Q}$ possédant des points d'ordre 2. C. R. Acad. Sci. Paris Sér. I **332**, 949–952 (1996)

167. Filipin, A., Fujita, Y.: The number of Diophantine quintuples II. Publ. Math. Debrecen **82**, 293–308 (2013)

168. Filipin, A., Jurasić, A.: A polynomial variant of a problem of Diophantus and its consequences. Glas. Mat. Ser. III **54**, 21–52 (2019)

169. Filipin, A., Szalay, L.: Triangular Diophantine tuples from $\{1, 2\}$. Rad Hrvat. Akad. Znan. Umjet. Mat. Znan. **27**, 55–70 (2023)

170. Fisher, T.: Higher descents on an elliptic curve with a rational 2-torsion point. Math. Comput. **86**, 2493–2518 (2017)

171. Fisher, T.: A formula for the Jacobian of a genus one curve of arbitrary degree. Algebra Number Theory **12**, 2123–2150 (2018)

172. Fisher, T.: On binary quartics and the Cassels-Tate pairing. Res. Number Theory **8**, Paper No. 74 (2022)

173. Franušić, Z.: Diophantine quadruples in $\mathbb{Z}[\sqrt{4k+3}]$. Ramanujan J. **17**, 77–88 (2008)

174. Franušić, Z.: A Diophantine problem in $\mathbb{Z}[(1+\sqrt{d})/2]$. Studia Sci. Math. Hungar. **46**, 103–112 (2009)

175. Franušić, Z.: Diophantine quadruples in the ring of integers of $\mathbb{Q}(\sqrt[3]{2})$. Miskolc Math. Notes **14**, 893–903 (2013)

176. Franušić, Z., Jadrijević, B.: $D(n)$-quadruples in the ring of integers of $\mathbb{Q}(\sqrt{2}, \sqrt{3})$. Math. Slovaca **69**, 1263–1278 (2019)

177. Franušić, Z., Soldo, I.: The problem of Diophantus for integers of $\mathbb{Q}(\sqrt{-3})$. Rad Hrvat. Akad. Znan. Umjet. Mat. Znan. **18**, 15–25 (2014)

178. Friedl, S.: An elementary proof of the group law for elliptic curves. Groups Complex. Cryptol. **9**, 117–123 (2017)

179. Fuchs, C., Heintze, S.: A polynomial variant of Diophantine triples in linear recurrences. Period. Math. Hungar. **86**, 289–299 (2023)

180. Fuchs, C., Hutle, C., Luca, F.: Diophantine triples in linear recurrence sequences of Pisot type. Res. Number Theory **4**, Paper No. 29 (2018)

181. Fuchs, C., Hutle, C., Luca, F., Szalay, L.: Diophantine triples and k-generalized Fibonacci sequences. Bull. Malays. Math. Sci. Soc. (2) **41**, 1449–1465 (2018)

182. Fuchs, C., Luca, F., Szalay, L.: Diophantine triples with values in binary recurrences. Ann. Sc. Norm. Super. Pisa Cl. Sci. (5) **7**, 579–608 (2008)

183. Fujita, Y.: The extensibility of $D(-1)$-triples $\{1, b, c\}$. Publ. Math. Debrecen **70**, 103–117 (2007)

184. Fujita, Y.: The extensibility of Diophantine pairs $\{k-1, k+1\}$. J. Number Theory **128**, 322–353 (2008)

185. Fujita, Y.: Any Diophantine quintuple contains a regular Diophantine quadruple. J. Number Theory **129**, 1678–1697 (2009)
186. Fujita, Y.: The number of Diophantine quintuples. Glas. Mat. Ser. III **45**, 15–29 (2010)
187. Fujita, Y., Miyazaki, T.: The regularity of Diophantine quadruples. Trans. Amer. Math. Soc. **370**, 3803–3831 (2018)
188. Gaál, I.: Diophantine Equations and Power Integral Bases. Theory and Algorithms. Birkhäuser, Boston (2019)
189. Gallagher, P.X.: A larger sieve. Acta Arith. **18**, 77–81 (1971)
190. Gardner, M.: Mathematical games. Sci. Am. **216** (1967), March 1967, p. 124; April 1967, p. 119.
191. Gardner, M.: Mathematical Magic Show, pp. 210, 221–222. Alfred Knopf, New York (1977)
192. Gebel, J., Pethő, A., Zimmer, H.G.: Computing integral points on elliptic curves. Acta Arith. **68**, 171–192 (1994)
193. Gibbs, P.E.: Comput. Bull. **17**, 16 (1978)
194. Gibbs, P.E.: A generalised Stern-Brocot tree from regular Diophantine quadruples. Preprint (1999) arXiv: math.NT/9903035.
195. Gibbs, P.E.: Some rational Diophantine sextuples. Glas. Mat. Ser. III **41**, 195–203 (2006)
196. Gibbs, P.E.: A survey of rational Diophantine sextuples of low height. Preprint (2016). https://doi.org/10.13140/RG.2.2.29253.65761
197. Gouvêa, F., Mazur, B.: The square-free sieve and the rank of elliptic curves. J. Amer. Math. Soc. **4**, 1–23 (1991)
198. Gross, B.H., Zagier, D.B.: Heegner points and derivatives of L-series. Invent. Math. **84** 225–320 (1986)
199. Gupta, S.: $D(-1)$ tuples in imaginary quadratic fields. Acta Math. Hungar. **164**, 556–569 (2021)
200. Gupta, H., Singh, K.: On k-triad sequences. Int. J. Math. Math. Sci. **5**, 799–804 (1985)
201. Gusić, I., Tadić, P.: A remark on the injectivity of the specialization homomorphism. Glas. Mat. Ser. III **47**, 265–275 (2012)
202. Gusić, I., Tadić, P.: Injectivity of the specialization homomorphism of elliptic curves. J. Number Theory **148**, 137–152 (2015)
203. Guy, R.K.: Unsolved Problems in Number Theory. Springer, New York (2004)
204. Gyarmati. K.: On a problem of Diophantus. Acta Arith. **97**, 53–65 (2001)
205. Gyarmati, K., Stewart, C.L.: On powers in shifted products. Glas. Mat. Ser. III **42**, 273–279 (2007)
206. Hajdu, L., Herendi, T.: Explicit bounds for the solutions of elliptic equations with rational coefficients. J. Symbol. Comput. **25**, 361–366 (1998)
207. Hammonds, T., Kim, S., Miller, S.J., Nigam, A., Onghai, K., Saikia, D., Sharma, L.M.: k-Diophantine m-tuples in finite fields. Int. J. Number Theory **19**, 891–912 (2023)
208. Hankerson, D., Menezes, A., Vanstone, S.: Guide to Elliptic Curve Cryptography. Springer, New York (2004)
209. He, B., Luca, F., Togbé, A.: Diophantine triples of Fibonacci numbers. Acta Arith. **175**, 57–70 (2016)
210. He, B., Togbé, A.: On the $D(-1)$-triple $\{1, k^2+1, k^2+2k+2\}$ and its unique $D(1)$-extension. J. Number Theory **131**, 120–137 (2011)
211. He, B., Togbé, A., Ziegler, V.: There is no Diophantine quintuple. Trans. Amer. Math. Soc. **371**, 6665–6709 (2019)
212. Heath, T.L.: Diophantus of Alexandria: A Study in the History of Greek Algebra. Powell's Bookstore/Martino Publishing/Mansfield Center, Chicago (2003)
213. Heichelheim, P.: The study of positive integers (a, b) such that $ab + 1$ is a square. Fibonacci Q. **17**, 269–274 (1979)
214. Herrmann, E., Pethő, A., Zimmer, H.G.: On Fermat's quadruple equations. Abh. Math. Sem. Univ. Hamburg **69**, 283–291 (1999)
215. Hindry, M., Silverman, J.H.: Diophantine Geometry. An Introduction. Springer, New York (2000)

216. Hoggatt, V.E., Jr.: Fibonacci and Lucas Numbers. The Fibonacci Association, Santa Clara (1979)
217. Hoggatt, V.E., Bergum, G.E.: A problem of Fermat and the Fibonacci sequence. Fibonacci Q. **15**, 323–330 (1977)
218. Horadam, A.F.: Generalization of a result of Morgado. Portugaliae Math. **44**, 131–136 (1987)
219. Hungerford, T.W.: Algebra. Springer, New York (1974)
220. Husemöller, D.: Elliptic Curves. Springer, New York (2004)
221. Irmak, N., Szalay, L.: Diophantine triples and reduced quadruples with the Lucas sequence of recurrence $u_n = Au_{n-1} - u_{n-2}$. Glas. Mat. Ser. III **49**, 303–312 (2014)
222. Jacobson, M.J., Jr., Williams, H.C.: Modular arithmetic on elements of small norm in quadratic fields. Des. Codes Cryptogr. **27**, 93–110 (2002)
223. Jacobson, M.J., Jr., Williams, H.C.: Solving the Pell Equation. Springer, New York (2009)
224. Jones, B.W.: A second variation on a problem of Diophantus and Davenport. Fibonacci Q. **16**, 155–165 (1978)
225. Jurasić, A.: Diophantine m-tuples for quadratic polynomials. Glas. Mat. Ser. III **46**, 283–309 (2011)
226. Kamienny, S.: Torsion points on elliptic curves and q-coefficients of modular forms. Invent. Math. **109**, 221–229 (1992)
227. Kanagasabapathy, P., Ponnudurai, T.: The simultaneous Diophantine equations $y^2 - 3x^2 = -2$ and $z^2 - 8x^2 = -7$. Q. J. Math. Oxf. Ser. (2) **26**, 275–278 (1975)
228. Kazalicki, M., Naskrecki, B. (with an appendix by Lasić, L.): Diophantine triples and $K3$ surfaces. J. Number Theory **236**, 41–70 (2022)
229. Kazalicki, M., Vlah, D.: Ranks of elliptic curves and deep neural networks. Res. Number Theory **9**, Article 53 (2023)
230. Kedlaya, K.S.: When is $(xy + 1)(yz + 1)(zx + 1)$ a square? Math. Mag. **71**, 61–63 (1998)
231. Kedlaya, K.S.: Solving constrained Pell equations. Math. Comput. **67**, 833–842 (1998)
232. Kenku, M.A., Momose, F.: Torsion points on elliptic curves defined over quadratic fields. Nagoya Math. J. **109**, 125–149 (1988)
233. Khinchin, A.Ya.: Continued Fractions. Dover, New York (1997)
234. Kihara, S.: On an elliptic curve over $\mathbb{Q}(t)$ of rank ≥ 14. Proc. Japan. Acad. Ser. A Math. Sci. **77**, 50–51 (2001)
235. Kihel, A., Kihel, O.: On the intersection and the extendability of P_t sets. Far East J. Math. Sci. **3**, 637–643 (2001)
236. Kim, S., Murty, M.R. (with an appendix by Sutherland, A.V.): From the Birch and Swinnerton-Dyer conjecture to Nagao's conjecture. Math. Comput. **92**, 385–408 (2023)
237. Kim, S., Yip, C.H., Yoo, S.: Diophantine tuples and multiplicative structure of shifted multiplicative subgroups. Preprint (2023). arXiv: 2309.09124
238. Klagsbrun, Z., Sherman, T., Weigandt, J.: The Elkies curve has rank 28 subject only to GRH. Math. Comput. **88**, 837–846 (2019)
239. Knapp, A.W.: Elliptic Curves. Princeton University Press, Princeton (1992)
240. Koblitz, N.: A Course in Number Theory and Cryptography. Springer, New York (1994)
241. Kolyvagin, V.: Finiteness of $E(\mathbb{Q})$ and CH$(E, \mathbb{Q})$ for a class of Weil curves. Math. USSR-Izv. **32**, 523–541 (1989)
242. Koshy, T.: Fibonacci and Lucas Numbers with Applications. Wiley, New York (2001)
243. Kulesz, L.: Families of elliptic curves of high rank with nontrivial torsion group over $\mathbb{Q}$. Acta Arith. **108**, 339–356 (2003)
244. Kwon, S.: Torsion subgroups of elliptic curves over quadratic extensions. J. Number Theory **62**, 144–162 (1997)
245. Lang, S.: Algebra. Springer-Verlag, New York (2002)
246. Laurent, M.: Linear forms in two logarithms and interpolation determinants II. Acta Arith. **133**, 25–348 (2008)
247. Laurent, M., Mignotte, M., Nesterenko, Y.: Formes linéaires en deux logarithmes et déterminants d'interpolation. J. Number Theory **55**, 285–321 (1995)

248. Le, M., Srinivasan, A.: A note on Dujella's unicity conjecture. Glas. Mat. Ser. III **58**, 59–65 (2023)

249. Lecacheux, O.: Rang de courbes elliptiques sur $\mathbb{Q}$ avec un groupe de torsion isomorphe á $\mathbb{Z}/5\mathbb{Z}$. C. R. Acad. Sci. Paris Sér. I Math. **332**, 1–6 (2001)

250. Lecacheux, O.: Rang de courbes elliptiques avec groupe de torsion non trivial, J. Théor. Nombres Bordeaux **15**, 231–247 (2003)

251. Lecacheux, O.: Rang de courbes elliptiques dont le groupe de torsion est non trivial. Ann. Sci. Math. Québec **28**, 145–151 (2004)

252. Lenstra, A.K., Lenstra, H.W. Jr., Lovász, L.: Factoring polynomials with rational coefficients. Math. Ann. **261**, 515–534 (1982)

253. Lenstra, H.W., Jr.: Factoring integers with elliptic curves. Ann. Math. **126**, 649–673 (1987)

254. LeVeque, W.J.: Topics in Number Theory I, II. Dover, New York (1984)

255. Ljunggren, W.: On the Diophantine equation $x^2 + 4 = Ay^4$. Norske Vid. Selsk. Forh. **24**, 82–84 (1951)

256. LMFDB Collaboration: The L-functions and modular forms database (2023). https://www.lmfdb.org

257. Long, C., Bergum, G.E.: On a problem of Diophantus. In: Philippou, A.N., Horadam, A.F., Bergum, G.E. (eds.) Application of Fibonacci Numbers, vol. 2, pp. 183–191. Kluwer, Dordrecht (1988)

258. Lozano-Robledo, Á.: Elliptic Curves, Modular Forms and Their L-Functions. American Mathematical Society, Providence (2011)

259. Luca, F.: On shifted products which are powers. Glas. Mat. Ser. III **40**, 13–20 (2005)

260. Luca, F., Fujita, Y.: On Diophantine quadruples of Fibonacci numbers. Glas. Mat. Ser. III **52**, 221–234 (2017)

261. Luca, F., Fujita, Y.: There are no Diophantine quadruples of Fibonacci numbers. Acta Arith. **185**, 19–38 (2018)

262. Luca, F., Osgood, C.F., Walsh, P.G.: Diophantine approximations and a problem from the 1988 IMO. Rocky Mountain J. Math. **36**, 637–648 (2006)

263. Luca, F., Szalay, L.: Fibonacci Diophantine triples. Glas. Mat. Ser. III **43**, 253–264 (2008)

264. Luca, F., Szalay, L.: Lucas Diophantine triples. Integers **9**, 441–457 (2009)

265. Mani, N., Rubinstein-Salzedo, S.: Diophantine tuples over $\mathbb{Z}_p$. Acta Arith. **197**, 331–351 (2021)

266. Martin, G., Sitar, S.: Erdős-Turán with a moving target, equidistribution of roots of reducible quadratics, and Diophantine quadruples. Mathematika **57**, 1–29 (2011)

267. Matthews, K.R., Robertson, J.P., White, J.: On a diophantine equation of Andrej Dujella. Glas. Mat. Ser. III **48**, 265–289 (2013)

268. Matveev, E.M.: An explicit lower bound for a homogeneous rational linear form in logarithms of algebraic numbers. II. Izv. Math. **64**, 1217–1269 (2000)

269. Mazur, B.: Rational points of abelian varieties with values in towers of number fields. Invent. Math. **18**, 183–266 (1972)

270. Mazur, B.: Modular curves and the Eisenstein ideal. Inst. Hautes Études Sci. Publ. Math. **47**, 33–186 (1977)

271. Mazur, B. (with an appendix by D. Goldfeld): Rational isogenies of prime degree. Invent. Math. **44**, 129–162 (1978)

272. Mestre, J.-F.: Construction d'une courbe elliptique de rang ≥ 12. C. R. Acad. Sci. Paris Sér. I **295**, 643–644 (1982)

273. Mestre, J.-F.: Formules explicites et minorations de conducteurs de variétés algébriques. Compos. Math. **58**, 209–232 (1986)

274. Mestre, J.-F.: Courbes elliptiques de rang ≥ 11 sur $\mathbb{Q}(t)$. C. R. Acad. Sci. Paris Sér. I **313**, 139–142 (1991)

275. Mestre, J.-F.: Courbes elliptiques de rang ≥ 12 sur $\mathbb{Q}(t)$. C. R. Acad. Sci. Paris Sér. I **313**, 171–174 (1991)

276. Mignotte, M.: A kit on linear forms in three logarithms. Preprint (2008)

277. Mignotte, M., Voutier, P. (with an appendix by Laurent, M.): A kit for linear forms in three logarithms. Math. Comp. **93**, 1903–1951 (2024)
278. Mikić, M.: On the Mordell-Weil group of elliptic curves induced by the families of Diophantine triples. Rocky Mountain J. Math. **45**, 1565–1589 (2015)
279. Milne, J.S.: Elliptic Curves. BookSurge Publishers, Charleston (2006)
280. Mohanty, S.P., Ramasamy, A.M.S.: On $P_{r,k}$ sequences. Fibonacci Q. **23**, 36–44 (1985)
281. Montgomery, P.L.: Speeding the Pollard and elliptic curve methods of factorization. Math. Comput. **48**, 243–264 (1987)
282. Mootha, V.K.: On the set of numbers {14, 22, 30, 42, 90}. Acta Arith. **71**, 259–263 (1995)
283. Mootha, V.K., Berzsenyi, G.: Characterizations and extendibility of P_t-sets. Fibonacci Q. **27**, 287–288 (1989)
284. Mordell, L.J.: Diophantine Equations. Academic Press, New York (1969)
285. Morgado, J.: Generalization of a result of Hoggatt and Bergum on Fibonacci numbers. Portugaliae Math. **42**, 441–445 (1983–1984)
286. Morgado, J.: Note on some results of A. F. Horadam and A. G. Shannon concerning a Catalan's identity on Fibonacci numbers. Portugaliae Math. **44**, 243–252 (1987)
287. Morgado, J.: Note on a Shannon's theorem concerning the Fibonacci numbers and Diophantine quadruples. Portugaliae Math. **48**, 429–439 (1991)
288. Morgado, J.: Note on the Chebyshev polynomials and applications to the Fibonacci numbers. Portugaliae Math. **52**, 363–378 (1995)
289. Nagao, K.: An example of elliptic curve over $\mathbb{Q}$ with rank $\geq$ 20. Proc. Japan Acad. Ser. A Math. Sci. **69**, 291–293 (1993)
290. Nagao, K.: An example of elliptic curve over $\mathbb{Q}(T)$ with rank $\geq$ 13. Proc. Japan Acad. Ser. A Math. Sci. **70**, 152–153 (1994)
291. Nagao, K.: Construction of high-rank elliptic curves with a nontrivial torsion point. Math. Comput. **66**, 411–415 (1997)
292. Nagell, T.: Introduction to Number Theory. Chelsea, New York (1981)
293. Najman, F.: Integer points on two families of elliptic curves. Publ. Math. Debrecen **75**, 401–418 (2009)
294. Najman, F.: Compact representation of quadratic integers and integer points on some elliptic curves. Rocky Mountain J. Math. **40**, 1979–2002 (2010)
295. Najman, F.: Torsion of rational elliptic curves over cubic fields and sporadic points on $X_1(n)$. Math. Res. Lett. **23**, 245–272 (2016)
296. Nguyen, P.Q., Stehlé, D.: An LLL algorithm with quadratic complexity. SIAM J. Comput. **39**, 874–903 (2009)
297. Nguyen, P.G., Vallee, B. (eds.): The LLL Algorithm. Survey and Applications. Springer, Berlin (2010)
298. Niven, I., Zuckerman, H.S., Montgomery, H.L.: An Introduction to the Theory of Numbers. Wiley, New York (1991)
299. Nowicki, A.: Liczby Kwadratowe. Podróże po Imperium Liczb 03. Olsztyn, Toruń (2011)
300. Ono, K.: Euler's concordant forms. Acta Arith. **78**, 101–123 (1996)
301. PARI Group: PARI/GP version 2.15.4. Bordeaux (2023). http://pari.math.u-bordeaux.fr/
302. Park, J., Poonen, B., Voight, J., Wood, M.M.: A heuristic for boundedness of ranks of elliptic curves. J. Eur. Math. Soc. (JEMS) **21**, 2859–2903 (2019)
303. Perron, O.: Die Lehre von den Kettenbruchen I, II. Teubner, Stuttgart (1954)
304. Petričević, V.: Personal communication (2019, 2022, 2023)
305. Piezas, T.: Extending rational Diophantine triples to sextuples. http://mathoverflow.net/questions/233538/extending-rational-diophantine-triples-to-sextuples
306. Rathbun, R.: Personal communication (2003)
307. Rickert, J.H.: Simultaneous rational approximations and related diophantine equations. Math. Proc. Camb. Philos. Soc. **113**, 461–472 (1993)
308. Rihane, S.E., Luca, F., Togbé, A.: There are no Diophantine quadruples of Pell numbers. Int. J. Number Theory **18**, 27–45 (2022)
309. Rockett, A.M., Szusz, P.: Continued Fractions. World Scientific, Singapore (1992)

310. Rosser, J.B., Schoenfeld, L.: Approximate formulas for some functions of prime numbers. Illinois J. Math. **6**, 64–94 (1962)

311. Sadek, M., El-Sissi, N.: On large F-Diophantine sets. Monatsh. Math. **186**, 703–710 (2018)

312. Sadek, M., Yesin, T:: Divisibility by 2 on quartic models of elliptic curves and rational Diophantine $D(q)$-quintuples. Rev. R. Acad. Cienc. Exactas Fis. Nat. Ser. A Math. RACSAM **116**, Article 139 (2022)

313. Sansone, G.: Il sistema diofanteo $N + 1 = x^2, 3N + 1 = y^2, 8N + 1 = z^2$. Ann. Mat. Pura Appl. (4) **111**, 125–151 (1976)

314. Schinzel, A.: Selected Topics on Polynomials. University of Michigan Press, Ann Arbor (1982)

315. Schinzel, A.: Polynomials with Special Regard to Reducibility. Cambridge University Press, Cambridge (2000)

316. Schmidt, W.M.: Integer points on curves of genus 1. Compos. Math. **81**, 33–59 (1992)

317. Schmidt, W.M.: Diophantine Approximation. Springer, Berlin (1996)

318. Schmidt, W.M.: Diophantine Approximation and Diophantine Equations. Springer, Berlin (1996)

319. Schmitt, S., Zimmer, H.G.: Elliptic Curves. A Computational Approach. de Gruyter, Berlin (2003)

320. Schütt, M., Shioda, T.: Mordell-Weil Lattices. Springer, Singapore (2019)

321. Serre, J.-P.: A Course in Arithmetic. Springer, New York (1996)

322. Shannon, A.G.: Fibonacci numbers and Diophantine quadruples: Generalizations of results of Morgado and Horadam. Portugaliae Math. **45**, 165–169 (1988)

323. Shockley, J.E.: Introduction to Number Theory. Holt, Rinehart and Winston, New York (1967)

324. Shorey, T.N.: Linear forms in the logarithms of algebraic numbers with small coefficients I, II. J. Indian Math. Soc. **38**, 271–292 (1974)

325. Shparlinski, I.E.: On the number of Diophantine m-tuples in finite fields. Finite Fields Appl. **90**, Paper No. 102241 (2023)

326. Silverman, J.H.: Computing heights on elliptic curves. Math. Comput. **51**, 339–358 (1988)

327. Silverman, J.H.: The difference between the Weil height and the canonical height on elliptic curves. Math. Comput. **55**, 723–743 (1990)

328. Silverman, J.H.: Advanced Topics in the Arithmetic of Elliptic Curves. Springer, New York (1994)

329. Silverman, J.H.: The Arithmetic of Elliptic Curves. Springer, Dordrecht (2009)

330. Silverman, J.H., Tate, J.: Rational Points on Elliptic Curves. Springer, Cham (2015)

331. Skolem, T: Diophantische Gleichungen. Chelsea, New York (1950)

332. Smart, N.P.: The Algorithmic Resolution of Diophantine Equations. Cambridge University Press, Cambridge (1998)

333. Soldo, I.: On the existence of Diophantine quadruples in $\mathbb{Z}[\sqrt{-2}]$. Miskolc Math. Notes **14**, 265–277 (2013)

334. Srinivasan, A.: $D(-1)$-quadruples and products of two primes. Glas. Mat. Ser. III **50**, 261–268 (2015)

335. Stewart, C.L.: On sets of integers whose shifted products are powers. J. Combin. Theory Ser. A **115**, 662–673 (2008)

336. Stichtenoth, H.: Algebraic Function Fields and Codes. Springer, Berlin (1993)

337. Stoll, M.: On the average number of rational points on curves of genus 2. Preprint (2009). arXiv: 0902.4165

338. Stoll, M.: Documentation for the ratpoints program. Preprint (2013). arXiv: 0803.3165

339. Stoll, M.: Diagonal genus 5 curves, elliptic curves over $\mathbb{Q}(t)$, and rational diophantine quintuples. Acta Arith. **190**, 239–261 (2019)

340. Stroeker, R.J., Tzanakis, N.: Solving elliptic Diophantine equations by estimating linear forms in elliptic logarithms. Acta Arith. **67**, 177–196 (1994)

341. Tijdeman, R.: Diophantine equations and Diophantine approximations. In: Number Theory and Applications, pp. 215–243. Kluwer, Dordrecht (1989)

342. Trudgian, T.: Bounds on the number of Diophantine quintuples. J. Number Theory **157**, 233–249 (2015)

343. Tzanakis, N.: Explicit solution of a class of quartic Thue equations. Acta Arith. **64**, 271–283 (1993)

344. Tzanakis, N.: Solving elliptic Diophantine equations by estimating linear forms in elliptic logarithms. The case of quartic equations. Acta Arith. **75**, 165–190 (1996)

345. Tzanakis, N.: Elliptic Diophantine Equations. A Concrete Approach Via the Elliptic Logarithm. de Gruyter, Berlin (2013)

346. Udrea, Gh.: A note on the sequence (W_n) of A. F. Horadam. Portugaliae Math. **53**, 143–155 (1996)

347. Udrea, Gh.: A problem of Diophantus-Fermat and Chebyshev polynomials of the first kind. Rev. Roumaine Math. Pures Appl. **45**, 531–535 (2000)

348. Vajda, S.: Fibonacci & Lucas Numbers, and the Golden Section. Theory and Applications. Ellis Horwood, Chichester (1989)

349. van Lint, J.H.: On a set of diophantine equations. T. H.-Report 68 - WSK-03, Department of Mathematics, Technological University Eindhoven, Eindhoven (1968)

350. van Luijk, R.: On Perfect Cuboids. Thesis, Universiteit Utrecht (2000)

351. Vellupillai, M.: The equations $z^2 - 3y^2 = -2$ and $z^2 - 6x^2 = -5$. In: Hoggatt, V.E., Bicknell-Johnson, M. (eds.) A Collection of Manuscripts Related to the Fibonacci Sequence, pp. 71–75. The Fibonacci Association, Santa Clara (1980)

352. Vinogradov, I.M.: Elements of Number Theory. Nauka, Moscow (1972) (in Russian)

353. Vorobiev, N.N.: Fibonacci Numbers. Birkhäuser, Basel (2002)

354. Voznyy, M.: Personal communication (2021, 2023)

355. Waldschmidt, M.: Open Diophantine problems. Moscow Math. J. **4**, 245–305 (2004)

356. Washington, L.C.: Elliptic Curves: Number Theory and Cryptography. CRC Press, Boca Raton (2008)

357. Weil, A: Number Theory: An Approach through History from Hammurapi to Legendre. Birkhäuser, Boston (1987)

358. Weintraub, S.H.: Factorization. Unique and Otherwise. CMS, Ottawa; A K Peters, Wellesley (2008)

359. Worley, R.T.: Estimating $|\alpha - p/q|$. Austral. Math. Soc. Ser. A **31**, 202–206 (1981)

360. Yip, C.H.: Multiplicatively reducible subsets of shifted perfect k-th powers and bipartite Diophantine tuples. Preprint (2023). arXiv: 2312.14450

361. Zannier, U.: Some applications of Diophantine Approximation to Diophantine Equations. Forum Editrice, Udine (2003)

362. Zhang, Y., Grossman, G.: On Diophantine triples and quadruples. Notes Number Theory Discrete Math. **21(4)**, 6–16 (2015)

363. Zwegers, S.: On the associativity of the addition on elliptic curves. Preprint (2024). arXiv: 2401.02346

Index

Printed in the USA
CPSIA information can be obtained
at www.ICGtesting.com
CBHW070318040924
13992CB00003B/35